CAMELOT'S COURT

ALSO BY ROBERT DALLEK

The Lost Peace: Leadership in a Time of Horror and Hope

Harry S. Truman

Nixon and Kissinger: Partners in Power

Lyndon B. Johnson: Portrait of a President

An Unfinished Life: John F. Kennedy, 1917–1963

Flawed Giant: Lyndon Johnson and His Times, 1961–1973

Hail to the Chief: The Making and Unmaking of American Presidents

Lone Star Rising: Lyndon Johnson and His Times, 1908–1960

Ronald Reagan: The Politics of Symbolism

The American Style of Foreign Policy: Cultural Politics and Foreign Affairs

Franklin D. Roosevelt and American Foreign Policy, 1932–1945

Democrat and Diplomat: The Life of William E. Dodd

CAMELOT'S COURT

Inside the Kennedy White House

ROBERT DALLEK

HARPER ● PERENNIAL

NEW YORK ● LONDON ● TORONTO ● SYDNEY ● NEW DELHI ● AUCKLAND

HARPER ● PERENNIAL

A hardcover edition of this book was published in 2013 by Harper, an imprint of HarperCollins Publishers.

CAMELOT'S COURT. Copyright © 2013 by Robert Dallek. All rights reserved. Printed in the United States of America. No part of this book may be used or reproduced in any manner whatsoever without written permission except in the case of brief quotations embodied in critical articles and reviews. For information, address HarperCollins Publishers, 195 Broadway, New York, NY 10007.

HarperCollins books may be purchased for educational, business, or sales promotional use. For information, please e-mail the Special Markets Department at SPsales@harpercollins.com.

All photos courtesy of the John F. Kennedy Presidential Library and Museum.

FIRST HARPER PERENNIAL EDITION PUBLISHED 2014.

Designed by Leah Carlson-Stanisic

The Library of Congress has catalogued the hardcover edition as follows:

Dallek, Robert.

Camelot's court : inside the Kennedy White House / Robert Dallek.— First edition.

pages cm

Includes bibliographical references and index.

ISBN 978-0-06-206584-1

1. Kennedy, John F. (John Fitzgerald), 1917–1963. 2. Kennedy, John F. (John Fitzgerald), 1917–1963—Friends and associates. 3. United States—Politics and government—1961–1963. 4. Cabinet officers—United States—History—20th century. 5. Political consultants—United States—History—20th century. 6. Political leadership—United States—History—20th century. I. Title.

E842.D269 2013

973.922092—dc23

2013012864

ISBN 978-0-06-206585-8 (pbk.)

14 15 16 17 18   OV/RRD   10 9 8 7 6 5 4 3 2 1

TO OUR GRANDCHILDREN

Hannah Bender
Ethan Bender
Sammy Sarathy-Dallek

Contents

o o o o

Introduction

o o o o

Some presidents hold an endless fascination for Americans: Washington, Lincoln, the two Roosevelts, and more recently John F. Kennedy and Ronald Reagan. The public's interest has a lot to do with its craving for heroes or, probably more important, its wish to understand and revel in what constitutes effective leadership. In a nation that often feels adrift in an uncertain world, where domestic and foreign crises repeatedly endanger the country's well-being, great presidents are a comfort—a sort of salve for the national psyche.

The affinity for presidential heroes goes far to explain a 2010 Gallup poll asking Americans to assess the last nine presidents from John F. Kennedy to George W. Bush. The survey gave Kennedy an astonishing 85 percent approval rating; only Reagan was in hailing distance of him, with 74 percent. In 2003, having published *An Unfinished Life*, a biography of Kennedy, I lived with the proposition that you write a book to forget a subject. But the poll rekindled my interest in Kennedy's leadership. In my first go-round on Kennedy's presidency, I saw ample reason for enthusiasm about parts of his performance, but 85 percent? Did the public have a better understanding of his leadership than I did? What was I missing? I understand that his assassination at the age of forty-six gives him a special hold on the country's sympathy. Moreover, his personal attributes and inspirational call for national commitments viewed in contrast to the flawed Lyndon Johnson, Richard Nixon, Gerald Ford, Jimmy Carter, and two Bush administrations have heightened Kennedy's attractiveness. Still, the public's conviction that he was so outstanding a chief on par with Washington, Lincoln, and the two Roosevelts is open to question.

Perhaps the best way to understand and assess Kennedy's presidential performance is through his interactions with the men whom Ted Sorensen, his speechwriter and political adviser, called his "Ministry of Talent." They were an extraordinary group of academics, businessmen, lawyers, foreign policy and national security experts, and career military officers who advised Kennedy in the many crises they confronted during his thousand-day presidency. Their focus was on the dangers to the country's safety posed by communist challenges in the Western Hemisphere, Europe, and Southeast Asia. They counseled Kennedy on the critical decisions that could make a difference between war and peace.

The backgrounds, aspirations, convictions, and judgments of the men around Kennedy are an essential part of understanding why and how they advised him.

Each in his own way was a combatant in a struggle to persuade the president that he had the best—if not always the right—answer to the various intimidating challenges they faced at home and particularly abroad, where the danger of a catastrophic war was constantly before them. They fought with each other—sometimes angrily, fearful that the opposing views of their colleagues could lead to disaster. Their passion was the product of genuine concern to serve the country and prevent the ultimate world catastrophe. But it also reflected the vanity of egotistical men who felt slighted by any rejection of their outlook when they were pushed aside, and several of them were. Kennedy took no pleasure in slighting them, but he saw the stakes as so high that he could not afford to tread lightly or put personal feelings ahead of hard decisions. He took comfort from knowing that every administration had its share of conflicts and unhappy advisers and that he had not forced any of them to assume the burdens of office.

Much of the White House tension sprang from the administration's focus on painfully difficult foreign policy questions. Yet Kennedy could not ignore domestic conflicts. They roiled the nation's stability and forced the president and Robert Kennedy, his brother and principal adviser, to devote attention to internal problems. The Kennedys were not indifferent to homegrown difficulties, especially the plight of African Americans. However privileged they were, as members of a minority group they despised the prejudice against blacks of segregation and wished to eliminate it. But they saw little chance to advance equal rights through Congress as long as southern congressmen and senators chaired crucial committees. By putting civil rights aside until 1963 or aiming for marginal gains through executive action, they hoped to pass other domestic reforms that could benefit all Americans, including African Americans.

Given existing conditions and the president's affinity for foreign affairs, however, the administration made domestic change a secondary priority. The main mission of the White House was to inhibit communist advance and avert a nuclear war. Immediately after the Cuban Bay of Pigs failure, in language the latter would understand, Kennedy told Richard Nixon, who shared his conviction about national priorities, "It really is true that foreign affairs is the only important issue for a president to handle, isn't it? I mean, who gives a shit if the minimum wage is $1.15 or $1.25, in comparison to something like this?" On more than one occasion, Kennedy said domestic politics can unseat you, but foreign dangers can kill you. The extent to which this outlook dominated Kennedy's thinking is reflected in the many, almost daily, meetings he held with officials from the State and Defense departments, the CIA, and Pentagon as compared to the occasional discussions he had with those responsible for managing domestic affairs.

The Kennedy who will emerge from the pages of this book is an astute judge of character and reasoned policy. He was an imperfect man whose foibles made him receptive to some bad advice that triggered misjudgments. Moreover, although I can imagine different outcomes in a second term, Cuba and Vietnam demonstrated his limited capacity to overcome all the foreign policy challenges of his thousand days. But these shortcomings were only a part of the story. His attribute as a quick learner helped make him an effective leader, particularly in restraining the actions of his military chiefs during crises that could have resulted in a nuclear war. His successes eclipsed the failings of his thousand days.

Kennedy's interactions with his Ministry of Talent not only enrich our understanding of his presidency; they also serve as useful cautionary tales for voters considering future aspirants to the Oval Office and judging those candidates' ability to meet the day-to-day problems of governing. Above all, the story of Kennedy and his advisers may remind us that the men and women we entrust with power are talented public servants who occasionally fall short of what we hope they will achieve but deserve our regard for assuming the heavy burdens of responsibility that come to every administration.

RD

Washington, D.C.

March 2013

CAMELOT'S COURT

John F. Kennedy: Prelude to a Presidency

Every presidency begins in a fog of uncertainty. The most ordinary of our chiefs, whose administrations left unremarkable legacies, never figured out how to make enduring contributions to the country. Even America's three most notable White House occupants—George Washington, Abraham Lincoln, and Franklin D. Roosevelt—initially puzzled, respectively, over how to launch, preserve, and rescue the ship of state.

Small wonder then that at the start of his presidency John F. Kennedy struggled to fulfill amorphous promises to secure the country from foreign dangers, restore prosperity, and end bitter racial divisions threatening public tranquility. Above all, he feared a crisis that could bring the world to the brink of a nuclear war. It cast a shadow over the realization that he would be the responsible official deciding the fate of millions everywhere. What would he do? He had no clear idea. After he heard that historian Arthur Schlesinger, Jr. wanted a White House appointment rather than an ambassadorship, Kennedy told him: " 'So, Arthur, I hear you are coming to the White House.' 'I am,' " Schlesinger replied. " 'What will I be doing there?' 'I don't know,' " Kennedy responded. " 'I don't know what I'll be doing there, but you can bet we will both be busy more than eight hours a day.' "

Like Lincoln in the midst of the Civil War, Kennedy instinctively met the burdens of office with humor that he hoped would insulate him from the anguish of potentially catastrophic decisions. But fourteen years of political activism—running twice for the House and twice for the Senate—had also imbued him with instrumental cynicism. The objective in his presidential campaign, for example, had been not to describe how he would fix the country's problems, but to win the election. He was following a well-developed tradition. The details of governing would have to come later.

Immediately after being elected, Kennedy was too tired to define how and where he would lead the nation. He was exhausted. His reach for the White House, which had begun in 1957 and consumed every waking hour during 1960, had drained his energies and left him ill-prepared for the arduous work ahead in the Oval Office. Health problems, including Addison's disease—a possibly fatal malfunctioning of the adrenal gland—chronic back pain that had led to major unsuccessful surgeries, spastic colitis that triggered occasional bouts of diarrhea, prostatitis, urethritis, and allergies, had added greatly to the normal strains of a nationwide campaign.

But voters knew little about Kennedy's lifelong illnesses, which had hospitalized him nine times for a total of forty-four days between 1955 and 1957. Only forty-three years old when he ran for the presidency, and masterful at giving the impression of youthful vigor, he had managed to mute questions about his capacity to meet the demands of governing. Yet during the campaign, political rivals had stirred suspicions about his health: Senate Majority Leader Lyndon B. Johnson, his principal challenger for the Democratic nomination, had encouraged journalists to inquire about his Addison's disease, describing Kennedy to one reporter as "a little scrawny fellow with rickets." Moreover, during the fall run against Republican Richard Nixon, break-ins of two Kennedy doctors' New York offices suggest that, like the 1972 Watergate burglary, the Nixon campaign was trying to steal medical records that could decide the election outcome. When a medical bag containing Jack's many medications went missing during the campaign, he was frantic to recover it—because of the political consequences rather than any threat to his health. "It would be murder," he told a political ally, if it got into the wrong hands.

Responding to Kennedy's frail appearance at a press conference the day after his election and unsubstantiated rumors about his health, a reporter asked whether talk about the president-elect's questionable fitness was true. Kennedy dismissed the inquiry with a wave of his hand and assurance that he was in "excellent" shape. Yet Ted Sorensen, his principal Senate aide and speechwriter, recalled that Kennedy's mind was neither "keen" nor "clear" two weeks after his election. He "still seemed tired then and reluctant to face up to the details of personnel and program selection." Kennedy aides felt compelled to follow up with public declarations that he was in "superb physical condition," assuring everyone that he was fully prepared to handle the demands of the presidency.

Kennedy echoed his doubts about satisfying the incessant demands from so many quarters with expressions of uncertainty about identifying and convincing the best people to serve in his cabinet and subcabinet. He told two of his aides: "For the last four years I spent so much time getting to know people who could help me get elected President that I didn't have any time to get to know people who could help me, after I was elected, to be a good President."

All thirty-five of Kennedy's presidential predecessors could have made the same complaint: The road to the White House was so uncertain, especially for the handful who ascended to the job from the vice presidency upon a president's death, that plans for how they would perform in office were never in the forefront of their thinking. Campaign rhetoric about managing current national and international problems was never more than that. Platforms and promises always told less about a president's coming agenda than calculations about appealing to majority sentiment.

Kennedy's route to the prize was as fortuitous as that of all who entered the office before him. True, his father's dream of seeing one of his sons in the White House and having the financial resources to make it happen certainly made Kennedy's ambitions more realizable than the presidential aspirations of men whose wealth and connections never matched his. Other conditions gave Jack an additional leg up. He was raised in a family that regularly breathed, talked, and consumed politics on a daily basis. His grandfathers were larger-than-life public figures who shadowed his early years and made him proud to be a Fitzgerald and a Kennedy. As important, their public visibility put a Kennedy entering Boston politics one step ahead of rivals.

Patrick Joseph Kennedy, Jack's paternal grandfather, was an upwardly mobile Boston Irishman who made it big. Although P.J., as he was affectionately known, lived only until 1929, when Jack was twelve, his accomplishments and affluence were family lore. Forced to work on the Boston docks at the age of fourteen as a stevedore to help support his widowed mother and three older sisters, P.J. used savings to forge a business career as the owner of three taverns and a whiskey importing company that made him a leading figure in the city's liquor trade. His standing as a successful East Boston businessman and a concern with the needs of the city's Irish population drew him into a political career as a five-term member of the Massachusetts lower house and a two-term state senator. As a prominent Boston Democrat, he was a member of the state's delegation to the 1884 national convention, where he gave a seconding speech for New York governor Grover Cleveland, the party's presidential nominee. Giving up elective office in 1895, P.J. spent the rest of his life as one of Boston's four principal Democratic Party ward bosses, choosing candidates for local and statewide offices and distributing patronage. As a part owner of a coal company and a bank, the Columbia Trust, P.J. established himself as one of the city's principal power brokers and wealthier members of what was locally called the "cut glass" set, or FIFs, "First Irish Families."

Jack's maternal grandfather, John F. Fitzgerald, was even more prominent than P.J. and was more instrumental in drawing his grandson into politics. Honey Fitz, as his followers lovingly called him, entered the political arena at age twenty-two as a Customs House clerk and secretary to one of Boston's leading Democratic bosses. He won his first election at age twenty-eight, in 1891 as a member of the Boston Common Council and simultaneously became ward boss of the North End when his mentor, the man he had been serving as secretary, died.

Fitz's meteoric rise in local politics rested on a natural affinity for the calling. An affable, charming character, with the gift of gab described as Fitzblarney, he loved people and center stage. "Fitzie," as the "dearos," the name he gave to his devoted supporters, also called him, was celebrated in a verse extolling his political virtues: "Honey Fitz can talk you blind / on any subject you can find / Fish and fishing, motor boats / Railroads, streetcars, getting votes." His oldest daughter said, "There was no one in the world like my father. Wherever he was, there was magic in the air."

His personal appeal translated into repeated victories at the polls. In 1894, after two years as a state senator, he began a six-year run as Massachusetts's only Democratic congressman and one of three Catholics in the House, where he established himself as the voice of an aggrieved Irish minority. The patrician Massachusetts senator Henry Cabot Lodge symbolized their sense of exclusion. Lodge's haughty manner and demeanor reminded them of the saying that up in Boston, "the Lowells speak only to the Cabots and the Cabots speak only to God." They delighted in the story of Fitzgerald's rebuke of Lodge when the senator lectured Fitzgerald on the corrupting influence of immigrants: "Do you think the Jews or the Italians have any right in this country?" Lodge asked. "As much right as your father or mine. It was only a difference of a few ships," Fitzgerald responded.

In 1905 Fitzgerald became Boston's mayor, signaling the emergence of the city's Irish as the principal political force and launching a personal dynasty lasting forty-five years. Although press stories about city hall corruption and rumors of an affair with Elizabeth "Toodles" Ryan, a beautiful cigarette girl at a local nightclub, marked his career as mayor, it little diminished Fitzie's hold on his Irish constituents, who loved him and his antics as a defiance of the city's imperious Brahmins. When Honey Fitz died at the age of eighty-seven in 1950, more than 3,500 people attended the church service. It impressed his grandson as a demonstration of "the extraordinary impact a politician can have on the emotions of ordinary people."

No one, however, contributed more to Kennedy's pre-presidential political career than his parents. His mother, Rose Fitzgerald Kennedy, was her father's favorite child. Her status as an attractive Boston debutante closely identified with her ethnic and religious roots made her a favorite of the city's aspiring Irish. They could imagine their sons and daughters sharing her rise to prominence that rivaled the standing of the town's Protestant elite. Her marriage to P.J.'s son, Joseph P. Kennedy, a brilliantly successful banker, and her visibility as the mother of nine sons and daughters gave her children instant fame that could open the way to a potentially stunning public career.

It was her husband, Joe, however, who was the engine of the family's special distinction that facilitated Jack's rise to power. Joe's middle name should have been ambition—for wealth, for status, for power. He grew up reading and identifying with the Horatio Alger rags-to-riches stories. Like so many other highly successful businessmen in his time, Joe enjoyed privileged beginnings. His family's economic and social standing gave him access to Boston Latin, the city's most famous public school, attended by its wealthiest residents. Despite an undistinguished academic record, Joe's athletic accomplishments on the baseball team, success as the captain of the drill team, and social skills that fostered his election as senior class president won him admission to Harvard College, where, again, he made a mark not as an outstanding student but as a budding politician and entrepreneur. He won election to student councils and the storied Hasty Pudding Club while also running a tour bus business that paid most of his college expenses and gave him a feel for moneymaking, which became his dominant focus after earning his B.A. in 1912.

Over the next twenty years, his talent for building successful businesses in banking, liquor, movies, stocks, and real estate made him one of the richest and most prominent men in America. Joe and his family, which had grown to nine children by 1932, enjoyed a standing that was the envy of the country's most famous figures—whether in Hollywood, sports, or politics. The onset of the Great Depression in the thirties convinced Joe, as he told his four sons, that the next generation of big men in America would not be in business, as when he came of age, but in government. And this is where Joe began investing his energies, and he expected Joseph, Jr., John, Robert, and Edward to do the same.

In 1934, Joe's financial contributions to Franklin Roosevelt's 1932 presidential campaign and reputation as a brilliant entrepreneur facilitated his appointment as chairman of the new Securities and Exchange Commission. Joe had been eager for a cabinet post, but public anger toward big business in the Depression precluded giving someone like Joe, who had a reputation for questionable financial dealings, a White House job. When asked why he had chosen a Wall Street insider to head the SEC, Roosevelt replied, "It would take a thief to catch a thief." In 1937, the president appointed Joe to head the new Maritime Commission, where he could draw on his World War I experience in shipbuilding to spur the growth of an American merchant fleet that FDR believed essential to the country's economic future and national defense in a likely European war.

Joe's reach for high public office culminated in a 1938 appointment as ambassador to Great Britain. Kennedy having established a reputation as an effective and evenhanded administrator at both the SEC and Maritime Commission, Roosevelt suggested he consider becoming secretary of commerce. But Joe saw the Court of St. James's, the most prestigious overseas diplomatic assignment, as better suited to his goals. He had thoughts of running for president, and a term as ambassador to Great Britain would school him in foreign affairs and supplement his credentials as a brilliantly successful businessman. White House insider Tommy Corcoran told Secretary of the Interior Harold Ickes, who puzzled over Kennedy's choice of London over an appointment to Roosevelt's cabinet, that the ambassadorship would open all doors to him. It wasn't just political ambition driving Kennedy's decision, Corcoran believed, but the chance to become America's first Irish Catholic ambassador to London. It gave him equal status with the country's most prominent Protestants.

FDR hoped that Kennedy's Irish roots would make him a critical observer of British prime minister Neville Chamberlain's conservative government and, specifically, his appeasement policy toward Hitler's Germany. As Kennedy was about to leave for London, Roosevelt privately described his selection of Kennedy as "a great joke, the greatest joke in the world," meaning that the British government would not be able to co-opt the Irishman, whom Roosevelt expected to make U.S. antagonism to Adolf Hitler clear. But Kennedy disappointed FDR's expectations. Wedded to American isolationist thinking and fearing a European war that could draw the United States into the fighting and risk the lives of his two oldest sons, Kennedy supported Chamberlain's soft line toward the Nazis and lobbied FDR to do the same.

Roosevelt, however, wanted no part of the disastrous Chamberlain-Kennedy indulgence of German aggression, which he saw threatening democratic nations everywhere. He shared Winston Churchill's observation that Chamberlain had a choice between dishonor and war. He chose dishonor and got war. "Who would have thought that the English could take into camp a red-headed Irishman?" Roosevelt said. The great majority of Americans shared FDR's outlook, and Kennedy's reputation as an appeaser and a closet anti-Semite partial to Hitler's persecution of Germany's Jews decisively ended his ambitions for high elective office.

The fall in public standing depressed Kennedy, and he began taking "solace . . . in his children's accomplishments." Kennedy shifted the focus of his political ambitions to his oldest son, Joe, Jr. And the young man was all too eager to meet his father's expectations. Like Joe, Sr., junior distinguished himself in prep school at Choate and at Harvard as an athlete and a young man on the make. In 1940, he entered Harvard Law School and simultaneously won election to the Massachusetts delegation to the Democratic National Convention in Chicago, where he favored party boss James Farley over FDR's ambition for a third term. In May 1941, after completing his second year in law school, Joe, anticipating U.S. involvement in the European war, which had begun in September 1939, enlisted in the Navy, winning wings as a naval aviator in May 1942. His commitment to military service rested on genuine concern about the nation's security but also on the conviction that a military record would be essential to anyone intent on a postwar political career.

Joe's hopes for his oldest son, who, in his twenties, already seemed marked out for extraordinary achievements by his ambition, family connections, and widely acknowledged charm, like that of Honey Fitz, collapsed in August 1944. Stationed in England, Joe volunteered for a risky mission aimed at German launch sites for their V-1 rockets on the coast of Belgium. The unmanned "buzz bombs," as they were called, were devastating London. Joe and one other Navy airman flew a PB4Y Liberator bomber armed with twenty-two thousand pounds of explosives, the largest concentration of dynamite on a plane prior to the atomic bombing of Hiroshima in 1945. Joe and his copilot were to bail out before crossing the English Channel, and the plane would continue to the target by remote control. But the plane, for unexplained reasons at the time, exploded while they were still aboard, killing both of them. In 2001, fifty-seven years after the accident, a World War II member of the British corps of Royal Electrical and Mechanical Engineers explained that a failure to inform British authorities to turn off ground-based radars in the south of England "upset the delicate radio controls" on Joe's plane and triggered the explosion.

Joe's death devastated his father, who told a friend, "You know how much I had tied my whole life up to his and what great things I saw in the future for him." To another friend, he said, "all my plans for my own future were all tied up with young Joe and that has gone to smash." Losing a child is torment enough for anyone, but for Joe it also meant the suspension of his dreams for a Kennedy in the highest reaches of American political power.

But not for long: In 1946, John, the second son, became the reluctant heir to Joe's political ambitions. Jack, as his family and friends called him, was less outgoing and more cerebral than his older brother. As a boy, his health problems compelled long stretches in bed or at least indoors: Three months before his third birthday, he had scarlet fever, a life-threatening disease in 1920 that hospitalized him for two months, followed by two weeks in a Maine sanatorium. He subsequently contracted all the illnesses— bronchitis, chicken pox, ear infections, German measles, measles, mumps, and whooping cough—that commonly afflicted school-age children and caused periods of isolation that he filled with reading adventure stories—everything from Sinbad the Sailor to *King Arthur and the Round Table.*

Other illnesses followed: At Choate, the exclusive Connecticut private school, which Jack began attending in 1931, when he was fourteen, he suffered from spastic colitis, an intestinal disorder that kept him from gaining weight and forced his hospitalization in 1933. He responded to his illness with a kind of wry humor that masked fears of a bleak future. He wrote his closest friend at Choate: "It seems that I was much sicker than I thought I was, and am supposed to be dead, so I am developing a limp and a hollow cough."

In June 1934, when Jack was seventeen, Joe sent him to the Mayo Brothers' clinic in Minnesota, where he spent a month while the doctors struggled to come up with a treatment for his intestinal malady. He wrote his schoolmate: "God what a beating I'm taking. I've lost 8 pounds and still going down. . . . Nobody able to figure what's wrong with me. All they do is talk about what an interesting case." They also subjected him to repeated rectal and intestinal exams: "I was a bit glad when they had their fill of that. My poor bedraggled rectum is looking at me very reproachfully these days. . . . The reason I'm here is that they may have to cut out my stomach—the latest news." He worried that his illness might limit his freedom to play competitive sports and blight his relationships with peers, especially girls, in whom he had developed a normal teenage boy's interest. "What will I say when someone asks me what I got?" he wondered.

His medical ordeal had just begun. In 1937, when Jack was twenty and a student at Harvard, his struggle with colitis led doctors to prescribe newly available corticosteroids, anti-inflammatory agents that were administered in the form of pellets implanted under the skin. Although the drugs reined in his colitis, they produced new health problems—stomach, back, and adrenal maladies that were apparently triggered by limited understanding of safe steroid dosages. Peptic ulcers, osteoporosis of the lumbar spine, or bone loss in the lower back, producing miserable backaches, and Addison's disease would afflict him for the rest of his life.

Kennedy's medical issues discouraged consideration of a political career—a vocation that involved exhausting campaigns that tested the endurance of the healthiest candidates. It was also a profession in which the appearance of any physical disability could be seen as disqualifying someone for high office. Franklin Roosevelt, mindful that his paralysis from polio could raise doubts about his capacity to perform effectively as president, discouraged public knowledge of his immobility, assuring that no photos of him in a wheelchair or on crutches ever became public during his twelve White House years. The Kennedys took every precaution to keep public disclosure of Jack's health issues to a minimum, even though there was no thought of having him run for office, at least as long as Joe, Jr. was expected to fulfill family aspirations for a Kennedy in high office. The assumption that no good could come of revealing Jack's medical problems made it something of a family secret. Even Ted, Jack's youngest sibling, had limited knowledge of his brother's medical history. He first learned of the extent of Jack's problems from reading this author's *An Unfinished Life*, published in 2003.

The physical limitations Jack seemed likely to face throughout his life encouraged him to think about a writing career—either of books or journalism. In 1940, he had published his Harvard senior honors thesis, *Why England Slept*, which commanded a popular audience interested in understanding how Britain had been so unprepared for the European war. In 1945, he had worked as a freelance journalist covering the founding of the United Nations in San Francisco. Moreover, his temperament seemed little suited to the glad-handing, backslapping, small talk, and favor-granting that Boston politicians used to keep constituents happy.

Having traveled extensively in Europe in the late thirties before the outbreak of the war, Jack had an affinity for international affairs that holding a local office or even a congressional seat would not satisfy. He was principally interested in why and how nations cooperated and opposed each other, and not in cozying up to residents of Boston worried about jobs, the cost of living, and educating their children. Unlike his brother Joe, he was not his grandfather's grandson. Becoming a foreign correspondent describing the rise and fall of foreign governments and monitoring international crises that he believed would agitate the postwar world were much closer to his heart's desire than any conceivable job as a politician—except maybe for president, and the odds of getting to the White House as a Catholic with serious health problems and a father who was something of a public liability seemed very long, if not insurmountable.

In 1945–46, however, Jack, under pressure from Joe, agreed to enter politics. "I never thought at school or college that I would ever run for office myself," Jack said in 1960. "One politician was enough in the family, and my brother Joe was obviously going to be that politician. I hadn't considered myself a political type, and he filled all the requirements for political success." But Joe, Sr. insisted that Jack replace his brother. "I got Jack into politics," Joe said in 1957. "I was the one. I told him Joe was dead and that it was therefore his responsibility to run for Congress." As Jack remembered it, "it was like being drafted. My father wanted his eldest son in politics. 'Wanted' isn't the right word. He *demanded* it. You know my father," Jack told a reporter.

Once committed to running for national office—first as a congressman, serving from 1946 to 1952, then as a senator, from 1953 to 1960—Kennedy acted as if his life depended on it. Winning, being on top, and staying there were the family's unspoken mottoes, and Jack was now the Kennedy standard-bearer. Yet winning elections to the House and the Senate did not necessarily translate into a satisfying career as a legislator. Jack despised being a congressman and took little more satisfaction from serving as a senator. His time in the Lower House persuaded him to endorse Mark Twain's snide observation: "Suppose you were an idiot. And suppose you were a member of Congress. But I repeat myself." He viewed the congressional leadership, which ranged in age from sixty-eight to eighty-three, as gray, stodgy, conservative, predictable, and unimaginative: men who worshipped at the altar of party regularity and lived by Texan Speaker of the House Sam Rayburn's adage—"to get along, go along." Kennedy saw most of his colleagues as time-servers: men who held safe seats, enjoyed the perks of the job—a degree of social standing, and a decent salary—and believed or convinced themselves that they were serving the national well-being.

By contrast, Jack thought that he and most of his colleagues accomplished little, if anything, of importance as congressmen. He told his close school friend, Lem Billings, that "most of his fellow congressmen [were] boring, preoccupied as they all seemed to be with their narrow political concerns." He hated "all the arcane rules and customs which prevented you from moving legislation quickly and forced you to jump a thousand hurdles before you could accomplish anything." He remained interested in ideas and unconventional thinking that challenged accepted norms, especially in international affairs. The House as an institution was decidedly unsuited for someone who thought as he did.

The Senate was not much more appealing to him. In his 1956 Pulitzer Prize–winning book, *Profiles in Courage*, Jack cited Daniel Webster's observation: "We have not fully recognized the difficulty facing a politician conscientiously desiring 'to push [his] skiff from the shore alone' into a hostile and turbulent sea." In Kennedy's view, as a group, senators were no more courageous or imaginative than their House counterparts. After a year in his seat, when asked, "What's it like to be a United States senator?" he replied: "It's the most corrupting job in the world." He saw senators as more intent on preserving their jobs than on boldly working for the national interest or addressing broad matters of national security. He complained that they were all too quick to cut deals and please campaign contributors to ensure their political futures.

Despite his complaints, Kennedy was never an especially independent voice in the Senate. In a 1960 tape recording he made for posterity, he described the life of a legislator as much less satisfying than that of a chief executive. He thought that effective leadership largely came from the top and one of a hundred senators lacked the power and influence exerted by a president. "The President today is the seat of all power. . . . The presidency is the place to be . . . if you want to get anything done," he said.

As a senator, he was as cautious as any of his colleagues in the service of his ambitions for the presidency. In 1954, when the Senate voted to condemn Joseph McCarthy for breaking Senate rules and abusing an Army general, Kennedy was the only Democrat not to cast a vote against him. True, Jack was in the hospital recovering from back surgery and said later that he was "in bad shape" and preoccupied with his health. Still, he acknowledged that he could have paired his vote with a senator favoring McCarthy to put himself on record in support of condemnation. Instead, he ducked the issue, mainly out of political expediency. He did not want to risk reelection to his Massachusetts seat, which he feared could result from a vote against McCarthy, who was popular with Massachusetts Catholics. In 1960, when Jack was running for president, Eleanor Roosevelt said about his abstention on the McCarthy vote, contrasting his action with his rhetoric and the argument in his 1956 book, "I wish he had shown more courage and less profile."

Journalists and party leaders questioned Kennedy's reach for the White House. His good looks and youth (after Eisenhower, who was leaving office at the age of seventy as the oldest man to have ever served as president) made Kennedy's appearance and age—forty-three in 1960—a distinct asset. "But what has all this to do with statesmanship?" a *New York Post* columnist asked. James Reston of the *New York Times* saw Kennedy's image of casualness and youthful energy as masterful in selling himself to the public, but all the emphasis "on how to win the presidency rather than how to run it" bothered him. Other journalists wondered whether someone so boyish-looking could possibly capture and then serve effectively in the White House. His campaign fashioned a song partly to counter the objection that he was too young to serve as president:

> *Do you want a man for president,*
> *Who is seasoned through and through?*
> *But not so doggone seasoned,*
> *That he won't try something new.*
> *A man who's old enough to know,*
> *And young enough to do.*

Many in the Democratic Party thought Kennedy would make an excellent choice for vice president and that he should wait his turn to run for president behind the sixty-year-old Adlai Stevenson, the party's two-time nominee, the fifty-nine-year-old Missouri senator Stuart Symington, an expert on national security in a dangerous world, and the fifty-two-year-old Lyndon Johnson, the Senate majority leader from Texas, whose prominence in the party and qualities as a force of nature seemed to make him a more deserving candidate than Jack. Stevenson, who hoped for a third nomination, privately dismissed Kennedy's ambition as a little foolish, telling a friend, "I don't think he'd be a good president. I do not feel that he's the right man for the job; I think he's too young; I don't think he fully understands the dimensions of the foreign affairs dilemmas that are coming up."

But Kennedy saw delay as a prescription for defeat. During the Wisconsin primary, when an elderly woman opposed to his candidacy told him, "You're too soon, my boy, too soon," Jack replied, "No this is my time. My time is now." It wasn't the eight years that he might have to wait should a Democrat win in 1960 that persuaded him to run at age forty-three, but the thought that at fifty-one, health problems might deprive him of the energy and stamina for a presidential campaign. Moreover, he was convinced that in eight years fresher faces would push him into the background or make him old news, with little ability to generate public excitement.

In running for the highest office, Kennedy saw himself as uniquely positioned to serve the country's well-being. He believed that the Republicans and most of his Democratic rivals for the nomination were locked into conventional thinking that would perpetuate the Cold War and endanger the peace. "The key thing for the country is a new foreign policy that will break out of the confines of the Cold War," he told a potential supporter. "Then we can build a decent relationship with developing nations and begin to respond to their needs. We can stop the vicious circle of the arms race and promote diversity and peaceful change within the Soviet bloc. We can get this country moving again on its domestic problems." Other Democratic aspirants for the highest office not only echoed Republican foreign policy ideas but also made the mistake of putting traditional welfare state assumptions—economic security and social programs—ahead of overseas challenges that could overshadow domestic concerns. Kennedy believed that his strongest claim on the presidency was an understanding that domestic issues had to take a backseat to national security dangers and that the next chief executive needed, above all, to assure long-term peace, because advances in destructive weapons made another all-out war impermissible.

The nomination and general election campaigns, however, offered limited opportunity for Kennedy to make a detailed case for how the country's direction in foreign affairs would change under his stewardship. Persuading party leaders and voters that he could lead them to victory in November and convincing a wider electorate that he would be a better national leader than his Republican rival moved him to focus on other matters than the substance of governing. Besides, he had no clear agenda for how he would achieve his larger designs, and since offering details of how he would proceed in office seemed likely to stimulate more opposition than support, he believed it just as well to let a program for governing remain unstated. As much to the point, he knew that the country's most effective presidents had never planned too far ahead. Circumstances were always changing, and any course of action was best designed in response to current events. As a student of history and the presidency in particular, he knew that the most influential recent presidents, from Theodore Roosevelt to Harry Truman, had shunned choosing a cabinet or White House staff or announcing precise policy choices in advance of their administrations. He took counsel from Abraham Lincoln's famous observation, "I claim not to have controlled events, but confess plainly that events have controlled me."

His fight for the nomination took him far afield from the substance of foreign policy making. During the West Virginia primary, for example, he felt compelled to address the state's struggle with economic problems. He described an agenda—increased unemployment benefits, expanded Social Security, food distribution for the needy, federal spending to stimulate coal production (the

state's biggest industry), and more defense investment—that echoed FDR's New Deal, which, like Roosevelt himself, was highly
popular in the state. But all this was muted alongside efforts to
convince the state's Protestants, who made up 96 percent of its
residents, that his Catholic religion would be of no consequence
in shaping a Kennedy presidency. "The Catholic question" was a
matter of vital concern to millions of American voters in 1960—
many, especially across the South, where anti-Catholic sentiment
was most pronounced, believed that a Catholic president would be
more loyal to the pope than to the United States. In confronting
the issue directly and effectively, Kennedy assured himself of an
essential electoral victory crucial to his nomination.

The question shadowed his campaign nonetheless, and in mid-
September, less than two weeks after the traditional Labor Day
start of the national contest, Kennedy felt compelled to defend his
religious affiliation before a meeting of Protestant, mainly Bap-
tist, ministers in Houston, Texas. Although some of his advisers
urged against speaking to what they described as a hostile, pro-
Republican group, Kennedy believed it essential to address the in-
nuendoes and outright distortions about the likely impact of his
religion on his capacity to serve as president. "I'm getting tired of
these people who think I want to replace the gold in Fort Knox
with a supply of holy water," he told two of his aides.

On September 12, before an audience of three hundred in the
ballroom of Houston's Rice Hotel, he respectfully dismissed concerns about his religion as a diversion from more essential considerations in the campaign. He emphasized his unqualified commitment
to the separation of church and state: "I am not the Catholic candidate for President," he famously declared. "I am the Democratic
Party's candidate for President, who happens also to be a Catholic.
I do not speak for my church on public matters—and the church
does not speak for me." He cautioned that 40 million Americans
should not lose their chance of being president on the day they were
baptized. If this were the case, he predicted, "the whole nation will
be the loser in the eyes of Catholics and non-Catholics around the

world, in the eyes of history, and in the eyes of our people." Although the religious issue by no means disappeared during the rest of the campaign, or stopped thousands of voters, especially across the South and in rural counties around the country, from casting anti-Catholic ballots, Kennedy's speech muted suspicions and disarmed some of the anti-Catholic hostility toward him.

It did not, however, make his campaign a model of constructive civic pronouncements on the substance of his future presidency. True, Kennedy felt compelled to offer generalizations about how he would get the country moving again, characterizing his future administration as leading the country on to "a New Frontier—the frontier of the 1960s—a frontier of unknown opportunity and perils—a frontier of unfulfilled hopes and threats." The Harvard economist John Kenneth Galbraith told him that the speech was an impressive rhetorical exercise, which it "had to be," safely negotiating "the delicate line that divides poetry from banality." But what did it mean? No one, including Kennedy himself, could say.

Kennedy and his advisers correctly believed that elections were generally won by voters coming to your side less out of convictions about how you would overcome current problems than from negative views of your opponent's character and record of flawed leadership. Kennedy remembered Franklin Roosevelt's campaign in 1932 against the hapless Herbert Hoover, whose failure to end the Depression spoke for itself, and Harry Truman's successful upset victory in 1948 against New York's Governor Thomas Dewey, whom Democrats characterized as the only man who could strut sitting down, and whom Truman tied to the "Do Nothing," "Good-for-Nothing" Republican-controlled Eightieth Congress. During the 1960 West Virginia primary, the Kennedy campaign promoted discussion of Hubert Humphrey's lack of World War II military service, implying that he had been a draft dodger, despite their understanding that medical problems had kept Humphrey out of the service.

In 1960, Vice President Richard M. Nixon perfectly fit the role of an opponent with a controversial political history that could be turned against him. Nixon had a reputation as a political assassin who had won U.S. House and Senate seats in California by falsely tarring opponents as communist fellow travelers. In 1952, he had attacked Adlai Stevenson, the Democratic nominee, as holding a Ph.D. from Secretary of State Dean Acheson's "cowardly college of Communist containment." Nixon's underhanded political tactics had also aligned him with Joseph McCarthy, who had been discredited by 1960. Liberals understandably despised Nixon. But his tactics had undermined him with voters more generally by making him appear sinister, untrustworthy, and not deserving of election to the presidency.

Nonetheless, Nixon also had his share of devoted supporters, who saw him as a leader who could effectively combat ruthless communists. Eight years as vice president under the still popular Dwight Eisenhower gave him impressive credentials as a seasoned foreign policy leader who had famously stood up to Soviet first secretary Nikita Khrushchev in what came to be known as the 1959 Moscow "kitchen debate." But his identification with the Eisenhower administration also carried liabilities: a so-called missile gap that Kennedy made much of during the campaign. In 1957, Moscow's launching of Sputnik, the first man-made satellite to orbit the earth, suggested that the Soviets had eclipsed the United States in capacity to deliver intercontinental ballistic missiles and had put the United States behind in the nuclear arms race. In addition, a series of economic downturns, including the continuing effects of a 1958 recession, gave Kennedy an advantage in emphasizing that he and the Democrats, who enjoyed higher standing as economic managers, would be better able to restore national prosperity.

The importance of negative images and impressions in defeating Nixon was most apparent in the results of a nationally televised debate, unprecedented in a presidential election, with Kennedy in September 1960. The debate attracted the largest audience ever to have watched two candidates battle each other. As a practiced debater confident of his ability to best any opponent, Nixon was receptive to the prospect of squaring off against his younger, less experienced opponent. Likewise, Kennedy was enthusiastic about the chance to demonstrate that he was as competent as Nixon in discussing the challenges facing the United States. Besides, Kennedy and his aides were confident that his more attractive personal attributes would create an appeals gap with the dour, humorless Nixon.

In their respective opening and closing statements, Kennedy ignored Nixon and spoke directly to the large viewing audience. Nixon, by contrast, tried to score points against Kennedy, reinforcing impressions of himself as a street fighter trying to win an election rather than demonstrate his qualities as a statesman. Moreover, Nixon, who had spent two weeks in a hospital for treatment of a knee infection suffered in an accident, was thin and pale and appeared scrawny and listless—almost cadaver-like. "My God," Chicago mayor Richard Daley said, "they've embalmed him before he even died." By contrast, Kennedy, the one with far greater physical problems than the vice president, came across as the picture of robust good health. The minority of the audience who heard the debate on the radio thought Nixon had won. But the great majority who watched it on television gave Kennedy the nod.

John Kenneth Galbraith, the liberal Harvard economist and campaign adviser, thought Kennedy was "simply superb." When he asked "the proprietor" of "a Negro shoe shine parlor" in San Diego, where he was at a conference on unemployment and Social Security, how he liked Kennedy's performance, the man replied: "So help me God, ah'm digging up two from the graveyard for that boy."

Kennedy won the election, but it was a close victory: a 118,574 popular vote margin, yielding 49.72 percent of the 68,837,000 total cast; it translated into an electoral count of 303 to 219. When Kennedy went to bed at 3:30 A.M. on election night, however, the contest still hung in the balance, with six states—California, Illinois, Michigan, Minnesota, Missouri, and Pennsylvania—too close to call. It wasn't until the next morning at a little after nine that he learned he had won, though Nixon's press secretary did not concede the result until after noon.

Kennedy puzzled over how to respond to the narrow margin. He could look back to Woodrow Wilson's 42 percent plurality in 1912 and take comfort from knowing that Wilson won a second term and became one of the most significant presidents of the century. But it did little to salve Kennedy's wounded pride and self-confidence, especially since everyone in his inner circle had been predicting a victory of between 53 and 57 percent: "How did I manage to beat a guy like this by only a hundred thousand votes?" he asked one of his aides. More important, it left him less room to maneuver; he would have to build an administration with greater regard for Republican sensitivities. His promise to adopt a fresh outlook on the Cold War, for example, gave immediate ground to decisions on choosing directors of the Central Intelligence Agency (CIA) and the Federal Bureau of Investigation (FBI). Liberals urged him to appoint new national security and law enforcement officials who could signal a change. Instead, two days after the election, Kennedy announced that Allen Dulles and J. Edgar Hoover would remain as heads of the two agencies.

The same week, Kennedy arranged a meeting with Nixon as a show of national unity. He privately acknowledged that he had nothing to say to his recent rival, but he thought it was important to give the impression that he would construct a bipartisan administration, though he would not offer him a job. After the meeting, in which Nixon did most of the talking, Kennedy privately remarked, "It was just as well for all of us that he didn't quite make it."

Two meetings with Eisenhower were more consequential. Kennedy wished to avoid any demonstration of antagonism, which had marked the transition from Truman to Eisenhower in 1952–53. Although Kennedy did not think well of Ike, seeing him as an "old fuddy-duddy" and calling him an "old asshole" who had lost control of his administration and become a "non-president," he understood that Eisenhower still enjoyed high public standing.

The first meeting at the White House in December 1960 focused on foreign policy problems. Eisenhower dominated the conversation; afterward he praised Kennedy as "a serious earnest seeker for information." He believed that Kennedy "will give full consideration to the facts and suggestions we presented," implying that despite party and campaign differences, Eisenhower foresaw continuity between their administrations. Kennedy kept his counsel largely because he didn't wish to reveal the limits of what he knew about the topics Ike had put before him or what he intended to do as president: "NATO nuclear sharing, Laos, the Congo, Algeria, Disarmament [and] Nuclear test suspension negotiations, Cuba and Latin America, U.S. balance of payments and the gold outflow."

In January, as Kennedy approached his inauguration, he asked for a second meeting. He was particularly worried about a civil war in Laos and the possibility that his first crisis would compel a decision on using military force to prevent a communist victory, which Eisenhower's advisers believed would pose a threat to all of Southeast Asia. Kennedy told an aide, "Whatever's going to happen in Laos, an American invasion, a Communist victory or whatever, I wish it would happen before we take over and get blamed for it." He feared a military action that went badly, diverted attention from other issues, and produced unfavorable contrasts with Ike. Comparison between him, a junior naval officer, and Eisenhower, the storied five-star World War II general, would clearly be disadvantageous at the start of Kennedy's term.

When he sat down with Eisenhower, Kennedy wanted to discuss administrative questions. In particular, he was keen to talk about "the present national security set up, organization within the White House . . . [and the] Pentagon." But Eisenhower put him off with the recommendation that he delay "any reorganization before he himself could become well acquainted with the problem." Ike's advice did not sit well with Kennedy, who believed that Eisenhower's affinity for a military command system had produced an overly cautious administration reluctant to act boldly and move in new directions. Kennedy gave Eisenhower the impression that he intended to set up a government that relied on having the right man in the right place. Eisenhower, who believed that successful administration depended more on smooth-running bureaucracies than on ambitious men pressing their personal agendas, considered Kennedy naïve in thinking that he could find miracle workers who would help him solve national and international problems.

Kennedy, however, had no precise plan for how he would organize his administration. He believed that it required considerable forethought and preparation. Consequently, after winning the nomination, he had invited Clark Clifford, Harry Truman's White House counsel and architect of his 1948 election victory, to discuss campaign politics. At the end of a breakfast meeting, Kennedy made "a request that had no precedent in American politics, one that was to set a pattern for future transfers of presidential power," Clifford recalled. Kennedy said, "Clark, I've been thinking about one matter where you could be of special help to me. If I win, I don't want to wake up on the morning of November 9 and say to myself, 'What do I do now?' I want to have a plan. I want someone to be planning for this between now and November 8." He asked Clifford to prepare a memorandum "outlining the main tasks of the new Administration." A week later, Kennedy told Clifford that a Brookings Institution group was studying past transitions and discussing ways to improve on them. He persuaded Clifford to be his representative on the committee.

At the same time, Kennedy invited Columbia University political scientist Richard Neustadt, who had just published a widely discussed book, *Presidential Power*, to write a transition plan for him as well. Neustadt, who knew that Kennedy had also directed Clifford to develop a strategy for taking control of the government, asked how he should coordinate his efforts with Clifford. Kennedy instructed him to ignore Clifford. "I can't afford to confine myself to one set of advisers," Kennedy told him. "If I did that, I would be on their leading strings."

Kennedy knew that the most effective presidents—Lincoln, Wilson, and the two Roosevelts—had consulted various advisers but at the end of the day had relied on their own counsel to make the most important decisions of their terms. As Harry Truman had said, "the buck stops here." It was the president who had the responsibility for choosing between the options available to him. Besides, for someone as young and inexperienced as he would be on entering office, Kennedy needed to insure against impressions of him as a cipher, a novice simply following the lead of subordinates who thought they knew better than their chief.

Neither Clifford nor Neustadt expressed an interest in becoming a part of the new administration, which pleased Kennedy. Tall, handsome, with the looks of a matinee idol and a reputation as a political miracle worker who had engineered Truman's 1948 upset victory, Clifford would be a competitor for center stage with any president who brought him into the White House. Moreover, Kennedy saw Clifford as someone whose ambition for control would provoke clashes with other advisers and create unwanted tensions in a new administration trying to develop policy initiatives. Kennedy joked that Clifford wanted nothing for his services "except the right to advertise the Clifford law firm on the back of the one-dollar bill." Kennedy had no interest in surrounding himself with yes-men, but he was determined not to be intimidated by veterans of earlier administrations, who found their way into the White House and believed themselves better prepared to lead the country than he was.

As for Neustadt, Kennedy had no plan to appoint him to some White House job that carried greater importance than the one he had held during the Truman presidency in the Bureau of the Budget. A personal encounter with Neustadt in December 1960 leads me to think that Kennedy's decision disappointed him. Neustadt spoke to a lunch meeting of Columbia College faculty, including myself, a new instructor in the history department. I have vivid memories of Neustadt speaking to us at the university's

Faculty Club from notes written on the back of an envelope about a recent meeting with the president-elect on plans for the transition. Listening to the professor, who seemed like a consummate Washington insider, I could not imagine Neustadt not wanting to become part of Kennedy's White House and a contributor to the young president's development of exciting New Frontier programs.

Kennedy saw Lyndon B. Johnson, the vice president–elect, as a prime example of someone convinced he had greater understanding than Kennedy of how to set the direction of the new administration. Vice presidents had traditionally been men of limited influence in the government. John Adams, the first vice president, described the position as "the most insignificant office that ever the invention of man contrived or his imagination conceived." Woodrow Wilson asserted, "In explaining how little there is to be said about it, one has evidently said all there is to say." During the 1960 campaign, when Eisenhower was asked to name a major idea of Vice President Nixon's that he had adopted as president, he replied, "If you give me a week, I might think of one. I don't remember." Eisenhower's comment spoke more about his reluctance to back Nixon's reach for the White House than Nixon's performance as vice president.

As the former Senate majority leader and a domineering personality who hated being anything less than top dog, Johnson arrived in the vice presidency determined to transform the office into something more important than it had been, though he was mindful of how Nixon had used the office to make himself into a credible presidential candidate. Johnson's twenty-seven years in Washington, first as a secretary to a Texas congressman, then as a congressman for eleven years and a senator for twelve, had been a case study in mastering the Capitol's congressional politics and making himself a prominent national figure. Many astute Washington insiders wondered why he would trade his powerful Senate post for the less consequential VP job. But Johnson believed that his days as a dominant majority leader were coming to an end:

If Nixon became president, he would be less cooperative with a Democratic-controlled Senate than Eisenhower had been; if Kennedy won the White House, Johnson assumed that he would be a secondary player with a Democrat as president. Better to be second fiddle to Kennedy as vice president than to be just one of several senators eclipsed by his party's new leader.

But presiding over the Senate and casting rare tie-breaking votes—a vice president's only constitutional duties—was not Johnson's idea of how he would serve in Kennedy's White House. Within days of becoming vice president, he asked Kennedy to sign an executive order giving him "general supervision" over a number of government agencies and directing cabinet secretaries to copy the vice president on all major documents sent to the president. Seeing Johnson's request as the opening wedge in a campaign to make himself a co-president, Kennedy simply dropped the memo in a drawer, where it was left to languish along with Johnson's ambitions for a larger role in the administration. In a 1964 interview with Arthur Schlesinger, Jackie Kennedy recalled Jack and Lyndon together vying or "fencing" with each other about "political things. And I always thought Lyndon was arguing with him or being rude, but Jack was sort of parrying with such amusement, and he always sort of bested him. Lyndon would give a big elephant-like grunt," grudgingly conceding that he was the subordinate in the relationship.

Kennedy's determination not to be the captive of any individual or set of advisers partly rested on a reading of Arthur Schlesinger's three-volume *The Age of Roosevelt*, a reconstruction of the first years of FDR's presidency. Schlesinger's history provided Kennedy with a useful model of how to manage advisers. Roosevelt had encouraged competition for influence among his closest associates. It was his way of compelling them to turn to him for final decisions on all the big issues of his presidency. Kennedy intended to do the same.

Moreover, he was determined to be an activist president, a chief executive who placed "himself in the very thick of the fight," a president unlike Harding and Coolidge in the twenties and now Eisenhower. Ike's contemporary reputation for passivity was over-drawn, but it was the conventional wisdom of 1960, and it was the sort of leadership that Kennedy believed current sentiment wished him to shun. The moment demanded a president more like the two Roosevelts and Truman—someone who would risk "incurring" the "momentary displeasure of the public" by exercising "the full-est powers of the office—all that are specified and some that are not." It was the picture of a president less interested in domestic affairs and day-to-day battles with congressmen and senators to pass legislation than in formulating and executing foreign policies to protect the nation from external threats and find ways to assure immediate and long-term peace.

But whatever Kennedy could take away from the experience of the Roosevelts and Truman to make himself a successful presi-dent, it was clear to him that there were no hard and fast formulas for presidential effectiveness, and that circumstances and his own temperament would determine his fate. George F. Kennan, the diplomat and historian who had designed Truman's containment policy, believed that Kennedy's personal attributes set him apart from other political leaders and gave him the wherewithal to be a great president. Kennan, who agreed to become ambassador to Yugoslavia after discussing the job with Kennedy, described him as "the best listener I've ever seen in high position anywhere." He was not a poseur or classic political glad-hander who loved to hear the sound of his own voice and craved the adulation that was expected in response. "He asked questions modestly, sensibly," Kennan re-called, "and listened very patiently to what you had to say and did not try, then, to tell jokes, to be laughed at, or to utter sententious statements himself to be admired." He did not "monopolize" a conversation but tried to learn from it—"a rare thing among men who have arisen to very exalted positions."

At the start of his term, Kennedy believed that most of those who would serve with him could make a significant difference in shaping his administration. He was determined to seek out the best and the brightest for the top White House jobs and then talk them into taking on the sometimes thankless work that carried risks to their reputations and peace of mind—not to mention the diminished public pay compared with what they could earn in the private sector. But to Kennedy and the people he brought into his administration, public service was a calling that gave them satisfaction and served the national well-being—at least that was the ideal that drew others to work for a president they believed was about to make a meaningful difference in the lives of millions of Americans and people everywhere.

Yet the strengths that Kennedy personally and the men advising him brought to the presidency provided no guarantee of a successful administration. Like all his White House predecessors, Kennedy faced uncertain events that could bedevil his time in office. But like the most successful of these men, Kennedy understood that presidential effectiveness required a capacity for imaginative thinking or flexibility that could help him master unforeseen challenges. He also appreciated that whether he could rise to that standard in every circumstance was an open question.

Yet however wise he might prove to be in response to unexpected events, he never fully faced up to personal limits that threatened to jeopardize his presidency. He was a compulsive womanizer. In the context of the times—a privileged young man growing up in the thirties, forties, and fifties—sexual escapades were not uncommon, especially among social lions in the country's great urban centers, and doubly so for someone as handsome and charming as Kennedy, who had enjoyed standing in Washington for years as perhaps the city's most desirable bachelor. Moreover, he was mindful of his father's reputation as a ladies' man, despite Rose Kennedy's strict religious belief in the sanctity of marriage vows. As Jack was about to marry Jacqueline Bouvier in 1953, his father told one of Jack's closest friends, "I am a bit concerned that he may get restless about the prospect of getting married. Most people do and he is more likely to do so than others."

Yet marrying Jackie, as she was called, was irresistible. From his first meeting with her in 1951, she impressed him as an ideal mate. She came from a Catholic *Social Register* family, and she was clearly beautiful, intelligent, delightfully charming, self-confident, and wonderfully poised. At twenty-two, she was thirteen years his junior and somewhat worshipful of the worldly-wise celebrity senator, who could fulfill whatever fantasies she may have had of a glamorous life with a Washington star. Their marriage in September 1953 at the Newport, Rhode Island, estate of Jackie's stepfather was described in the press as the social event of the year.

More than love drew Jack to marriage, however. As an ambitious politician who had his eye on higher office, Kennedy believed that he had to marry, however much he enjoyed his bachelorhood and freedom to sleep around. Yet he did not see marriage as a deterrent to multiple partners. It had not ended his father's womanizing.

As Joe foresaw, his son's philandering did not subside. He remained as promiscuous as ever. Lem Billings, Jack's closest friend, recalled the "humiliation" Jackie "would suffer when she found herself stranded at parties while Jack would suddenly disappear with some pretty young girl." Priscilla Johnson, an attractive young woman who worked on Kennedy's Senate staff in the fifties and resisted his overtures, described him as "a very naughty boy." Her rejection of his advances made him more respectful of her, and moved him to speak openly to her about women in general and his reckless behavior in particular. "I once asked him," she said, "why he was doing it—why he was acting like his father, why he was avoiding real relationships, why he was taking a chance on getting caught in a scandal at the same time he was trying to make his career take off. He took a while trying to formulate an answer. Finally, he shrugged and said, 'I don't know really. I guess I just can't help it.' He had this sad expression on his face. He looked like a little boy about to cry."

Kennedy's response speaks loudly about the sources of his actions. His frenetic need for conquests was not the behavior of a sexual athlete. It was not the sex act that seemed to drive his pursuit of so many women, but the constant need for reaffirmation, or a desire for affection and approval, however transitory, from his casual trysts. It is easy to imagine that Jack was principally responding to feelings of childhood emptiness stemming from a detached mother and an absent father. As the mother of nine children, including a disabled daughter who followed Jack's birth by only a little more than a year, Rose struggled to attend to her two oldest sons. Busy building his fortune and compelled by business demands to travel widely, Joe was more a family patriarch than a hands-on father closely interacting with his children.

Kennedy's affinity for womanizing found an extended outlet in an eighteen-month affair with a young White House intern beginning in the summer of 1962. The publication in 2012 of *Once Upon a Secret*, a recounting by Mimi Beardsley Alford of her relationship with Kennedy, when she was just nineteen and twenty, provides the most revealing details ever into his sexual escapades. A companion on summer trips with him and on occasional weekends at the White House when Jackie was away, Mimi offered him a reliable retreat from the demands of his duties.

His time with Mimi also appealed to his attraction to risk-taking. Unprotected sex once led to an unrealized scare of pregnancy. Not to mention that Mimi's nights at the White House and presence on trips suddenly made her visible to White House insiders and journalists, which created risks that could have politically touched off a ruinous scandal. Given the assumptions of the time that the mainstream press would not write about a president's sex life, Kennedy was confident that he could avoid any public attack on his character. But he could not be sure. And he was mindful that as president he could be more than embarrassed by accusations of philandering. He either knew or at least understood James Monroe's observation that "national honor is the national property

of the highest value" and that every president is the temporary
custodian of that property.

Alford remembers Kennedy's affair with her as "a reckless desire
for sex." But, according to her account, something else was at work:
Their relationship wasn't "romantic. It was sexual, it was intimate,
it was passionate. But there was always a layer of reserve between
us, which may explain why we never once kissed. . . . In fact I don't
remember the President *ever* kissing me—not hello, not goodbye,
not even during sex." Her function, as she recalls, was to provide
"good company . . . because he hated to be alone but also because
he found a change of pace in someone like me—young full of energy, willing to play along with whatever he wanted."

And what he wanted occasionally was to give expression to "his
demons and . . . his more sinister side." During a visit with him
to Bing Crosby's house in Palm Springs, California, where a rau-
cous Hollywood party was in full swing, Alford refused to try
a "popper," an amyl nitrate capsule that "purportedly enhanced
sex." Unwilling to take no for an answer, Kennedy "popped the
capsule and held it under my nose." She "panicked and ran crying from the room" when the drug caused her heart to race and
hands to tremble. Sometime after, Kennedy was "guilty of an even
more callous and unforgivable episode at the White House pool,"
involving long-time aide Dave Powers: He challenged Mimi "to
give Powers oral sex" while he watched. Although she was "deeply
embarrassed afterward" and Kennedy apologized to both her and
Powers, it did not stop him from asking her at a later date to do
the same thing with his younger brother Ted Kennedy. She angrily
rejected the suggestion. "You've got to be kidding," she told him.
"Absolutely not."

What Alford didn't quite understand was how dependent Kennedy had become on her and how these callous actions expressed
the anger he fixed on her as the object of his dependence. In some
unspecified way, she filled a vacuum in his need for mothering,
affection, and attention as well as for a release from presidential

obligations. "He always asked . . . about her social life." When she told him that she had begun dating someone she thought was "really nice," he said, "Ah Mimi, you're not going to *leave* me, are you?" It was true words spoken in jest. At their last meeting in November 1963, Kennedy embraced her and said, "I wish you were coming with me to Texas," which was ruled out by Jackie's presence on the trip. "I'll call you when I get back," he added. Reminding him that she was getting married, Alford recalled him saying, " 'I know that,' . . . and shrugged. 'But I'll call you anyway.' " He seemed to need her more than she needed him—however sad she was at ending their remarkable affair.

As he entered the presidency, Kennedy had to be mindful of the risks to his public reputation from revelations about his closely guarded health problems and his affinity for extramarital affairs. Attempting to hide these personal weaknesses became an additional challenge to the daunting problems that now awaited him in the Oval Office.

Robert Kennedy: Adviser-in-Chief

As Kennedy searched for the men who would become his closest White House associates, his campaigns and House and Senate service gave him some feel for what he could expect from interactions with advisers. His experience in the military and as a keen observer of public policy had made him more than a little cynical about so-called experts. He was conversant with Irish leader Charles Parnell's counsel: "Get the advice of everybody whose advice is worth having—they are very few—and then do what you think best yourself."

Kennedy saw decision-making and governing as a high-risk business and the advice of the experts as essential, but always to be viewed with measured skepticism. He never doubted that advisers were eager to help and had his best interests at heart, but he also understood that they were competitors for his attention and might be intent on ingratiating themselves with him. Only members of his family could be fully trusted to act in the unselfish interest of their son or brother. It was a caution his father had preached repeatedly to his sons and daughters: Only your closest relatives would put your needs and ambitions first or ahead of theirs.

When Jack first entered politics in 1946 to run for a House seat from Boston, his principal adviser was his father, who financed the campaign and presided over all the major decisions on Jack's candidacy. As Jack put it to a friend at the start of his reach for a House seat, "There goes the old man! There he goes figuring out

the next step. I'm in it now, you know. It's my turn. I've got to perform." Joe believed that if Jack were to get anywhere in politics, he
needed his help. Joe saw the twenty-nine-year-old Jack as "rather
shy, withdrawn and quiet." Joe worried that he lacked his drive and
that of Joe, Jr., who "used to talk about being President." But Joe's
concerns were overdrawn: Winning and being the best were family
watchwords that Jack had fully absorbed. He played competitive
sports—football and yacht racing—with a fierce determination.
He told a reporter in 1960, "The fascination about politics is that
it is so competitive. There's always that exciting challenge of competition."

Yet at the start of his political journey, Jack had little confidence
in his ability to attract voters. He had limited affinity for the false
camaraderie common among Boston politicos. "Backslapping with
the politicians," he acknowledged, was not his idea of a good time.
"I think I'd rather go somewhere with my familiars or sit alone
somewhere and read a book." Mingling with an audience after giving a speech was nothing he relished. And speechmaking was an
ordeal. In 1946, his delivery was stiff and wooden, with no trace
of humor to lighten his remarks. He read his speeches, reluctant to
depart from the text for fear he would lose his place and embarrass himself by unconvincing off-the-cuff reflections. He was like
a novice teacher reluctant to make eye contact with students whose
boredom would be all too apparent in their looks.

But Joe believed that he could create the conditions that would
allow Jack to overcome his natural limitations as a conventional
politician. Just as he had succeeded in building a fortune for himself from small beginnings, so Joe assumed that he could turn Jack
into a winning candidate for high office. But first he had to repair
his own damaged public standing: His poor judgment on British
dealings with Hitler, his affinity for American isolationism, and
his alleged anti-Semitism were liabilities opponents would try to
attach to his son. In 1945–46, he improved his reputation in Massachusetts by promoting well-publicized programs of economic
expansion for Boston and the state.

At the same time, he worked quietly to launch Jack's political career: He persuaded James Curley to give up his Eleventh Congressional District seat to return to his old job as Boston's mayor by offering to finance the campaign and help him pay legal fees incurred fighting fraud charges. When one of Jack's potential opponents for the Democratic Party's congressional nomination in the district and his supporters offered to help Jack run for some undesignated office in the future if Joe would agree to keep him out of the race, Joe dismissed their demand as "crazy," saying, "My son will be President in 1960."

Whatever Jack's limitations—and they included attacks on him as an interloper or carpetbagger who had only recently taken up residence in the congressional district—he had the advantage of being a war hero, which was a compelling asset only months after the end of the war. Polls Joe commissioned advised Jack to emphasize his wartime service. They also indicated that Jack's identification with the storied Kennedys and Fitzgeralds would be valuable in the campaign. The importance of the family connection was not lost on Jack. "My biggest help . . . getting started," he recalled in 1960, was "my father having been known. . . . That's a far greater advantage to me, I think, than the financial. Coming from a politically active family was really the major advantage."

Nonetheless, Joe's investment of more money than normally went into a Boston congressional race gave Jack high visibility among voters. Joe paid for one hundred thousand reprints of a *Reader's Digest* summary of a *New Yorker* article about Jack's wartime heroism in rescuing the crew of his torpedo boat, which was cut in half by a Japanese destroyer. The spending on the offprint was a small part of the $250,000 to $300,000 Joe is supposed to have put into the campaign. It was an unprecedented sum for a congressional primary. "With what I'm spending I could elect my chauffeur," Joe joked. Two political journalists compared the lavish flow of money to "an elephant squashing a peanut."

A campaign aide and veteran of Boston politics compared the nomination fight to a war that required three things to win: "The first is money and the second is money and the third is money." California Democratic Party boss Jesse Unruh later echoed the point in the sixties: "money is the mother's milk of politics," he said. Joe's cash allowed the campaign to hire a public relations firm that saturated the district with billboard, newspaper, radio, and subway ads; flattering pictures of Jack, the war veteran prepared to fight for the needs of district voters, became a familiar sight to its residents. Jack threw himself into the campaign with dawn-to-dusk appearances punctuated by appeals to voter patriotism that warned about the growing communist threat, which a Navy veteran like Jack would know how to fight. The election outcome was all Joe could have hoped for: Jack beat his closest primary rival by a two-to-one margin and defeated his Republican opponent in November by nearly three to one.

Once in office, like many congressmen from safe districts, Jack seemed to have secured a lifetime job. But it wasn't the job he wanted. From the first, he saw the work of a congressman as unrewarding. "We were just worms in the House—nobody paid much attention to us nationally," he complained. He saw service there as a stepping-stone; when the time and circumstances seemed right, he would run for a Senate seat, which in turn would be a prelude to a race for the presidency.

As he reached for the higher rungs on the political ladder, Jack needed more help and someone he could rely on to discuss campaign plans and strategies. His father would supply the money and give him access to people who could promote his candidacy first in Massachusetts and then around the nation. But Joe, already sixty-four when Jack ran for the Senate in 1952, could not be constantly at his side, and he remained enough of a controversial figure that his help was best given behind the scenes.

The alternative was Jack's younger brother, Bobby, who initially did not seem like an appropriate choice as campaign manager and confidant. Born in 1925, he was eight years younger than Jack, and the seventh Kennedy child. Four sisters had followed

Joe, Jr. and Jack. Joe, Sr. was not very invested in Bobby, as the boy was called. Before Bobby could hypothetically run for president in 1964, Joe would be seventy-six. If one of his sons were to reach for the White House in his lifetime, Joe assumed that it would have to be Joe, Jr. or Jack, and Joe, Jr.'s death made Jack the focus of Joe, Sr.'s ambition.

Besides, Bobby was Rose's son. His diminutive size alongside his more robust brothers stirred Rose's maternal protectiveness, as it did with Rosemary, the first daughter and third child, who suffered from birth defects that compelled special attention to her needs. Unlike the somewhat rebellious Jack and Kathleen, the second sister, who married an English nobleman in 1944 outside the faith, Bobby was the obedient, observant child, most devoted to his mother's insistence on fidelity to their Catholicism. He became an altar boy and shared his mother's affinity for regular church attendance. He stammered as a child and seemed most in need of parental reassurance, which he reciprocated with attentiveness to his parents' demands for putting family first. No one was more faithful to Joe's dictates about family loyalty and winning every contest than Bobby. He became a crusader of sorts for the Kennedy reputation. Combative and intolerant of any criticism of his family or of opposition to his father's and brother's ambitions, Bobby became the principal and devoted manager of the family's political campaigns.

In 1952, when Jack ran for the Senate, Bobby was not very close to his older brother. While Bobby was growing up, Jack was already off at Choate and then Harvard. During the war, as Jack served in the Pacific, Bobby attended prep school at Milton Academy. After graduation in 1944, he enlisted in the Navy, where he served until 1946. Bobby joined Jack's congressional campaign that year, but he was not welcomed with open arms. "It's damn nice of Bobby wanting to help," Jack wrote a mutual friend, "but I can't see that sober, silent face breathing new vigor into the ranks." To the contrary, Jack worried that Bobby would do more to antagonize than attract voters. Bobby had a reputation as "kind of a nasty, brutal, humorless little fellow," who was "tough on himself and tough on the people around him." Jack assigned him to East Cambridge, an area unfriendly to Jack's candidacy, where Bobby would be largely out of the way and couldn't do much harm. But his campaign work impressed Jack and the professional party operatives Joe paid to secure Jack's election: Though Jack lost the East Cambridge wards to his opponents, the vote was closer than anticipated, for which Bobby, who had disarmed some of the hostility to the Kennedys by playing softball with local teenagers in a park, got the credit.

While Jack served six years in the House, Bobby was at Harvard, mainly playing football and arguing about politics, the topic of most interest to him after athletics. After graduation in 1948, Bobby followed the Kennedy tradition of foreign travel, and in Jack's footsteps as a correspondent for the *Boston Post*. Eager to go where history was being made, he traveled to the Middle East: Cairo, Jerusalem, and Lebanon. He wrote several dispatches about the emerging state of Israel, which birth he witnessed firsthand as the British prepared to leave Palestine and Arabs and Jews prepared for war. The grand tour took him to Italy and through Belgium, Holland, and Germany. During the trip, the death of his oldest sister, Kathleen, in a plane crash greatly distressed him, as did talk of a war with Russia, which seemed to be imminent, according to the diplomats and military men he spoke with in Vienna. He saw such a conflict, which could well include the use of atomic bombs, as too horrifying to contemplate. But at the end of the year, after he had returned to the States, the arrest and trial of Hungary's Catholic prelate, Cardinal Mindszenty, moved Bobby to advocate "forceful action."

In September 1948, Bobby entered the University of Virginia Law School, where he assembled a respectable record, graduating in June 1951 in the middle of his class. It was a major improvement over Harvard, where his poor academic record had made his Virginia application a near failure. In June 1950, while in law school, he married Ethel Skakel, the daughter of a Chicago coal industry millionaire, and in July 1951, the first of their eleven children was born.

In the fall, he joined his congressman brother on a seven-week trip to the Middle East and Asia. Joe had to talk Jack into inviting Bobby, whom Jack saw as "moody, taciturn, brusque, and combative," and seemed likely to be "a pain in the ass." But family ties trumped personal tensions; Jack felt obliged to put up with his younger brother's irritating qualities. Still, Jack was never entirely happy about his father's directives, whether about familial relations or politics. "I guess Dad has decided that he's going to be the ventriloquist," Jack told a friend about Joe's pressure on him to cast a congressional vote, "so I guess that leaves me the role of dummy." At the same time, Jack never lost sight of how Joe's fame and money had been so instrumental in facilitating his rise in politics. As Jack said later about his career, Joe made it happen.

During the trip, Jack for the first time took a shine to his younger brother, who charmed him with his sense of humor and playfulness by teasing people. As important, they shared a sense of how the United States needed to deal with the emerging Asian countries they visited. They agreed on "the importance of associating ourselves with the people rather than just the governments, which might be transitional." In Indochina, where the French were fighting to hang on to their colonies, Bobby and Jack saw it as a losing cause that ran counter to the will of the masses. They believed that Western nations, including the United States, were putting themselves at a disadvantage in competing with communism by not identifying themselves with the aspirations of the majority of Asians for freedom from colonial control. They took away from the trip a mutual affinity for rescuing emerging nations from the grip of communism. Bobby's religious orthodoxy made him more doctrinaire than Jack, who was more skeptical about church teachings and a little cynical about all institutional affiliations.

Nonetheless, they found enough in common to imagine working together on future political issues.

The moment came in 1952 when Jack ran for the Senate from Massachusetts. His candidacy was something of a long shot; he aimed to unseat the storied Henry Cabot Lodge, Jr., whose family—grandfather and grandson—had held the seat for forty-five of the last sixty years. Moreover, Jack had hardly distinguished himself as a congressman and was reaching for the Senate office when the Republicans seemed likely to win the 1952 presidential campaign. President Truman and the Democrats were in poor standing over the stalemated Korean War, and the Republican nominee, General Dwight D. Eisenhower, an architect of victory in World War II, enjoyed considerable popularity in the state, where supporters sported "I like Ike" buttons.

Although Bobby persuaded Kenneth O'Donnell, his Harvard classmate, to join the campaign, Bobby preferred to remain at the Justice Department, where he was enjoying his work as a prosecutor. Besides, Bobby saw no place for himself alongside his father, who had established himself as the major domo of the operation. Working largely behind the scenes, Joe supervised how his large contributions to the campaign should be spent on publicity and monitored the content of Jack's speeches and campaign messages he considered essential to attract voters. He bypassed campaign finance laws by setting up statewide committees supposedly dedicated to advancing the state's shoe, fish, and textile industries, but which in fact were subterfuges for advancing Jack's candidacy. He lent the publisher of the *Boston Post* $500,000 to keep his paper afloat and to assure an endorsement of Jack, which could attract as many as forty thousand votes. As Jack later told a reporter, "We had to buy that fucking paper."

After the campaign, Lodge complained that he was overwhelmed by Joe's spending. But during the campaign, O'Donnell believed that despite all Joe's money, they were headed for an "absolute catastrophic disaster." O'Donnell saw Joe as out of touch with current state interests and popular ideas and as too strong-minded or dogmatic to see the error of his ways. Joe was a brilliant businessman, but he consistently misread the state of public affairs. His views on foreign policy were particularly out of sync with current majority sentiment. He advocated a return to isolationism in the early Cold War years at a time when the country was receptive to a new internationalism to beat back communism. He also misread the country's mood in the presidential election, believing that Illinois governor Adlai Stevenson, the Democratic candidate, would decisively defeat Eisenhower in 1952.

O'Donnell argued that the only one who could rescue Jack from defeat in the Senate race was Bobby, whose family influence and visceral feel for Massachusetts politics would make the necessary difference. Not only could he bring a required discipline to the campaign, but he was also the only one who could rein in his father and persuade him to support a separate Kennedy operation rather than rely on the state's traditional Democratic Party apparatus.

Bobby initially resisted suggestions that he take over the direction of the campaign, but the role offered an irresistible opportunity to prove himself to Joe and Jack: His mastery of the challenge would show his father and brother that he deserved their regard as one of the family's leading lights. After taking on the assignment, Bobby threw himself into the fight with uncommon energy, working eighteen-hour days and creating an incomparable organization that set up separate offices from the local party ones in every city and town across the state. He blitzed voters with 1,200,000 brochures, which landed on every doorstep in Massachusetts. Two journalists monitoring the campaign described Bobby's organization as "the most methodical, the most disciplined and smoothly working state-wide campaign in Massachusetts history—and possibly anywhere else."

Jack gained an astonishing victory. In a year when the Republicans won all the principal races in the state—a 200,000-vote victory for Eisenhower and a 15,000-vote margin against the sitting Democratic governor—Jack defeated Lodge by 70,000 ballots. Although Jack's personal attributes made the greatest difference in a contest with an opponent who largely shared his views on most foreign and domestic questions, a considerable part of the success belonged to Bobby. It was during this campaign that Jack and Joe realized, as a mutual friend of Jack's and Bobby's said, that Bobby "had all this ability." Jack was greatly impressed by Bobby's achievement, and suddenly Joe discovered that "he had another able son."

In January 1953, Joe used ties to Wisconsin senator Joseph McCarthy, the new chairman of the Permanent Subcommittee on Investigations of the Senate Government Operations Committee, to make Bobby a minority counsel. Although Bobby would take considerable heat for serving with McCarthy, whose ill-founded attacks on political opponents as national security risks enraged many Democrats, Bobby's work stood apart from the senator's. Where McCarthy's probes played fast and loose with the facts and questioned the loyalty of those being scrutinized, Bobby established a reputation as scrupulous about the evidence cited in reports and as loath to accuse anyone of disloyalty to the country. And though he resigned after six months out of disgust with McCarthy's methods, he shared Joe's anxiety about an internal communist threat to the United States. It had been his and Joe's explanation for why Bobby chose to work with a senator who was under such fierce criticism for reckless, unwarranted accusations against Americans with no alleged communist connections.

Bobby's identification with McCarthy added to an already negative picture of Joe and Rose's third son as a carbon copy of his father—difficult and arrogant. And truth be told, he was a "very cross, unhappy, angry young man." Often during evening social engagements at someone's dinner table, he would provoke quarrels with anyone who disagreed with him. Ted Sorensen remembered him as "militant, aggressive, intolerant, opinionated, and somewhat shallow in his convictions . . . more like his father than his brother."

Perhaps Bobby's most distinguishing feature was his indifference to negative opinion about him. In April 1954, when McCarthy began attacking the Department of the Army as infiltrated by communists and his Senate subcommittee investigated the charges, Bobby signed on as the lead counsel for the Democratic minority. It provoked sharp criticism from McCarthy's allies that Bobby was a tool of those who wished to smear the country's best defender against internal subversion. During the hearings, after a heated argument and near fistfight with Roy Cohn, the Republican majority's chief counsel, Bobby wrote the subcommittee's minority report, which roundly condemned McCarthy's accusations and tactics. A Senate censure vote of McCarthy in December 1954 vindicated Bobby and the Democrats who had recommended the reprimand. It signaled the collapse of McCarthy's influence and won Bobby praise for his integrity. With the Democrats having gained control of the Senate in the November elections, Bobby was rewarded with an appointment as the chief counsel of the Investigations Subcommittee.

The McCarthy episode included a striking bit of irony. Democratic senator Lyndon B. Johnson of Texas, who would become a principal antagonist of Bobby in the late fifties and sixties, was an unacknowledged ally of Bobby's in battling McCarthy. As Senate minority leader in 1953–54, Johnson had been under considerable pressure to strike at McCarthy. But he shrewdly cautioned Senate liberals to wait until McCarthy began attacking conservatives and their favored institutions. Consequently, when McCarthy and his top aides hit out at Protestant clergymen and the Army, Johnson moved against them, arranging to have the Army-McCarthy hearings televised in the expectation that they would reveal McCarthy's sinister character and unsavory methods and would undermine his public standing. Moreover, when it came time to appoint a Senate committee to consider McCarthy's censure, Johnson persuaded conservative Democrats and Republicans to serve. It was an effective strategy that made Senate and national audiences receptive to Bobby's report.

Although Jack and Bobby both worked in the Senate, and saw each other on a regular basis—socially, if not professionally—they operated in separate spheres: Bobby focused on domestic corruption as the subcommittee's chief counsel and Jack increasingly concerned himself with foreign affairs. Because Bobby had not spent time abroad, except for his 1948 and 1951 excursions, Joe insisted that he travel to the Soviet Union with associate Supreme Court justice William O. Douglas, who was a compulsive traveler to remote lands. Bobby spent much of the 1955 trip arguing with his Russian hosts until Douglas told him, "You can never argue with these fellows, so why don't we just forget about it," and not spend evenings "trying to convert some guy who will never be converted." By the end of the trip in September, Bobby had become sympathetic to the Soviet masses, especially the various ethnic groups in Central Asia he viewed as victims of communist exploitation that was the equal of anything European nations had done to Asian and Middle Eastern peoples under colonial rule. His sympathy for Soviet citizens, however, did not reduce his distrust of the Kremlin: On his return, he publicly warned against being fooled into concessions to Moscow without reciprocal commitments.

In 1956, he and Jack came together again on trying to make Jack a potential presidential candidate. It was a considerable reach: A first-term thirty-nine-year-old senator with no visible credentials as a national political leader, Kennedy needed to expand his public profile as an attractive personality with whom millions of people could identify and as someone capable of dealing with the communist threat abroad and the racial divide at home.

Jack and Bobby saw the Democratic Party's vice presidential nomination as a giant step toward the Oval Office. Traditionally, the vice presidency had been a burying ground of political ambitions. Vice presidents had come and gone without much public notice of who they were or what they had accomplished. By contrast, Vice Presidents Theodore Roosevelt and Harry Truman had performed admirably after succeeding to the presidency on the deaths of William McKinley in 1901 and Franklin Roosevelt in 1945, respectively, and winning elections in their own rights, in 1904 and 1948. They had significantly increased public regard for vice presidents and how service in the office could prepare someone for the presidency. Moreover, Richard Nixon's three years as an active vice president under Eisenhower from 1953 to 1956 had added to the view that the office counted for something, especially since Nixon seemed likely to run for the higher office in 1960 at the end of a second Eisenhower term or if Ike lost his bid for reelection.

When Jack and Bobby told Joe, who was out of the country, that Jack was about to try for the vice presidential nomination at the Democratic convention, Joe exploded in anger. Initially, he had been ambivalent about the idea. In October 1955, when Eisenhower was recovering from a heart attack and speculation abounded that he would not run again, Democrats believed that they might recapture the White House the following year. In these circumstances, Joe agreed that the vice presidential nomination was worth fighting for. At least, he believed it worthwhile to have Tommy Corcoran, a prominent Washington fixer and friend, approach Lyndon Johnson about making Jack the VP candidate. It was accepted wisdom that Johnson, the Senate majority leader, was running for president, and that he had a better chance than Adlai Stevenson, who had lost to Ike in 1952, to win the White House. Joe told Johnson that if he would publicly announce his candidacy and privately commit to taking Jack as his running mate, Joe would finance his campaign.

But Johnson was reluctant to make a commitment before he was certain that Ike was not running. In addition, he believed it a mistake to get out front and become the object of a stop-Lyndon campaign. No southerner had won the presidency since before the Civil War and Johnson's identification as a Texas segregationist

would make it difficult enough for him to get his party's nomination and win the White House without the additional burden
of having the first Catholic running mate. Memories of Catholic governor Al Smith's losing 1928 campaign suggested that any
Catholic on the ticket could be toxic.

Johnson's rejection of Joe's proposal infuriated Bobby. "He believed it was unforgivably discourteous to turn down his father's
generous offer," Corcoran recalled. Johnson's response was perfectly understandable. He saw Joe's suggestion as more helpful to
Jack's ambitions than his own and he had no interest in being a
stalking horse for Joe's wish to put a son in the White House.
Bobby was so focused on the family's ambitions that he could not
see Johnson's side of the issue. It also did not help that Johnson
would acknowledge Bobby when they passed in the corridors of
the Senate Office Building with a patronizing "Hi, sonny!" Jack
was less frustrated by Johnson's decision, accepting it as nothing
more than self-serving politics.

By the time Jack decided to get in the race for vice president anyway, it was clear that Ike was running and likely to win a second
term and Stevenson was the most likely Democratic nominee. Joe
opposed Jack's decision because he thought that Stevenson would
be badly beaten and that the defeat would partly be blamed on
Jack's Catholicism, which would then damage his chances for a
future presidential nomination.

But Jack and Bobby believed that a vice presidential nomination would give Jack the sort of national visibility that would
propel him into the presidential nomination in 1960. Stevenson,
however, was not sold on Jack as a running mate; he thought he
needed a southerner or a border-state senator. To avoid alienating
the Kennedys, who could be an important source of campaign
financing, Stevenson refused to pick a vice president. Instead, he
told the convention to decide for him. It produced a sharp contest in which Jack ran second to Tennessee senator Estes Kefauver
on the first ballot. Conservative southerners, including Johnson's

Texas delegation, backed Kennedy against Kefauver, who had antagonized them with his support of civil rights. Kennedy took a second ballot lead over Kefauver, 648 to 551½, just 38 short of nomination. Liberals, however, irritated by Jack's failure to vote for McCarthy's censure, fearful that Jack's Catholicism would hurt the ticket, and appreciative of a border-state senator's backing of civil rights, rallied behind Kefauver, who won the nomination on the third ballot.

Commentators agreed that Jack had done himself nothing but good as a national political figure by his performance at the convention, where his attributes as a young war hero and attractive personality impressed many of the delegates. The defeat, however, frustrated Jack and Bobby. Jack was depressed, saying, "This morning all of you were telling me to get into this thing. And now you're telling me I should feel happy because I lost it." At first Bobby was inconsolable, complaining, "They should have won and somebody had pulled something fishy." He lamented their ignorance of convention procedures that could have made a difference, but he filed away the lessons for the future. Moreover, he took solace from the belief that Jack was now "better off," and Stevenson was "not going to win and you're going to be the candidate next time," he told Jack.

Despite his assessment of Stevenson's chances, Bobby accepted an invitation to travel with the candidate and work in the campaign. For Stevenson, it was a way to mend fences with the Kennedys, who resented Stevenson's failure to take Jack as his running mate. From the start, Bobby thought the whole operation was a disaster. Arthur Schlesinger remembered him in the campaign as "an alien presence, sullen and rather ominous, saying little, looking grim and exuding an atmosphere of bleak disapproval." For Bobby, it was a chance to learn how to run a national campaign or, more to the point, how not to run a campaign. Bobby thought Stevenson's style of speaking was terrible, always reading speeches when he should have been speaking extemporaneously. He came across as insincere or too cerebral, too focused on obscure issues instead of people and more devoted to words than actions.

Meanwhile, Jack had also signed on to the campaign, but less out of an interest in advancing Stevenson's candidacy than in becoming better known around the country. Instead of confining himself to Massachusetts and a few of the big swing states, as Stevenson's advisers asked, Jack went into twenty-four states, where he gave more than 150 speeches and charmed everyone with his wit and good looks. He endeared himself to audiences with the observation on his lost fight for the vice presidential nomination: "Socrates once said that it was the duty of a man of real principle to avoid high national office, and evidently the delegates at Chicago recognized my principles even before I did."

Both brothers were becoming nationally recognized figures. In 1957, mass-market magazines featured them in articles: *Look* published a photographic spread about "The Rise of the Brothers Kennedy," and the *Saturday Evening Post* led its September issue with "The Amazing Kennedys." The *Post* saw "the flowering of another great political family" like "the Adamses, the Lodges and the La Follettes." Amazingly, the article predicted that Jack would become president, Bobby the U.S. attorney general, and the youngest brother, Ted, a senator from Massachusetts.

The 1956 ventures were schooling for Jack's and Bobby's 1960 reach for the White House. The campaign began as soon as Stevenson lost to Eisenhower in November 1956, leaving the Democratic nomination for 1960 wide open. Jack broached the subject with Joe on Thanksgiving Day, raising questions about the viability of his candidacy. Joe, ever confident that his son could become president, brushed aside Jack's doubts with assurances that millions of second-generation Americans were waiting for the chance to put one of their own in the White House. Jack didn't need much persuading; he was eager to run and said, "Well, Dad, I guess there's only one question left. When do we start?" He began courting all the party's leaders and all its factions, while denying that he was a candidate, for fear he would stimulate a "stop-Kennedy" counter-campaign.

With no formal organization operating on Jack's behalf, Bobby returned to his job as counsel for the Senate Permanent Subcommittee on Investigations. In September 1959, after more than two and a half years investigating labor corruption, Bobby resigned as counsel to the subcommittee to write a book on the subject, *The Enemy Within* (1960).

By October 1959, however, he was caught up in Jack's campaign. Bobby, Ted Sorensen said, was Jack's first and only choice for campaign manager. He trusted Bobby to "say 'no' more emphatically and speak for the candidate more authoritatively than any professional politician." Bobby at once made clear that he would be a driving force in the operation. He convened a meeting of seventeen principals who were close to Jack and would be at the center of the nominating and national campaigns. At his Cape Cod home, on a beautiful fall day, he pressed everyone to say what was being done to ensure Jack's success. When no one could provide decisive answers, Bobby chided Jack: "How do you expect to run a successful campaign if you don't get started. . . . It's ridiculous that more work hasn't been done already!" Appreciating Bobby's tough-minded realism and signaling the group to prepare themselves to be pushed hard by Bobby, Jack joked: "How would you like looking forward to that voice blasting in your ear for the next six months?" To reassure Bobby and the rest of the gathering that he had been busy laying the groundwork for the campaign, Jack spent the next three hours describing in detail the political challenges they faced in every part of the country to his securing the nomination and winning the election.

Bobby's initial field assignment in November was to sound out Lyndon Johnson on his intentions. Although Johnson denied his interest in running, telling fellow senators that a southerner couldn't get the nomination or be elected, few Washington insiders believed him. Adlai Stevenson, who was angling for a third nomination and refused to acknowledge his own ambition, assumed Johnson was in the chase and Jack thought he was "running very hard." Bobby went to Johnson's ranch in the Texas Hill Country—a show of deference by the thirty-four-year-old Bobby for the fifty-one-year-old majority leader, who saw Jack's candidacy as a premature attempt to bypass senior, more accomplished, and more deserving members of the Senate and party. Johnson began reminding party bosses that the young man had little to show for his thirteen years in the House and Senate. "That kid," as Johnson derisively called him, "needs a little gray in his hair."

During Bobby's visit to his ranch, Johnson denied that he was running and refused to endorse Jack or anyone else, but he did say he opposed a third Stevenson nomination. Eager to put Bobby and his brother in their place, Johnson insisted that Bobby join him in a deer hunt. Johnson correctly assumed that Bobby would be out of his element and forced to take instruction from him. Bobby was, indeed, a reluctant participant. He was knocked to the ground and suffered a cut above the brow from the recoil of a shotgun Johnson had insisted he use. With thinly disguised disdain, Johnson said, as he helped Bobby to his feet, "Son, you got to learn to handle a gun like a man." Bobby understandably took Johnson's remark as a slap in the face—not only to himself but also to his brother, who was daring to oppose what Johnson saw as his greater claim on the presidency. Johnson's behavior reminded Bobby of his earlier refusal to take up his father's "generous offer" and gave birth to a feud that would color all future relations between Johnson and the Kennedys, but especially Bobby.

Johnson's response convinced Bobby and Jack that Johnson was in the hunt and strengthened their determination, as was typical whenever they faced opposition, to pull out all stops in the nomination fight. Bobby, like his father, took any defeat as not only a personal insult but also a demonstration of inadequacy. Any loss was proof of incompetence, of the larger society's view of Irish Catholic inferiority. In 1960, an Irish Catholic running for president was a challenge to the unacknowledged hierarchy of white Protestant America. Many in the country saw the Irish as only one cut above African Americans, whose inferiority was written into law across the South.

But as Johnson understood, in 1960, southern whites, like Catholics, were also unwelcome participants in the reach for the White House. True, Harry Truman, with his border-state twang and indelible middle American qualities—the bow tie, hair parted in the middle, and blue serge suits—had diminished some of the prejudices about who deserved to hold the highest office. But Johnson's candidacy, like Kennedy's, was a call to reshuffle the accepted standards for access to the Oval Office. They were implicit allies in trying to break down old barriers. But until they sorted out who would take the lead in redefining the country's political standards, they were bitter rivals.

And for Bobby Kennedy, with Jack's tacit approval, so was anyone who stood in the way of his brother's White House campaign. The first demonstration of their hardball approach to winning the nomination came in Wisconsin, where they competed with Minnesota senator Hubert H. Humphrey in the Democratic primary. Primaries in 1960 were no surefire route to the party's nomination. There were too few of the state contests to ensure anyone the prize. But for Jack, they were an essential demonstration to party bosses and convention delegates that he could win sufficient Protestant votes to become a viable national candidate.

Understanding the importance of the primary in Wisconsin—a state with a large number of Protestant as well as Catholic voters—Bobby spared no effort to win. He gave a demonstration of what was ahead in Wisconsin when at the end of 1959 he pressed Governor Michael DiSalle of Ohio to be the first governor to come out for Kennedy. When DiSalle resisted, Bobby gave him what John Bailey, the Democratic National chairman, who attended a meeting between them, called "a going-over," the likes of which shocked Bailey. It consisted of bare-knuckle threats to DiSalle's political future. But it worked, and DiSalle endorsed Jack's candidacy in January 1960 as the campaign for Wisconsin began.

Bobby's successful hard-nosed tactics encouraged him to remain aggressive. As campaign manager, he blitzed the state with Kennedy operatives; family members and hired guns seemed to be everywhere, talking up Jack to anyone who would listen. Humphrey said he felt like "an independent merchant competing against a chain store." Bobby brought Paul Corbin into the campaign—a slick political Houdini whose mantra was winning, regardless of how it was done. Humphrey complained about Bobby's "ruthlessness and toughness"—specifically, what Bobby encouraged Corbin to do: anti-Catholic tracts sent to Catholic households that were calculated to anger Catholics and stimulate them to vote for Kennedy. Corbin also spread rumors that the corrupt Teamsters union was campaigning for Humphrey.

Although Jack would carry the state, it was a Pyrrhic victory. Jack's margin was too small to be considered decisive, and commentators immediately described his success as the result of Republican Catholics coming to Jack's rescue. Jack muttered: "Damn religious thing." One of his sisters asked: "What does it all mean?" Jack replied: "We've got to go to West Virginia and do it all over again."

West Virginia, only 3 percent Catholic, became an acid test of whether Jack could win Protestant votes. The primary would make or break Jack's candidacy. The Kennedys spared neither money nor scruples to win: West Virginia was notorious for vote-buying, and relying on Joe's advice and money, Jack's campaign paid top dollar to the party's county bosses to ensure strong majorities for him. Humphrey didn't blink at the local requirement for vote-buying, either. But "our highest possible contribution was peanuts compared to what they [the county chairmen] received from the Kennedy organization," he said. He was right. Where Humphrey spent a total of about $25,000 on his campaign, Jack's TV ads alone came to $34,000.

Because they couldn't be sure that the local party operatives could be relied on to produce promised votes, the Kennedy campaign also assumed that it had to motivate voters to go to the polls for Jack. Joe, Bobby, and West Virginia Democrats more familiar with local attitudes urged the strongest possible identification with Franklin Roosevelt's memory and the New Deal. Because the state still struggled with pockets of poverty and prided itself on traditional values and patriotism, the campaign declared itself for "food, family and flag."

Jack's campaign between April 5 and May 10, the date of the primary, emphasized his determination to bring West Virginia families out of poverty with federal programs, promising to put West Virginia "on the top of my agenda at the White House." The campaign also made sure that voters would know about Jack's heroic Navy service and his brother's sacrifice in a suicide mission. Nor did Jack neglect the religious issue, which posed a serious threat to his election. He implored audiences not to make religion a consideration in their choice of a candidate. After all, he said, no one asked his or his brother Joe's religion when they risked their lives in combat. A beautifully crafted television documentary about Jack and his family informed voters about his merits as a candidate and heightened his appeal as someone deserving of their support for the presidency. It was an early use of the TV medium as a vehicle for reaching lots of people who normally paid scant attention to politics.

Still, the campaign, led by Bobby, who wished to ensure a landslide, believed it essential not only to broaden and deepen Jack's appeal, but also to give voters reasons to vote against Humphrey. Bobby saw an opening in allegations about Humphrey's lack of a military record. Partly responding to reminders to voters of Joe Kennedy's sympathy for Britain's prewar appeasement policy and attacks on Jack as a rich man's son who was buying the election, Bobby pressured Franklin Roosevelt, Jr., who had come into West Virginia to identify Jack with FDR's New Deal, to alert voters to

Humphrey's absence from the war against the Axis. The Humphrey campaign protested against implicit allegations of draft-dodging. In fact, a 4-F classification had deterred Humphrey from serving. Jack repeatedly decried the use of such tactics, but more to remind voters that Humphrey had not served than to discourage them from taking it into account. Humphrey said later, they "never shut FDR, Jr. up, as they easily could have."

As far as Bobby was concerned, dirty tricks were a justifiable response to dirty tricks. The Humphrey people were playing the religion card and so Bobby had no problem with Paul Corbin's recruitment of Catholic priests to knock on doors in Catholic areas to get out the vote. They convinced seminarian volunteers helping in the campaign to dispose of their frocks when visiting Protestant households to solicit Kennedy votes.

Jack decisively won in West Virginia, 60.8 to 39.2 percent, and Humphrey announced his withdrawal from the nomination fight. As a goodwill gesture that could soften Humphrey's sense of loss and deter him from throwing his support in the convention to Lyndon Johnson or any other potential rival, and a bow toward party liberals, who were Humphrey's strongest backers, Bobby went to see Humphrey and his wife in their hotel room to praise him and his contributions to the party's domestic record. It was also a calculated step toward persuading Humphrey delegates to back Jack instead of Adlai Stevenson, the liberal alternative.

Jack and Bobby went to the convention in Los Angeles in July hopeful but uncertain about gaining the prize. Johnson had emerged as their principal rival. After their encounter with him in 1956 and Bobby's humiliating visit to his ranch in 1959, the Kennedys were angry at Johnson and privately denounced him as a "chronic liar" whose pronouncements on his noncandidacy were nothing more than a political ploy.

The antagonism intensified when Texas governor John Connally, Johnson's campaign manager, used a press conference to describe Kennedy as suffering from Addison's disease and unfit to serve as president. Johnson himself attacked Joe Kennedy as a "Chamberlain umbrella policy man." Before traveling to Los Angeles, Johnson conferred with Eisenhower at the White House. He urged the president to come out against Kennedy's nomination, describing him as "a dangerous man" whose lack of foreign policy experience could jeopardize the country's security. When Bobby learned of Johnson's attacks on his brother, he exploded at Bobby Baker, Johnson's Senate aide: "Lyndon Johnson has compared my father to the Nazis, and John Connally . . . lied by saying that my brother was dying of Addison's disease. You Johnson people are running a stinking damned campaign and you'll get yours when the time comes."

It was an empty threat. As soon as Jack won the nomination, he decided to invite Johnson to join the ticket. Of all the several possible running mates, Johnson seemed the most likely to help Jack get elected. Moreover, he was every bit as qualified as Jack or any of the other senators—Humphrey, Stuart Symington of Missouri, or Henry "Scoop" Jackson of Washington—to step in as president should that unlikely development occur. The South was Jack's weakest electoral region: His Catholicism and identification with civil rights causes, which he had encouraged to ensure support from the party's liberal base, made him vulnerable in the old Confederacy. Johnson seemed able to help hold some, if not all, of the southern states, especially Texas with its largest number of southern electoral votes. Finally, as a very seasoned politician with twenty-five years in Washington as a congressional aide and House and Senate member, Johnson had the potential to become a valuable Kennedy adviser.

Bobby was unhappy about the choice. He had been assuring liberals at the convention that Johnson was not in the running for the second spot. It may have been good pre-nomination politics to keep the liberals, who saw Johnson as too conservative, in line. But Bobby's personal antagonism toward Johnson gave a ring of truth to his assurances. Moreover, Bobby's later recollections of how

Johnson joined the ticket echoed his resistance to the selection. He remembered that Jack offered Johnson the nomination in the belief that he would turn it down. According to Bobby, it was Jack's way of stroking Johnson's outsized ego and winning his goodwill for the national campaign.

As word spread at the convention that Jack was offering Johnson the vice presidency, Bobby told liberals that they were shocked when Johnson accepted. But Jack and Bobby may have invented this story to appease Johnson's opponents, who threatened to fight Johnson's selection on the floor of the convention. To soften liberal resistance, Jack and Bobby argued that Johnson as vice president would make him easier to control than having him as majority leader, where he could play havoc with a liberal legislative program. The Kennedys also told liberals that once they learned how resistant they were to Johnson's selection they tried to persuade him to serve as the party's national chairman instead. But Bobby reported that when he made Johnson the offer, he seemed on the verge of tears and refused. He declared himself ready to fight for the VP nomination if Jack would support him. Bobby said that he then acknowledged Jack's willingness to stay with his decision to put Johnson on the ticket.

The truth of exactly what happened will never be known. It is plausible that Bobby pressed the case against Johnson with his brother, but in a show of determination to follow his own counsel and set a pattern for how he would function as president, Jack rejected Bobby's advice. It was no measure of Jack's continuing regard for Bobby's importance in his campaign and future influence at the White House.

Once Jack had the nomination, it was a foregone conclusion that Bobby would run the campaign. Bobby's operation in 1952 had made "every politician in Massachusetts . . . mad at Bobby," Jack said, "but we had the best organization in history." Bobby intended to replicate the experience in 1960. His technique was to demand unrelenting effort from everyone without regard for their needs: "Gentlemen, I don't give a damn if the state and county organizations survive after November, and I don't give a damn if you survive," he told New York's feuding Democrats. "I want to elect John F. Kennedy." He had "all the patience of a vulture," a journalist said. He sent Paul Corbin, a political aide, to upstate New York to bring the warring factions into line. Like Bobby, he had few scruples about winning: He promised the ambassadorship in Quito, Ecuador, to three people until someone told him: "There were nineteen republics down there, and he could at least spread his offers around."

Bobby showed no deference to anyone: Governors, mayors, congressmen—they were all subjected to the same insistent pressure for results in advancing Jack's candidacy. Adlai Stevenson privately called Bobby the "Black Prince." Eisenhower, who saw Jack as the self-indulgent son of a rich man, called Jack "Little Boy Blue." He would leave the dirty work of the campaign to his brother, just as Ike, who knew something about relying on a subordinate to sling mud and bring people in line, had left it to Nixon. Eisenhower described Bobby as "that little shit." Former president Truman told Bobby to moderate his behavior, and that being a "son of a bitch" made people angry and might be doing more to undermine than serve the campaign.

Bobby was not convinced. He continued to be a son of a bitch, mindful that Truman had succeeded by being the same. "I'm not running a popularity contest," he protested. Getting people mad at him went with the job of being campaign manager. "It doesn't matter if they like me or not. Jack can be nice to them," Bobby said. A journalist called it a "sweet-and-sour brother act, Jack uses his charm and waves the carrot and then Bobby wades in with the big stick."

The contrast between the two brothers was summed up by Eleanor Roosevelt's evolving view of Jack. He traveled to Hyde Park, New York, the Roosevelt home, to disarm her doubts about his liberalism. His charm campaign worked. She came away from the meeting convinced that he was someone capable of learning. "I liked him better than I ever had before," she wrote a friend, "because he seemed so little cock-sure, and I think he has a mind that is open to new ideas." She did not hold Bobby in similar regard. But he stayed away from her, and like Truman, she muted whatever doubts she had about the Kennedys, because the idea of Nixon as president was more than either of them could bear. "I never liked Kennedy," Truman said. "I hate his father. Kennedy wasn't so great as a Senator. . . . However, that no-good son-of-a-bitch Dick Nixon called me a Communist and I'll do anything to beat him." Mrs. Roosevelt felt the same way.

For all his toughness, Bobby had a deeply ingrained sense of morality that occasionally trumped his ruthless side or eclipsed his political calculations. Segregation offended him. Although he understood that they had to tread lightly in dealing with the white South, he considered its treatment of blacks morally and legally unjust and reprehensible. He insisted on a strong civil rights plank in the party's platform, and when he visited Georgia during the campaign, he refused to attend a dinner unless blacks were included. Political considerations were not absent from Bobby's actions. In taking a stand on civil rights, he appreciated that most southerners would find it difficult to vote for a Republican president and that black votes in northern states could make a difference in places like Illinois, Ohio, and Pennsylvania.

Yet political assessments could not trump visceral antagonism to racial abuse. The imprisonment of Martin Luther King in October with a sentence of four months' hard labor for a minor traffic infraction confronted Jack and Bobby with a moral dilemma. Mrs. Coretta Scott King, who was five months pregnant and terrified that her husband would be killed in prison, asked Harris Wofford, Jack's civil rights representative, to intervene. When Jack called Mrs. King to sympathize with her and send an indirect message to King's jailers, Bobby told the campaign aides he thought responsible for Jack's gesture that three southern governors would now probably support Nixon and that would probably cost them the election. When he learned about the trumped-up charge justifying King's sentence, however, Bobby called the responsible judge to complain, telling him as a lawyer who knew something about the rights of defendants that it was "disgraceful" to deny King the right to make bond. The judge released King the next day, with the unpredictable result that black voters moved decisively into Jack's column. More was at work here, however, than Bobby's indignation. Jack had made the initial contact to the judge through Georgia's governor, Ernest Vandiver, who was eager to see Kennedy become president. After a mutual friend of the judge and the governor called the judge, he agreed to release King if Jack or Bobby gave him cover with a phone call.

Not everything worked out as the Kennedys had hoped, though, and Jack and Bobby had their differences in the course of the campaign. When a reporter on the campaign trail who provoked Bobby's wrath complained, "That brother of yours has no manners," Jack replied, "Ignore him." Despite the problems, at the end of the day they won, however close the margin. And Jack knew that Bobby, who had given unstintingly to the campaign, was a driving force in their success. "He's the hardest worker. He's the greatest organizer," Jack said later. "Bobby's easily the best man I've ever seen."

With Jack exhausted after the election and taking a little time to recuperate, Bobby became the principal manager of the transition. "It is Bobby . . . who will be the new man-to-see in Washington," *Newsweek* reported on November 21, sixteen days after the election. But just what place Bobby would hold in the new administration became a topic of speculation. He seemed a logical choice for chief of staff, but two considerations worked against it: Bobby did not want to work directly under Jack; nor did it appeal to Jack. As Jack was well aware, Bobby could be abrasive, and so close a working relationship seemed bound to produce tensions and arguments that both of them wished to avoid. Second, making Bobby the chief meant bypassing other aides who felt they had a claim on the job. So it seemed best to name no one and let Jack be his own chief and use the people around him on a day-to-day basis as he saw fit.

On November 19, Jack casually mentioned to a *New York Times* reporter the possibility that he might ask Bobby to become attorney general. It was a calculated conversation, an attempt to sound out press and public reaction, or more to the point, to prepare the press and public for Bobby's elevation to a central place in the administration. The response was swift and uniformly negative: The *Times* dismissed the idea as a politicization of an office that should be above politics. Bobby had been Jack's "political manager" and "no matter how bright or how young or how personally loyal" he was to his brother, the attorney general's office "ought to be kept completely out of the political arena." Others warned that Bobby's appointment would provoke charges of nepotism and complaints about making someone so young—Bobby was only thirty-five— and inexperienced in the practice of law the chief legal officer of the United States. Of course, every high administration appointment partly rested on rewarding political supporters. But the need for the appearance of impartiality at the Justice Department made Bobby's selection a serious political liability.

In sounding out opinion, Jack was doing more than testing the waters; he was preparing the ground for Bobby's appointment. Bobby encouraged speculation that he was interested in getting out of Washington and Jack's shadow by considering a run for Massachusetts governor or becoming a university president. Or if he remained in government, he thought it would be best to steer clear of the attorney general's job: He said it "would be a very bad mistake." As attorney general he would become the administration's advocate on civil rights, which would enrage southerners and turn them against Jack, who inevitably would be identified with his brother. If he remained in Washington, Bobby thought it should be as an undersecretary of defense or an assistant secretary of state. But his credentials for either of those jobs were even less than those for becoming attorney general.

All this, however, was posturing on Bobby's part. After his central role in getting Jack to the White House, it was inconceivable that he was going to take his distance from the Oval Office—either by leaving town or by serving in a subcabinet position at Defense or State. As Connecticut governor Abe Ribicoff told Jack, "I have now watched you Kennedy brothers for five solid years and I notice that every time you face a crisis, you automatically turn to Bobby. You're out of the same womb. There's empathy. You understand one another. You're not going to be able to be President without using Bobby all the time." The talk about Bobby leaving Washington or becoming a subordinate national security official reflected Jack's sensitivity to making a close relative a key member of the administration. It would be an unwelcome start to his administration.

To disarm criticism of Bobby's appointment, Jack leaked stories that he had offered the attorney general's post to Ribicoff and Adlai Stevenson. But it was done with the understanding that neither of them would take it. To further blunt complaints about Bobby's appointment, Jack and Bobby encouraged the belief that their father forced the decision on them. They even went so far as to ask Clark Clifford to see Joe in New York to persuade him to drop the idea. The request amazed Clifford, who thought it "truly a strange assignment." Joe, who was in on the ploy, played his part to the hilt, telling Clifford that he would not back down: "I am going to see to it that Bobby gets the same chance that we gave to Jack."

Bobby played his part as well, telling a friend that he was going to see Jack and tell him that he wouldn't take the job. "This will kill my father," Bobby said. Jack and Bobby met over breakfast and Bobby brought a friend, the reporter John Seigenthaler, along. Seigenthaler witnessed the little drama Bobby and Jack had prepared. Jack insisted that Bobby accept the appointment, explaining that he needed someone around who would tell him "the unvarnished truth, no matter what." It echoed what Jack had already told former secretary of state Dean Acheson, that his cabinet members would be strangers to him and he needed "someone whom he knew very well and trusted completely with whom he could just sort of put his feet up and talk things over." Finally, to demonstrate how reluctant he and Jack were, Bobby recounted how Jack joked that Bobby shouldn't "smile too much" when they announced his selection or else the press would "think we are happy about the appointment." Jack then told his friend Ben Bradlee, the editor of *Newsweek*, that he thought about announcing his decision by opening the front door of his house at two in the morning, when no one would be outside, and whispering to the empty street, "It's Bobby."

Was any of this to be believed? Or, to put it another way, were Jack and Bobby capable of so elaborate a cover-up? Unquestionably. They had effectively muted public knowledge of Jack's health problems, which had required a more elaborate deception than disguising their determination to make Bobby attorney general. With the exception of Joe, Rose, and Bobby, no one knew all of Jack's medical history—not even his various doctors, who were consulted about individual problems yet never the whole array of difficulties.

The reality of how Jack and Bobby were massaging the truth is evident in a note Bobby sent to the columnist Drew Pearson on December 15, one day before the breakfast meeting at which Jack supposedly persuaded Bobby to head the Justice Department. The letter, which is in Bobby's attorney general papers at the John F. Kennedy Library, was meant to disarm Pearson's doubts about him taking the job. Pearson had warned Bobby that he would be forced to deal with "so many controversial questions with such vigor that your brother in the White House would be in hot water all the time." Bobby's letter states: "I made up my mind today and Jack and I take the plunge tomorrow. For many reasons I believe it was the only thing I could do—I shall do my best and hope that it turns out well." What Seigenthaler witnessed the next day was a charade meant to have him reveal the anguish Jack and Bobby supposedly suffered in installing Bobby at Justice. Even without the letter to Pearson, one can hardly believe that Jack and Bobby would have included a journalist in their deliberations if they were actually settling so vital a question as Bobby's future role in the administration.

Bobby's appointment put him at the center of the new administration, where he would become the president's leading adviser on every major question. "Who would you say . . . the President depended and relied upon most?" Secretary of the Treasury C. Douglas Dillon was asked. "Obviously the person he depended upon the most was the Attorney General," Dillon replied. "The Attorney General was in and out of the White House a great deal. He talked with the President all the time, and the President relied heavily on him in all sorts of matters, from high policy to personal matters."

o o o o

"A Ministry of Talent"

The day after the November election, the exhilaration Kennedy, Bobby, and those closest to them felt at having won the country's greatest political prize gave way to the hard work of building an administration. Kennedy saw himself as a bit at sea. He knew he wanted Bobby at his side in some capacity, but everything beyond that was pretty much a blank slate. During a post-election vacation in Palm Beach, Florida, he complained to his father, "Jesus Christ, this one wants that, that one wants this. Goddamn it, you can't satisfy any of these people. I don't know what I'm going to do." Joe was not very sympathetic: "Jack," he said, "if you don't want the job, you don't have to take it. They're still counting votes up in Cook County."

Kennedy's highest priority was choosing a national security team. The Cold War with the Soviet Union and fears that the United States had fallen behind in the arms race with Moscow, especially in the development of intercontinental ballistic missiles, a state the Kennedy campaign described as the "missile gap," had made initiatives that might reduce chances of armed conflict with Moscow Kennedy's greatest concern. A conflict between pro-communist and pro-Western forces in Laos, Soviet talk of expelling the West from Berlin, and a Soviet foothold in Cuba threatening expanded communist control in the hemisphere heightened Kennedy's fear of a crisis that could provoke a war. His worst nightmare as president was a nuclear conflict that would kill millions of people and scar parts of the earth for as far into the future as anyone could see.

Compounding Kennedy's worries about communist challenges that could lead to armed conflict was his limited confidence in America's military chiefs and uncertainty about finding wise national security advisers. His memories of the military in the southwest Pacific during World War II and pressure to appoint militant anticommunists to Defense and State, who might be more inclined to use nuclear weapons than he would, troubled him. For example, he remembered General Douglas MacArthur's reputation as a brilliant Pacific commander as overblown. He thought that MacArthur's island-hopping campaign had cost too many lives and had prolonged the Pacific fighting. He shared the average GI's contempt for MacArthur as "Dug-out-Doug," the general who re-fused to emerge from his "dug-out in Australia." The professional Army and Navy officers—West Point and Annapolis graduates—gave him great pause. He recalled "ferrying quite a lot of gener-als around," who thought they could advance further through the ranks by being seen in a PT boat—a symbol of courageous deter-mination after a squadron of the crafts had carried MacArthur to safety from the Philippines. He saw these armchair warriors as the architects of what he called "this heaving puffing war machine of ours." He had also been critical of the average fighting man. He became "cynical" about them during his time on the front line, describing them as prone to excessive "bellyaching and laying off."

The Soviets might have taken satisfaction, or might have been understandably frightened, to know that Kennedy distrusted America's military establishment as being too enamored of nuclear weapons and readiness to use them. As worrisome, General Lyman Lemnitzer, the chairman of the Joint Chiefs of Staff, reciprocated the president-elect's doubts: Could so young a man with such lim-ited military experience serve as commander in chief?

Lemnitzer was a West Point graduate who had made his way up the ranks as a member of Eisenhower's World War II staff, helping plan the successful invasions of North Africa and Sicily. He commanded an infantry division in Korea, became Army chief of staff in 1957, and was Eisenhower's choice for head of the Joint Chiefs in 1960. The sixty-one-year-old general was little known outside of military circles, but his fellow soldiers remembered him as standing six foot two and two hundred pounds, with a large "bear-like" frame, "booming voice," and "a deep infectious laugh." His passion for golf, notable for smashing "a golf ball 250 yards down a fairway," had partly made him an Eisenhower favorite. More important, he mimicked Ike's talent for maneuvering through Army and Washington politics. Like Eisenhower he was not bookish or particularly drawn to grand strategy or big-picture thinking. He was the nuts-and-bolts sort of general who made his mark managing day-to-day problems.

Kennedy knew Lemnitzer only from congressional hearings, but since Lemnitzer was a career officer and the leading military official left over from the Eisenhower administration, Kennedy saw him as the representative not only of conventional Army thinking but also the embodiment of Ike's affinity for reliance on massive retaliation with nuclear weapons, which Kennedy believed could engage the United States in a suicidal conflict known as MAD, mutually assured destruction. Schlesinger, who was close enough to Kennedy to know what he thought of Lemnitzer, privately described the Joint Chiefs chairman as "that sweet but dopey man." Eisenhower had every confidence that he could count on Lemnitzer's deference, but Kennedy doubted that Lemnitzer would be as respectful toward him.

Kennedy was right to be suspicious of Lemnitzer and all of Ike's chiefs. After Kennedy's victory, Lemnitzer's briefing on military affairs deepened the mutual skepticism about their respective capabilities and good judgment. Lemnitzer questioned the new president's qualifications to manage the country's national defense. Privately, he lamented the fact that Eisenhower's departure meant there would no longer be "a Pres with mil exp available to guide

JCS." He later said of Kennedy, "Here was a president with no military experience at all, sort of a patrol boat skipper in World War II." Lemnitzer took pains to fill the vacuum with a detailed briefing about national emergency procedures or how the president should respond to a foreign threat. Kennedy's preoccupation in that meeting with worries about having to make "a snap decision" on launching a nuclear response to a Soviet first strike added to Lemnitzer's belief that Kennedy didn't sufficiently understand the challenges before him.

The real issue between them was not Kennedy's inexperience and limited understanding of how to ensure the country's safety but Kennedy's doubts about the wisdom of using nuclear arms and the military's excessive reliance on them as a deterrent against communist aggression. The affinity for nuclear weapons rested on recent experience in World War II and Korea. Victory over Germany and Japan had been the product of total war—an all-out use of America's industrial and military might against enemies whose defeat could only come with unconditional surrender. By contrast, the limited war in Korea had produced a stalemate that had left the military and the larger American public frustrated. Eisenhower's doctrine of massive retaliation rested on the view that success in combat required the utilization of all the country's power or at least the need to deter adversaries from acts of aggression, by encouraging the belief that risking war with the United States would bring the sort of destruction that had befallen the Axis powers. Kennedy's reluctance to risk the carnage a nuclear war would bring by threatening all-out conflict put him at odds with the Chiefs he had inherited.

As Lemnitzer would soon find out, he was now part of an administration that had diminished regard for the military's judgment on defense questions. It was not long before newspaper stories began describing how the White House bypassed and ignored the Chiefs in reaching decisions on national security matters formerly under their principal control. One critic of Kennedy's greater concentration of military issues in civilian hands than had been the case with Eisenhower called this a "yo-yo form of government." But Kennedy was less interested in how critics described his management of the Chiefs than in holding them in check.

Admiral Arleigh Burke, the fifty-nine-year-old chief of naval operations, a Naval Academy graduate with thirty-seven years of service, was a combat veteran of World War II and Korea. He was an early problem for Kennedy. As an anti-Soviet hawk, he believed that U.S. military officials needed to intimidate Moscow with threatening rhetoric. Burke "pushed his black-and-white views of international affairs with bluff naval persistence," Schlesinger said. Kennedy had barely settled into the Oval Office in January when Burke proposed publicly, in the words of Arthur Sylvester, Kennedy's new Defense Department press officer, to attack "the Soviet Union from hell to breakfast." When Sylvester brought the speech to Kennedy's attention, he ordered Burke to back off. Although Burke promised to write a new speech, he called Sylvester an "old son-of-a-bitch" and leaked the story to the *New York Times*, which provoked Senate hawks on the Armed Services Committee to criticize the White House for muzzling an admiral who was trying to warn Moscow against reckless acts of aggression.

Burke didn't stop there. He went to see Secretary of Defense Robert McNamara. "Mr. Secretary, I'm quite a bit older than you are," he began. "I've been in this jungle of Washington for a good many years. I would like to explain some things that you aren't going to like. But I'd like to have you listen, and as a matter of fact, you must listen. I've got to get this off my chest." Burke warned him against destroying what others had spent lifetimes building. He urged him not to focus on day-to-day actions but to think about the long term and to plan for what U.S. military power would look like in a decade. "Was he receptive to this?" an interviewer asked him. "Yes, yes," Burke answered. "And thereafter I went down to see him two or three times a week in the early morning, or he came up to see me, until I left." New to his job and eager not to antagonize the Chiefs, McNamara was being polite. But he wasn't very receptive to Burke's advice. And Kennedy wanted no part of any suggestion that the military would have free rein to say and act as they thought best in either the short or the long run. Soon after he heard about Burke's speech, Kennedy ordered all active-duty officers to clear speeches with the White House, and told Sylvester, "The greatest thing that's happened in the first three months of my administration was your stopping the Burke speech."

Kennedy's biggest worry about the military was the freedom of field commanders to launch nuclear weapons without explicit White House permission; it risked the devastation of Western Europe and cities in the United States. At the end of January, ten days after becoming president, Kennedy learned that "a subordinate commander faced with a substantial Russian military action could start the thermonuclear holocaust on his own initiative." A top Kennedy aide recalled that "we became increasingly horrified over how little positive control the President really had over the use of this great arsenal of nuclear weapons." Kennedy pushed the military to replace Eisenhower's strategy of "massive retaliation" with what he called "flexible response"—a strategy of calibrated force other than strict reliance on nuclear weapons; the strategy had been described by General Maxwell Taylor in a 1960 book, The Uncertain Trumpet. But the stalemate in the Korean War frustrated military chiefs. They preferred the use of atomic bombs, as General Douglas MacArthur had proposed, to win a decisive victory. They were reluctant to have Kennedy, a president with so limited military experience, assume exclusive control over deciding when a first nuclear strike would be appropriate.

The NATO commander, General Lauris Norstad, and two Air Force generals, Thomas Power and Curtis LeMay, offered stubborn opposition to White House directives reducing their choice of when to go nuclear. The fifty-four-year-old Norstad had a reputation for being fiercely independent. During a visit to NATO headquarters by the secretaries of state and defense, they asked him to whom he had a primary obligation—the United States or the European alliance members. He saw the question as "challenging my loyalty. My first instinct was to hit him," McNamara, Norstad recalled. Instead he "just stood there and . . . tried to smile and cool off a bit. . . . And I said, 'Well gentlemen, I think that ends this meeting.' Whereupon I walked out and slammed the door." Norstad made his reluctance to concede Kennedy's ultimate authority so clear that the president's national security adviser urged Kennedy to tell Norstad that the president "is boss."

Power confided to a journalist his worries over civilian control of his freedom to use America's ultimate weapons. The fifty-five-year-old Power had joined the Army Air Corps after high school in 1928 and worked his way up through the ranks. During World War II, he had served in North Africa, Italy, and the Pacific, where he led the first firebombing raid on Tokyo in March 1945. In 1957 he became head of the Strategic Air Command. He had unqualified faith in the use of airpower and was contemptuous of anyone urging restraint in a war with Russia. "Why are you so concerned with saving their lives?" he asked the authors of a RAND Corporation study counseling against attacks on Soviet cities at the start of a conflict. "The whole idea is to kill the bastards. At the end of the war if there are two Americans and one Russian left alive, we win." LeMay, his superior, described Power as a "sadist" and "not stable."

The fifty-four-year-old LeMay, who had the nickname "Old Iron Pants," was not much different. He shared Power's faith in the untrammeled use of airpower to defend the national security. The child of a working-class Ohio family, whose father was a harsh taskmaster, LeMay imbibed his father's insistence on strict discipline and the value of dealing harshly with opponents. After earning a degree in civil engineering at Ohio State University, LeMay joined the Air Corps in 1928 and, like Power, became a pioneer in developing an Army air wing that mounted all-out assaults on Germany and Japan during World War II. The burly, cigar-chomping LeMay believed that the United States had no choice but to bomb its foes into submission. He had no qualms about striking at enemy cities, where civilian populations would pay the price for their governments' misjudgments in fighting the United States. He was the principal architect of the incendiary attacks on Tokyo by B-29 heavy bombers that destroyed most of the city and killed more than two hundred thousand Japanese. He was convinced that the air raids shortened the war.

During the Cold War, LeMay was prepared to inflict greater damage on the Russians with nuclear bombs in a preemptive first strike. He dismissed civilian control of his decision-making, complaining that the White House had a phobia about nuclear weapons and privately asking, "Would things be much worse if Khrushchev were Secretary of Defense?" LeMay was the inspiration for "General Jack D. Ripper" in the 1964 film farce *Dr. Strangelove: How I Learned to Stop Worrying and Love the Bomb,* a harrowing satire of a paranoid U.S. commander who believes that the Russians have poisoned the U.S. water supply and orders a nuclear first strike on the Soviet Union. Ted Sorensen called LeMay "my least favorite human being."

When Kennedy's national security adviser, McGeorge Bundy, asked the staff director of the Joint Chiefs for a copy of the Joint Strategic Capabilities Plan (JSCP), the blueprint for nuclear war, the general at the other end of the line said, "We never release that." An exasperated Bundy explained, "I don't think you understand. I'm calling for the president and he wants to see the JSCP." Small wonder that the Pentagon was reluctant to let even the president read a plan that threatened to strike Moscow with 170 atomic and hydrogen bombs. Every major Soviet, Chinese, and East European city was slated for destruction, including the annihilation of hundreds of millions of people in the communist bloc. When Kennedy received a formal briefing on the war plan, he was sickened by what he heard. Turning to a high administration official, he said, "And we call ourselves the human race."

At the time, Schlesinger reflected on how some people questioned Kennedy's willingness to retain "Chiefs who occasionally seemed so much out of sympathy with his policy. The reason was that, in his view, their job was not policy but soldiering, and he admired them as soldiers." Kennedy said that "it's good to have men like Curt LeMay and Arleigh Burke commanding troops once you decide to go in. But these men aren't the only ones you should listen to when you decide whether to go in or not," he told *Time* columnist Hugh Sidey. Schlesinger also noted Kennedy's sensitivity "to the soldier's role—dangerous in war and thankless in peace." Kennedy was fond of quoting a poem:

> *God and the soldier all men adore,*
> *In Time of trouble and no more;*
> *For when War is over and all things righted,*
> *God is neglected—the old soldier slighted.*

Yet it wasn't simply regard for career military men dictating Kennedy's willingness to keep quarrelsome chiefs like Burke and LeMay in place. As long as they confined their differences with him to the conference room and did not embarrass him with public protests, he was reluctant to stir up anything resembling the political criticism Harry Truman had caused with Douglas MacArthur's dismissal in 1951. The closeness of Kennedy's election victory influenced not only the choice of men he appointed to high administration positions but also the minimal freedom he believed he had to provoke controversies that could erode his limited political capital.

Lemnitzer was not entirely off the mark in wondering about Kennedy's readiness to deal with national security questions. For someone who saw foreign affairs and threats to the peace as the central issues of his presidency, Kennedy had entered the transition period with astonishingly little certainty about his choices for secretaries of defense and state. His highest priority, however, was finding appointees who satisfied domestic political considerations. True, Kennedy wanted men he saw as impressively bright and tough-minded—"a ministry of talent," he said—but the closeness of the election convinced him that he had to choose appointees who would be considered bipartisan or would not intensify national political divisions. These same political considerations had led him to reappoint Allen Dulles as head of the CIA and J. Edgar Hoover as director of the FBI, signaling that continuity rather than radical innovation would be the initial hallmark of the administration. A reluctance to antagonize Hoover, who had files he could use against any officeholder who might try to unseat him, and, specifically, embarrassing material about Kennedy's womanizing, may also have influenced the decision.

Feeling initially surrounded by the Eisenhower military and security officials he felt compelled to keep on—at least for a time—Kennedy moved quickly to ensure that in addition to Bobby he would have some familiar faces close to him. Ted Sorensen and Arthur Schlesinger topped the list.

No one on his staff had been closer to Kennedy during his ten Senate years than Sorensen. Reporters described him as "the president's 'intellectual alter ego,' and 'a lobe of Kennedy's mind.'"

Sorensen was born in Nebraska in 1928, the son of the Danish American state's attorney general, who was a progressive Theodore Roosevelt Republican, and a Russian Jewish mother, who was a pacifist advocate of women's rights. Sorensen was a superb student at the University of Nebraska, where he graduated first in his class from the law school in 1951, at the age of twenty-three. As a high school, college, and law student, he had been an activist on behalf of equal rights for African Americans, urging a fair employment law upon the Nebraska legislature and testing local segregationist practices in places of public accommodation like restaurants, the municipal swimming pool, and the university's dormitories. He was also an avowed pacifist who, when registering for the draft, requested classification for noncombat service. In the late 1940s, when the country was awash in anticommunist agitation, being a proponent of left-wing causes took some courage. It was a demonstration of Sorensen's affinity for high-minded missions and self-righteousness, which throughout his years as a Senate and White House staffer would become a source of conflict with anyone who did not share his views.

After graduating from law school, Sorensen headed to Washington without a job, an "immense gamble," as he saw it, but he hoped to find some kind of public service that could "make a difference," which was shorthand for saying he was looking for a calling that would not only change the world but also make him famous or at least a public figure. He had been drawn to Washington by what had attracted so many others through the generations—a chance to do good things and become someone important.

It took several days of walking the city's steaming July pavements before he found an opening at the Federal Security Agency, part of the Social Security system as well as a forerunner of the Department of Health, Education, and Welfare. He stayed there only for a little over a year, however; a chance to write a report on the Railroad Retirement System ended his service at the FSA, where the likelihood of budget cuts seemed certain to eliminate his job. But writing reports that seemed likely to have little, if any, impact did not measure up to what had brought him to the capital.

Sorensen resumed the hunt for a government position that fulfilled his larger ambitions. With Eisenhower about to become president and executive branch openings in short supply, he looked for an appointment on Capitol Hill. He was not impressed with what he had seen of Congress from a seat in the Senate gallery or at one of its committee hearings: The public posturing and intellectual and moral strength of the men he observed "disillusioned" him. In time, however, he came to believe that congressmen and senators were a mixture of good and bad: "an ambitious young idealist can realize his highest ambitions and a greedy demagogue can exercise his worst traits," he wrote later.

When opportunities for appointments with newly elected senators Henry Jackson from Washington and John F. Kennedy from Massachusetts developed, Sorensen chose Kennedy. He liked Kennedy: He was "a good guy" who didn't try to impress him with his pedigree or status; moreover, he offered Sorensen "a more challenging, exciting, and promising" assignment, developing an economic plan of renewal for New England, than anything Jackson proposed. But Sorensen also saw Kennedy as a senator on the make. He sensed from the first that Joe Kennedy's son had higher aspirations than being one of ninety-six legislators. For someone as devoted to liberal causes as Sorensen, joining Kennedy was more the result of ambition to tie himself to a rising star than to be a true believer in bold social reforms. If he had followed Kennedy's Senate campaign, Sorensen knew that Kennedy was essentially a budding Cold Warrior rather than an advocate of progressive domestic advance.

Political considerations more than any assessment of Sorensen's abilities had dictated the Jackson and Kennedy job offers. Jackson believed that a Sorensen on his staff would help his political standing with Scandinavian constituents. Kennedy didn't mention his political conviction that Sorensen's outspoken liberalism would help him with left-of-center Democrats suspicious of Kennedy's commitment to their causes.

Nor did Sorensen mention in his 2008 memoir, *Counselor*, the political gain Kennedy saw from hiring him. Although Sorensen complained in his book about "the idolaters who have almost buried the memory of the real man under a Camelot myth too heroic to be human," Sorensen's recollections contribute to the romanticized portrait of a great president. He wrote: "Despite the exaggerated attention and speculation, some malicious, some merely mindless focused on allegations about his private life, and despite the revisionist detractors, these hindsight distortions of his life and record have not lessened his hold on America's affectionate memory."

Like Kennedy's critics, who aimed to undermine Kennedy's historical profile, Sorensen's recollections suffer from distortions aimed at maintaining and improving it. For one, he was unwilling to acknowledge that Kennedy's invitation to him was partly based on political considerations about JFK's shaky relations with liberals. Moreover, my revelations about Kennedy's medical history and his efforts to hide the true state of his health from the public prior to and during his thousand days in the White House provoked complaints from Sorensen that I was mistaken: "There was no cover-up," he told me several times, which of course there was.

And in *Counselor,* Sorensen acknowledged that he and the "Kennedy team . . . to some extent" downplayed, especially in the 1960 campaign, "the true state of JFK's health." They "may have obscured the stark truth," he adds, but he couldn't bring himself to describe fully the lengths to which Kennedy and those closest to him hid his health problems. It may be that Sorensen did not know the full extent of Kennedy's medical troubles until the records were opened. He was surprised by what they revealed and regretted having agreed to open them.

Arthur Schlesinger certainly didn't know. In July 1959, when he asked Kennedy about rumors that he had Addison's disease and was taking cortisone to control it, Kennedy replied that he had an associated problem from malaria that had been brought under control and that "[n]o one who has the real Addison's disease should run for the Presidency, but I do not have it," which, of course, was untrue. The Kennedys understood that being more forthcoming might have cost him the election. They were surely correct: Given the closeness of the final tally and the problems Kennedy had convincing voters that his Catholicism and youth or inexperience should not be a bar to his election, adding a discussion of his medical history would probably have put his victory out of reach.

During his lifetime, Kennedy had reciprocated Sorensen's regard. From the first, Kennedy relied on him for a variety of jobs: "My work at the office varies 100% from week to week," Sorensen wrote his father in the spring of 1956. "One week it may be a commencement address . . . another week it may be a legislative matter . . . and another week politics." Sorensen considered it "the most wonderful job in the world . . . for a wonderful, responsible guy . . . with whom I get along excellently." He took exceptional satisfaction from having helped Kennedy write his Pulitzer Prize–winning *Profiles in Courage*, though he always downplayed his contribution to the book. He lived by the traditional speechwriter's and ghostwriter's creed of never laying claim to the words that came from his boss's lips or pen. Yet he was protective of his turf, jealously guarding against other speechwriters altering his handiwork. In 1962, when Schlesinger persuaded Kennedy to add a paragraph to his State of the Union message that Sorensen had written, Kennedy said, "Ted certainly doesn't go for additions to his speeches." When the *New York Times* used words from Schlesinger's paragraph as its "Quotation of the Day," Kennedy told Schlesinger, "Ted will die when he sees that."

Sorensen's importance to Kennedy grew as the latter reached for the presidency. And the 1960 campaign was something of "a turning point." Sorensen recalled with great satisfaction that Kennedy referred to him as "indispensable . . . one of his very key men who got the work done." As Kennedy traveled the country and made campaign decisions, Sorensen was someone he "trust[ed] implicitly." Not surprisingly, Kennedy decided to make the thirty-two-year-old Sorensen his "principal adviser on domestic policy and programs—his source of ideas, his draftsman of speeches and messages, the formulator of presidential legislative and administrative programs turning campaign promises into feasible action." Sorensen was "overjoyed" when Kennedy announced at the family compound in Hyannis Port, Massachusetts, the afternoon after the election that he would be the president's "policy-program" adviser, with the title of "special counsel to the president."

As Sorensen would eventually find out, his White House post was a source of satisfaction and conflict. He came to see the truth of the adage that the White House is, as Thomas Jefferson said, "a splendid misery." The chance to make a difference in shaping the country's current and future affairs exists side by side with the tensions and divisions that eventually develop in every administration. The competition for a president's attention and regard provokes rivalries among ambitious men and women with exaggerated views of their own wisdom and importance.

Sorensen's difficulties were principally with O'Donnell and other members of the so-called Irish Mafia who were unconvinced that Sorensen's ideas and judgments were always Kennedy's best options. Sorensen never entirely understood what antagonized O'Donnell, who was decidedly hostile to him. Sorensen wondered whether it was his "reticent reserve or inability to schmooze" or what some saw as his abrasive personality that offended O'Donnell. Sorensen also speculated that it may have been his status as "an outsider among the Irish-Catholic politicians from Massachusetts who thought they had a proprietary stake" in Kennedy's career and presidency. He never found out. But the division was a reality that played a part in shaping Kennedy's White House.

Arthur Schlesinger, Jr. was a partner of sorts in Sorensen's handling of ideas for policies and programs, though less of a Kennedy acolyte. Schlesinger had become a Kennedy supporter in 1959, when Adlai Stevenson remained a possible candidate and JFK badly needed liberal backing for the nomination. Eager to be part of a presidential administration, Schlesinger concluded that Kennedy was a much better bet to reach the White House than Stevenson, a two-time losing nominee. A public statement by liberal politicians, academics, and journalists endorsing Kennedy in June 1960, including Schlesinger and Galbraith, put Schlesinger firmly in JFK's camp. A book, *Kennedy or Nixon: Does It Make Any Difference?*, which praised JFK and pilloried Nixon as an opportunist devoid of ideas for advancing the national well-being, boosted Kennedy's appeal. In addition, help with debate preparation and speechwriting during the campaign gave Schlesinger a claim on the White House role he craved. Kennedy's initial offer to him of an ambassadorship and Galbraith's posting to the New Delhi embassy were calculated to put Kennedy's most identifiable liberal advisers at a distance from Washington; after the closeness of the election, it was calculated to mute conservative criticism.

Bobby Kennedy recalled that Jack liked Arthur but "thought he was a little bit of a nut sometimes," and was a little ambivalent about having him so close at hand. Still, Schlesinger was a stimulating "gadfly" who could generate ideas that could make a difference, and the presence of the Eisenhower men made putting him at the White House more appealing. By 1961, the forty-three-year-old Schlesinger had established himself, to use the term of the time, as a public intellectual whose writings on Andrew Jackson and Franklin Roosevelt made him the country's most identifiable Democratic historian. He was also recognized as a Harvard academic notable for his horn-rimmed glasses, receding hairline, prominent brow, and colorful bow ties. His presence in the White House echoed FDR's appointment of Columbia University professors—his brain trust—advancing new ideas to overcome a national crisis.

Schlesinger's appointment realized his hope of translating his academic prominence into a policymaker's influence. But he hated suggestions that he had compromised his academic independence in order to join the president's court. When the literary critic Alfred Kazin "juxtaposed" him against Richard Hofstadter, the distinguished Columbia University professor and Pulitzer Prize–winning historian, Schlesinger bristled at the comparison of "the power-loving stable mate of statesmen as against the pure, dispassionate, incorruptible scholar." Yet as events in the White House unfolded, Schlesinger found that he could not get so close to power without occasional compromises of his integrity that no courtier can resist.

When Kennedy agreed to have Schlesinger in Washington as a special assistant to the president instead of in some remote embassy, he gave him an office in the East Wing, where various secondary aides to the president and first lady worked. Except for Schlesinger, Sorensen said, these folks were "inhabitants of another world." It was symbolic, if not the reality, of the distance Kennedy wanted people to believe Schlesinger would have from the center of authority. When Schlesinger asked Kennedy whether his appointment was firm enough for him to request a leave from Harvard, where he was teaching, Kennedy said, "Yes, but we won't say anything about this until Chester Bowles is confirmed. I don't want the Senate to think that I am bringing down the whole ADA [Americans for Democratic Action]."

Kennedy's caution about liberals' visibility in his administration made them unhappy. When Schlesinger mentioned this to Kennedy, he replied, "Yes, I know. . . . But they shouldn't worry. What matters is the program. We are going down the line on the program." Schlesinger defined it for him as "an administration of conservative men and liberal measures." Kennedy agreed, and said that after a year or so, he planned "to bring in some new people." But then he "reflectively" acknowledged: "It may be hard to get rid of these people once they are in."

While Sorensen and Schlesinger might be a kind of intellectual blood bank that provided progressive ideas and tempered liberal criticism, Kennedy needed to make the difficult decisions about who would hold the administration's highest offices. Defense and Treasury were first on his list of crucial selections. The men he considered for Defense were also seen as suitable for Treasury, indicating that he saw them as pillars of Wall Street or corporate America who functioned comfortably in the boardrooms of industry. His first choice for either Defense or Treasury, and the State Department as well, if he preferred it, was the sixty-five-year-old Robert A. Lovett, a scion of the northeastern establishment. A prominent executive at the Brown Brothers Harriman banking house, Lovett had served in the country's fledgling naval air arm during World War I and directed the expansion of air forces in the War Department during World War II. He had become undersecretary of state under Dean Acheson in the Truman administration and then Truman's secretary of defense, succeeding the storied General George C. Marshall. Joe Kennedy, ever mindful of the millions of Americans who had voted for Nixon partly out of concern about having a Catholic as president, urged his son to appoint Lovett to a high cabinet post. Implicit in his recommendation was the belief that Lovett could help Jack disarm the fears about so young a man of such different background from past occupants of the White House.

In a December meeting at Kennedy's three-story redbrick townhouse on Georgetown's N Street, Lovett charmed the young president-elect, candidly explaining that he had not voted for him and considered liberal Harvard economist John Kenneth Galbraith, whom Kennedy would shortly make ambassador to India, "a fine novelist." "No doubt Lovett's urbane realism was a relief from the liberal idealists, like myself," Schlesinger said, "who were assailing the President-elect with virtuous opinions and nominations." In questionable health from chronic ulcers, and probably reluctant to tie himself to a president whose values seemed removed from his and about whom he had serious doubts, Lovett declined Kennedy's offer of any high cabinet job. Lovett, however, had recommendations for Treasury, State, and Defense that included Douglas Dillon, the sitting Treasury secretary, and Dean Rusk and Robert McNamara for State and Defense, respectively. He did not want to serve, but he was eager to get people in place who he believed could insulate the country from the liberals Lovett feared might shape Kennedy's policies.

When Kennedy escorted Lovett to the front door of his house, where reporters waited in the cold to have news of administration appointments, Kennedy told them that he had offered Mr. Lovett his choice of cabinet posts. If he could not get Lovett to join his government, he could still have the advantage of letting the world know that he wanted men like Lovett to serve with him and was seeking his counsel about who would be excellent choices for top cabinet positions. He was telling Wall Street, I'm no radical, but he was also heightening suspicions in the Kremlin that he was just another front man for the capitalists like his father who aimed to destroy communism.

With Lovett out of the picture, Kennedy was more concerned than ever to convince establishment Republicans to join his administration. He immediately had Bobby Kennedy call Robert McNamara in Detroit, where he had become president of the Ford Motor Company. The forty-four-year-old McNamara had graduated in 1937 from the University of California, Berkeley, where he earned a degree in economics and was elected to Phi Beta Kappa. His energy and drive to succeed matched his brilliance as a student. Reared in a lower-middle-class Irish family in Oakland, California, during the Great Depression, McNamara was determined to make a mark in the world of commerce and secure himself and his family from the financial hardships he saw everywhere in the 1930s. After earning an MBA in 1939 at Harvard Business School, where he impressed himself on his professors as an exceptional student, he worked for a year at the accounting firm of Price Waterhouse in San Francisco. He hated it. For someone excited by innovative solutions to business productivity and opportunities to build a reputation as a manager, the day-to-day grind of accounting bored him. In 1940, at the age of twenty-four, he seized the chance to become the youngest assistant professor in the history of the Harvard Business School, where he could explore ideas about the efficiency of large enterprises. With the United States at war beginning in 1941, he received temporary draft deferments to instruct Army Air Forces officers in statistical analysis for tracking the resources and capability of the country's air arm.

In 1943, he entered the Air Force as a captain serving in the Office of Statistical Control, first in England and then India, where he computed everything from fuel consumption to means of saving planes and crews from various hazards when flying over the Himalayan Mountains between India and China. Promoted to the rank of lieutenant colonel and transferred to the Pacific, the theater of principal combat against Japan, he worked for Curtis LeMay's bomber command, helping design the devastating B-29 firebombings of Tokyo, which killed thousands of Japanese civilians at limited cost to U.S. airpower.

After leaving the Air Force, McNamara accepted a job at the Ford Motor Company rather than return to Harvard. Married, with two small children and a wife who had been afflicted with polio, he felt the need for a more substantial income than anything the university could offer him. Colonel Charles Thornton, his commanding officer, who had pioneered much of the statistical work the Air Forces relied on in the war, persuaded McNamara to join him and other "Whiz Kids," as they called themselves, in using their statistical methods to restore the auto company to prosperity. Ford's revival under the leadership of Thornton and McNamara gave McNamara a reputation as a brilliant manager of a large corporation. His innovative practices, including the design of popular Ford models, made him not only Ford president in November 1960 but also something of an industrial celebrity.

Bobby Kennedy told McNamara that his brother, the president-elect, wanted him to meet with their brother-in-law, Sargent Shriver. Shriver had married Eunice Kennedy, Jack's younger sister, and had worked for Joe in Chicago, managing his many properties. Shriver had helped organize Jack's Wisconsin and West Virginia primary campaigns and had become the head of a committee working to fill high administration positions. McNamara, who hadn't made the connection that Bobby was Jack's brother, agreed to see Shriver the following week. But Bobby, making clear that he was speaking for the president-elect, insisted it be that day. When

Shriver showed up that afternoon, he explained that Kennedy wanted McNamara to be secretary of the Treasury. The bewildered but flattered McNamara replied, "You're out of your mind. I'm not remotely qualified for that." Shriver countered, "Well then, I'm authorized to say, he wishes you to serve as secretary of defense." McNamara said, "This is absurd. I'm not qualified." Shriver answered, "Well the president-elect at least hopes you will give him the courtesy of agreeing to meet with him tomorrow in Washington." McNamara agreed.

His reluctance to acknowledge his competence to assume a cabinet post was false modesty. What gave him pause was the abandonment, even temporarily, of his proven talent for managing one of the country's largest corporations and the diminished compensation of a government job. He was his own boss and accepting a cabinet post would make him a presidential subordinate. This was nothing he was willing to rush into. And yet going to Washington from Detroit would put him on a world stage. It was at least something he should consider, and being courted by the new young president-to-be had irresistible appeal.

Once McNamara had been ushered into Kennedy's home through the back door, the president-elect repeated Shriver's offer of the Defense Department post. And McNamara echoed his want of qualifications to serve in the cabinet job, underplaying his war work and management of Ford. "Who is?" Kennedy asked rhetorically. He explained that there are no schools for defense secretaries or presidents, either. But Kennedy had taken an instant liking to the clearly dapper, brilliant forty-four-year-old McNamara and convinced him to meet again the following Monday, after McNamara had had a chance to think over the offer. Kennedy's appeal dissolved McNamara's doubts. The man and the mission of serving the national defense excited McNamara's ambitions to do big things; it was an opportunity to make an indelible mark on history. After conversations with his wife and children, McNamara decided to accept, on the condition that he was free to select his deputies and not be required to frequent the Washington social scene. At their next meeting, McNamara handed the president a letter setting forth his conditions. Jack and Bobby immediately agreed to McNamara's terms and the appointment was announced to the press waiting outside the house.

The Kennedys didn't care much, if at all, about McNamara's demands. His appointment was serving their political purposes, and besides, they didn't think he would make much difference in the administration. They accepted what Lovett had told Jack when he asked, "What makes a good secretary of defense?" Good values and a good president, Lovett had replied, adding "and he can't do much damage. Not that he can do much good, but he can't do that much damage."

It was an astonishing exchange. At the height of the Cold War, when national security had become the country's greatest concern, the most important official after the president in assuring the nation's safety was to be someone whom Kennedy didn't know and who thought himself a total novice at the job. Kennedy's casualness about the appointment spoke volumes about his assumption that McNamara would be of small consequence in controlling the national defense. McNamara, however, who could not imagine becoming an ornament, accepted the post on the assumption that he could make a difference. He had mastered every professional challenge he had faced in his life and he believed that this would be no different.

Initially, McNamara was just another new face in Washington. The first time Schlesinger met him, at a Georgetown party a few days before the inauguration, he was "a quiet agreeable man with rimless glasses looking like a college professor." Schlesinger failed to catch his name. "That's Bob McNamara," Steve Smith, Kennedy's brother-in-law, told him. In time, everybody would come to know a man most everyone saw as fiercely ambitious, aloof, calculating, and combative. He would become a much more influential figure in the administration than Kennedy assumed, but for the moment he would be off in the Defense Department managing an impenetrable bureaucracy.

Kennedy was more focused on choosing his national security adviser, who would work at the White House and have more direct access to and interaction with the president. His choice was McGeorge Bundy, another Republican, and the youngest Harvard College dean in the university's history. Bundy's identity as a Republican also served Kennedy's initial need for bipartisanship, but Kennedy saw him as likely to be a larger part of national security and foreign policy discussions than any of the cabinet officers. Kennedy considered cabinet meetings a formality that wasted valuable time. But he relished conversations with someone as smart, accomplished, and realistic about the world as Bundy seemed to be.

Kennedy knew Bundy's history, or perhaps *pedigree* is a better word. He was a descendant of the Boston Lowells. Kennedy could not help but imagine his father's satisfaction at knowing that a Brahmin would now work for an offspring of Boston's Irish. The irony of that relationship, however, was of distinctly less consequence in bringing Bundy to the White House than Kennedy's regard for him as one of the brightest men he had ever known. Bundy's reputation for brilliance—notable as the applicant who had the highest score in history on Yale's entrance exam, a Harvard Junior Fellow, someone too gifted to bother with a traditional Ph.D., and the Harvard dean who had mastered the university's temperamental prima donnas—had found confirmation for Kennedy when he interacted with him as a Harvard trustee in the fifties. At five foot ten and 160 pounds, with clear plastic-frame glasses, a receding hairline, and round face with steely blue-gray eyes, Bundy was hardly a physically imposing figure. But no one who met him could dismiss him as some ordinary bookish academic too cerebral to make much of a mark on the world. His brilliance, sharp wit, precise thought, ability to think on his feet, and talent for cutting through rhetorical nonsense from politicians, journalists, and fellow academics made him a formidable adversary and an extraordinary colleague. All who knew him may have feared or loved him, but above all, they found him unforgettable.

At Harvard, Bundy had seen himself as one cut above his faculty, and in government, he would see himself as a kind of circus master, disdainful of congressmen and senators, the many know-nothings from the hinterlands who he thought were best ignored in the making of foreign policy, and the army of bureaucrats who could be troublesome and needed to be circumvented. Bundy quickly developed a reputation as someone who, in the familiar phrase, did not suffer fools gladly. When a Defense Department official provided a too long-winded, somewhat self-serving account of how he had uncovered a Joint Chiefs war plan hidden from the White House, Bundy snapped, "Is this a briefing or is it a confessional?" He snidely called a national security colleague he saw as too philosophical "the theologian." He was no more patient with the press corps: "A communiqué should say nothing in such a way as to feed the press without deceiving them," he advised Kennedy's press secretary. Bundy's arrogance would leave a trail of angry Washington colleagues and commentators who would later dish out verbal payback.

Kennedy's eagerness to have so intelligent a man at his side had led him to consider asking Bundy to become secretary of state, but having so young a secretary—Bundy was only forty-one—seemed likely to trouble people at home and abroad; they were already on edge about a forty-three-year-old president, and so Kennedy dropped the idea. Kennedy then suggested that Bundy become undersecretary of state for political affairs, but he withdrew that proposal when his choice for secretary objected. Kennedy then asked Bundy to become undersecretary for administrative affairs, but Bundy thought it would be less interesting than running Harvard College. Serving as special assistant for national security affairs, however, was irresistible; it presented an opportunity to make a significant difference in an administration that would be primarily focused on foreign affairs. One Harvard colleague, however, doubted the wisdom of putting Bundy so much at the center of power. Sociologist David Riesman thought that the "arrogance and hubris" that had made Bundy so effective as Harvard's "perfect dean . . . might be very dangerous" for the nation. Decisions about war and peace were best left to humbler men.

Because Kennedy intended to maintain the closest possible control over defense and foreign policy, Bundy was slated to carry a heavy load of responsibilities. To help him deal with the extensive daily challenges Kennedy envisioned for his office, he invented the job of Deputy Special Assistant for National Security Affairs, or Deputy National Security Adviser.

Kennedy chose Walt W. Rostow to fill the post. The forty-four-year-old Rostow was another one of the brilliant academics Kennedy had become acquainted with as senator from Massachusetts and a Harvard overseer. He was the offspring of a Russian Jewish immigrant family. His parents were socialists who named the second of their three sons Eugene V. Debs Rostow after the radical leader of the Industrial Workers of the World union. Also enamored of their adopted country, they named their other two sons Ralph Waldo Emerson Rostow and Walt Whitman Rostow. Rostow earned a Yale B.A. by the age of nineteen, was a Rhodes Scholar at Oxford in 1935–36, and completed a Ph.D. in economics in 1940, when he was only twenty-three. He served in the Office of Strategic Services (OSS), predecessor of the CIA, during World War II, and after the war became an economic adviser to the State Department, where he helped develop the significant reconstruction and relief impetus of the Marshall Plan to defend Western Europe against communist subversion. Between 1950 and 1961, Rostow was a professor of economic history at the Massachusetts Institute of Technology, a speechwriter for President Eisenhower, and a counselor to Kennedy on international affairs. In 1960, he published *Stages of Economic Growth: A Non-Communist Manifesto*, which outlined a path to national prosperity for developing countries. It was an early statement of Rostow's commitment to winning the international competition against communism. "I was glad I lived long enough to see the demise of communism," he said in 1992.

During Kennedy's 1960 campaign, Rostow, who was a prolific writer, provided JFK with memos on everything from foreign aid to arms control, space, policy toward Asia and Africa, and the United Nations. He helped Kennedy coin the phrase "Let's get this country moving again" and ingratiated himself with Kennedy and his backers by warning, "If the Republicans win, this country will have gone round a corner from which there may be no return." It was the sort of hyperbole that campaigns feed on, but overstated rhetoric could be a problem in responding to overseas threats, as Kennedy would see when Rostow later counseled him on Vietnam. Nonetheless, Kennedy was greatly impressed with Rostow's capacity to write so extensively about so many different topics. Kennedy told him once, "Walt, you write books faster than I can read them."

Kennedy had initially hoped to put Rostow in the State Department, to convert a stodgy bureaucracy into a more productive center of fresh foreign policy thinking. But Rostow was unwelcome in a department with so many cautious bureaucrats; he was too full of himself and grand theories they thought were probably unworkable and best avoided. Kennedy then decided to bring him to the White House, where Bundy, with Kennedy's enthusiastic approval, set up a National Security Council, which was more like a college faculty than a government bureaucracy. Bundy was not eager to hear Rostow's endless sermons on how to combat communism, but if Kennedy wanted him, Bundy thought it best not to resist. After all, Rostow did meet Kennedy's standard of an independent thinker who had never been a bureaucratic yes-man uncritically endorsing what higher-ups wanted to hear.

Nonetheless, Bundy and Rostow were strikingly different personalities with little affinity for each other: Bundy, the Brahmin with a birthright to dominate and govern other men; Rostow, the ambitious ethnic with the talent to make his way in a competitive world. The self-confident Bundy expected deference but admired intellectual independence and relished the give-and-take of contested ideas. "Goddammit, Mac, I've been arguing with you about this all week long," an exasperated Kennedy would explode at him during a tense period in the White House. But neither Kennedy nor Bundy would really mind. Nor did Rostow's affinity for theorizing and exuberance for what Bundy sometimes saw as bad ideas greatly trouble the latter. At one level, he was still the dean arguing with the tenured professor whose ego eclipsed his better judgment.

Bundy recruited other smart academics, chiefly from Harvard, who aimed to reorient the country's external dealings away from Eisenhower's brinksmanship and massive retaliation to General Maxwell Taylor's "flexible response." The premium was on reducing Soviet-American tensions, inhibiting the arms race, and avoiding a blowup over Berlin and Germany and possible brushfire wars in Asia. Carl Kaysen led Bundy's list of potential NSC colleagues. A professor of economics who had also been a Junior Fellow at Harvard, Kaysen was invited to focus on trade policy but was free to offer advice on arms control discussions and negotiations, as well.

When another colleague wrote Bundy about a successful conference in Geneva, Switzerland, he replied: "Your description of Geneva makes it sound like the opposite of Washington. There you have serious discussions in an atmosphere of unconcern." That was the exception here, he complained. But "I think perhaps we are moving toward a period in which we shall be able to take serious decisions, some of them even based on thought." As far as Kennedy was concerned, however, the State Department and Dean Rusk, the man he chose to head it, did little to advance the administration toward fresh, rational foreign policies.

Adlai Stevenson had wanted the job of secretary of state. There was some precedent for a twice-defeated presidential candidate to become the lead cabinet officer: In 1913, Woodrow Wilson had made William Jennings Bryan, the Democratic Party's three-time losing White House nominee, his secretary of state. But Wilson owed Bryan: He had made the difference in helping Wilson become the Democratic nominee on the convention's forty-sixth ballot. By contrast, Stevenson had stood in the way of Kennedy's nomination by allowing supporters to unreservedly contest Kennedy's candidacy. Neither Jack nor Bobby thought all that well of Stevenson. True, he had managed to excite liberal enthusiasm, but they saw him as rather prissy and ineffective. He never met their standard of tough-mindedness, on which they put a high premium for service in their administration. Also, they worried that he might forget who was president and who was secretary. But more important, they couldn't forgive his refusal to support Jack for the nomination, even when they had sent word that they were willing to give him the secretary's post in return for his early backing. "Fuck him," Kennedy told an all-out Stevenson supporter after he won the presidency. "I'm not going to give him anything."

Because Kennedy had asked him to write a foreign policy report during the campaign, Stevenson had some expectation that Kennedy would invite him to head the State Department. Stevenson's hopes also rested on the conviction that Kennedy needed the backing of party liberals, who remained loyal to Stevenson and wanted to see him at the center of a new Democratic administration. In addition, Stevenson believed that his international standing as a prominent exponent of improved relations with Moscow might carry some weight with Kennedy.

In the end, although Jack and Bobby would have been just as happy to freeze Stevenson out of the administration, they felt compelled to offer him something; they could not ignore his continuing influence with party liberals. Kennedy offered him a choice of becoming ambassador to Britain, attorney general (this was before he turned to Bobby), or ambassador to the United Nations. Stevenson was cool to all these proposals. He unequivocally rejected the London assignment and command of the Justice Department.

Despite the prospect of limited impact on foreign affairs, the U.N. ambassadorship at least had the appeal of joining a list that included Eleanor Roosevelt, America's first representative at the United Nations. Stevenson tested the limits of Kennedy's patience by saying that he would need to see who was secretary of state before he accepted the U.N. appointment.

Bobby Kennedy remembered the process with Stevenson as "so unpleasant. . . . The President really disliked him. He was so concerned about what he was going to do and what his role was going to be and whether he'd take the position or not that the President almost withdrew it."

Although Kennedy knew that he didn't want Stevenson as secretary of state, finding a secretary proved to be more of a problem than Kennedy had anticipated. Part of the difficulty was Kennedy's ambivalence about the department. He considered it something of a dinosaur, a sort of prehistoric beast that lumbered along with no discernible contribution to the national well-being. Kennedy's consideration of nominees for the department's leadership revealed how torn he was between trying to find either someone who could turn it into a useful engine of fresh thinking about overseas problems or else a caretaker who would simply keep it in line while the president and his national security advisers managed the serious work of policymaking.

Kennedy's impulse to give the department new life registered in his arrangement during the campaign to have Stevenson write a foreign policy report in preparation for the day he would become responsible for the country's external affairs. At the same time, he asked John Sharon, a Stevenson associate who helped him prepare the report, to give him a "shit list," as Kennedy described it, "of people in the State Department who ought to be fired." His inclination to place Bundy and Rostow in the department also suggested his leaning toward making it into a more productive source of constructive foreign policy proposals.

Kennedy's thoughts of reforming the department found full-est expression in his consideration of Arkansas senator J. William Fulbright as secretary. Kennedy had developed a cordial relationship with Fulbright during their service on the Senate Foreign Relations Committee, which Fulbright chaired. Kennedy viewed Fulbright as a match for Bundy and Rostow. A star football player at the University of Arkansas, a Rhodes Scholar at Oxford, and president of the University of Arkansas from 1939 to 1941, Fulbright was a combination of brains and athletic skill that Kennedy admired. He also shared Fulbright's affinity for internationalism, which had made Fulbright an early supporter of U.S. commitment to the United Nations and a program of international student exchange that, beginning in 1946, enjoyed institutional standing in the United States as the Fulbright Fellowship Program.

When Schlesinger talked to Kennedy on December 1 about the State Department, "it was clear that his thoughts were turning more and more to Fulbright. He liked Fulbright, the play of his civilized mind, the bite of his language and the direction of his thinking on foreign affairs." Bobby Kennedy had the same impression: "The President was quite taken with having Fulbright. . . . [He] had worked with Fulbright and thought he had some brains and some sense and some judgment. . . . He was the only person mentioned as Secretary of State whom he knew."

But Bobby talked Jack out of appointing him, even if it was only after heated arguments. Having signed a southern manifesto opposing the Supreme Court's 1954 *Brown v. Board of Education* decision desegregating schools, backed a 1957 filibuster against the first major civil rights law since Reconstruction, and signed a friend-of-the-court brief supporting Governor Orval Faubus's opposition to integrating Little Rock's Central High School, Fulbright had openly identified himself as an unequivocal supporter of segregation. Bobby convinced Jack that Fulbright would be a liability in dealing with Africa and Third World nations generally. It was no small consideration: Winning the contest with the communists for the hearts and minds of Third World peoples was a Kennedy priority. His appointment of Rostow partly rested on Rostow's identification as an economist aiming to draw developing countries into the Western camp. A secretary of state who rejected equality for people of color would give Moscow and Peking an advantage in emerging nations deciding between East and West. In addition, Fulbright's opposition to "an all-out anti-Nasser policy," implying a degree of sympathy for Egypt and other Arab nations, also made Jewish supporters of Israel distrustful of Fulbright and potential vocal opponents of his appointment as secretary of state.

With Fulbright eliminated from the competition, the job fell to Dean Rusk as a kind of consolation prize, though Rusk never saw it that way. Bobby Kennedy may have best captured the spirit in which Rusk won the appointment when he said later, "It finally had come down to where everybody had been eliminated—and Rusk was left. . . . Time was running out. We had to get somebody. . . . So the President—he had never met him—invited him down to Florida and asked him right away. So Rusk was selected, not for any great enthusiasm about him as such, although people spoke highly of him."

It was not as if Rusk was without credentials. Born in rural Georgia in 1909, the fifty-one-year-old Rusk was a classic example of the self-made man. A graduate of Atlanta's public schools, Rusk worked his way through Davidson College in North Carolina, where he played basketball, commanded his Reserve Officers' Training Corps battalion, and graduated at the top of his class. Selected as a Rhodes Scholar in 1931, he spent the next two years at Oxford before studying in Germany during the first year of Hitler's regime. His exposure to Nazism deepened an affinity for Wilsonian pacifism and moved him to write an essay on British relations with the League of Nations, which won him the Cecil Peace Prize. From 1934 to 1940 he taught at Mills, a women's college in Oakland, California, while also attending the University of California, Berkeley's Boalt Hall Law School. In 1940, foreseeing American involvement in World War II, Rusk joined the U.S.

Army, where he won the rank of colonel as a staff officer to General
Joseph W. Stilwell in the China-Burma-India Theater. Posted to
the operations division of the general staff in Washington in 1945,
Rusk helped identify the 38th parallel as the dividing line between
U.S. and Soviet forces in Korea. At the close of his military service in 1946, Rusk became a State Department official focused on
United Nations affairs. Proving himself a master of the department's bureaucratic ins and outs, he rose to deputy undersecretary
of state, becoming a favorite of Secretaries of State George Marshall and Dean Acheson.

Rusk endeared himself to Acheson in 1950 when he accepted
appointment as assistant secretary for Far Eastern affairs. The
communist victory in China in 1949 had opened the department,
especially its Asian specialists, to attacks from right-wing politicians, led by Wisconsin senator Joseph McCarthy, for having
"lost" China. Rusk spent the next twenty-one months echoing the
Truman administration's defense of its China policy published in
the State Department's *China White Paper*; rationalizing the White
House decision to expand the fighting in Korea above the 38th
parallel, which trapped the United States in a stalemated war; and
helping negotiate the 1951 Japanese Peace Treaty. Resigning from
the department in December 1951, Rusk became president of the
Rockefeller Foundation, where he was serving when Kennedy invited him to become secretary of state.

When Kennedy offered Rusk the job in mid-December, it wasn't
simply that he was the last man standing, though Kennedy had
pretty well exhausted the list of candidates. Rusk in fact satisfied
Kennedy's vision of what a secretary of state in his administration
should be. After their initial meeting at Kennedy's Georgetown
house, where, according to Rusk, Kennedy never raised the prospect of his becoming secretary, Rusk told a friend, "We couldn't
communicate. If the idea of my being Secretary of State ever entered his mind, it's dead now. We couldn't talk to each other. It's
all over." From Kennedy's vantage point, however, Rusk's passive

or low-key style was desirable. Kennedy described Rusk after their meeting as "lucid, competent, and self-effacing," hardly the sort of enthusiastic endorsement a new president usually provides for a high-level appointee.

When Kennedy called the next day to offer him the job, Rusk asked that they meet again before either of them made a final decision. Kennedy agreed and invited him to fly to West Palm Beach, where he had gone for a vacation. As Rusk sat in Kennedy's living room, waiting to see the president-elect, he noticed a copy of the *Washington Post* sitting prominently on a coffee table—it announced Rusk as secretary of state. When Kennedy entered and saw the headline, he "blew his top," asking Rusk if he was the source of the leak. Told no, Kennedy called *Post* publisher Philip Graham to chide him for printing the story. After Graham explained that Kennedy was the one who had told him, Kennedy said, "But that was off the record." Hardly, since it was exactly what Kennedy wanted: Kennedy had no interest in giving Rusk a choice of accepting; he was compelling him to take the job, and by forcing the issue, was also making clear that Rusk was now under the president's command.

As Kennedy already understood, Rusk was the sort of man who would take orders without complaint and do the president's bidding. Indeed, it was Rusk's diffidence that especially appealed to Kennedy. With Fulbright, Bundy, and Rostow eliminated from service in the department, Kennedy planned to concentrate control of foreign policy strictly in the White House, specifically with Bundy's emerging team of high-powered national security advisers. Kennedy had read an article Rusk had published in the journal *Foreign Affairs*, titled "The President." It argued for a return to presidential dominance of foreign policy making, a shift away from what had allegedly been the arrangement between Eisenhower and Secretary of State John Foster Dulles. Moreover, the journalist Walter Lippmann's description to Kennedy of Rusk as a "profound conformist" who "would never deviate from what he considered the official view" was additional confirmation for Kennedy of what he now wanted in the State Department. As a friend of Schlesinger's told him, Rusk was "the lowest common denominator," meaning he would be the least controversial and most compliant of the several men Kennedy had considered.

Once in office, Rusk was promptly seen as the gray eminence in an administration of scintillating figures. Mindful of the image he had taken on at the Kennedy White House, Rusk would jokingly say that in this crowd of dazzling characters "he looked like the friendly neighborhood bartender." When Warren Christopher, another understated personality, served as secretary of state under President Bill Clinton, critics would joke that Christopher was another Dean Rusk, without his charisma.

Rusk's caution was the consequence not just of his persona but also of a conviction that a secretary of state is obliged to stand in a president's shadow. As Rusk's son, Richard Rusk, said, his father "believed that a secretary of state should never show any blue sky with his president, that policy differences between them must remain confidential, and that failure to do so weakens an administration." Rusk was obsessed with maintaining confidentiality: He resisted making records of phone conversations with Kennedy or having secretaries prepare memos of conversations. He had regular sweeps of his home and office for "bugs," lest any government agency be recording his conversations. "His passion for secrecy was so strong," his son adds, "that after leaving office, he went back to the State Department, pulled out his copies of telephone memos of conversations with his two presidents, and threw them away."

Rusk's prudence made him the butt of some hostile Washington humor spread by McGeorge Bundy. During a White House meeting in the Oval Office between only Kennedy and Rusk, when the president asked his opinion, Rusk is supposed to have whispered: "There's still too many people here, Mr. President." While Rusk compulsively deferred to the president, he was less accommodating toward others in the administration, especially competitors for the president's ear on foreign policy. He would have his share of differences with Bundy and others in the national security bureaucracy who saw Rusk's restraint as amounting to a State Department foreign policy vacuum that they had no choice but to fill.

Yet Rusk was never as passive and self-effacing as he pretended to be. He had quietly lobbied for his appointment. The publication of his *Foreign Affairs* article in the spring of 1960 was no accident. It was meant to send a message to any Democrat who might get the nomination. Moreover, letters to Kennedy recommending Rusk for the post were part of an orchestrated campaign. His silence in the face of the not-so-quiet Washington gossip about his meekness angered him; he once told a colleague that "it isn't worth being secretary of state" when a president gives so much preference to his White House national security team. But it wasn't just White House competitors who ignored Rusk; some in the State Department, unhappy with his caution, soon saw fit to bypass him, graphically belittling his habit of protecting his private parts.

Once he made Rusk secretary of state, Kennedy did not believe the department would be the source of much fresh thinking about foreign affairs. He held Rusk at arm's length, never addressing him as other than "Mr. Rusk." Kennedy was content then to satisfy domestic political obligations by making Chester Bowles undersecretary, the department's second in command, rather than appoint someone likely to stimulate innovative foreign policy discussions.

The fifty-nine-year-old Bowles was a devoted Stevenson supporter and a spokesman for liberal Democrats whom Kennedy had courted as essential in his reach for the White House. Notable as the architect of the Benton & Bowles advertising firm, which had made him wealthy; as governor of Connecticut from 1948 to 1950; as Truman's ambassador to India from 1951 to 1953; as a one-term congressman at the end of the fifties; and as Kennedy's foreign policy adviser during the campaign, Bowles had a record of public service that made him a reasonable choice for a top State Department post. Given Kennedy's decision to deny the secretary's post to Stevenson and give him a distinctly secondary appointment as U.N. ambassador; the selection of Lyndon Johnson as his running mate, which survived furious objections from liberal labor leaders; his reappointment of conservatives J. Edgar Hoover and Allan Dulles, respectively, as FBI and CIA directors; and the choice of McNamara and Bundy, nominal Republicans, as leading national security officials, Kennedy was acting out of political considerations more than foreign policy ones when he chose Bowles.

Kennedy appointed him in spite of personal tensions between them. Bowles had offended Kennedy when he would not campaign for him against Humphrey in the Wisconsin primary. He also angered Kennedy when he rejected Kennedy's advice to run again for the House seat he had won in 1958; it would have relieved Kennedy of having to give him a job in the administration. They were temperamentally incompatible. Bowles was an idealist who offended Kennedy's affinity for practical, realistic solutions.

Yet their backgrounds and politics were close enough. As a graduate of Choate and Yale, with a sense that privileged Americans like himself should help the needy at home and abroad and that attention to Third World countries was vital in defeating communist ambitions for world control, Bowles shared with Kennedy concerns about domestic change and international challenges that promised to make him a credible member of Kennedy's State Department.

Bowles's call for aid to India and other emerging Asian nations as a way to counter Soviet appeals to follow their lead toward state socialism particularly resonated with Kennedy. They also shared a belief in the need for America's identity with anticolonialism, or the right of Asian, African, and Middle Eastern peoples to self-determination. But their mutual concerns could not bridge a fundamental divide: For Bowles, democracy or independence for former colonies was a moral imperative, a matter of *principle*; for Kennedy, it was a means to a self-serving American end—a way to ensure that the resources of Asia, Africa, and the Middle East, especially oil supplies, were available in the West, and that the strategic areas of these regions could serve as potential bases for the United States and its European allies to contain communism.

"Annoyed" is probably too kind a description of how Kennedy responded to his undersecretary. His view of Bowles was closer to contempt for someone who, holding high office, seems very detached from the hard realities a responsible official needs to confront. And Bowles, in Kennedy's view, was the kind of soft-minded intellectual who thought that hunger and poverty were greater dangers to the United States than a Kremlin laying plans to control Cuba and throw the West out of Berlin. Kennedy had been mindful of former secretary of state Dean Acheson's view of Bowles as a "garrulous windbag and an ineffectual do-gooder." But politics dictated that he include him in the ranks of administration appointees. Besides, putting him in the State Department seemed like less of a problem once he had appointed Rusk and concluded that the department would be a sort of fifth wheel in the crucial decisions affecting foreign affairs.

None of this is to suggest that Kennedy wouldn't have been receptive to initiatives from the State Department that offered fresh ideas about the country's many foreign problems. But he was not very hopeful, once Fulbright was out of the picture, that Rusk and Bowles had the wherewithal to capture his attention with the kinds of imaginative proposals he expected to see from his national security team.

Kennedy's appointment of George Ball as undersecretary for economic affairs deepened rather than eased his doubts about the department's likely contribution to foreign policy. Trained as a lawyer, the fifty-one-year-old Ball had been an associate general counsel in the Lend-Lease Administration and a counsel to the European Economic Community. As someone who had written reports for Kennedy during the campaign on foreign economic policy, Ball certainly had a claim on that post. But Ball was another strong Stevenson supporter who had come late to Kennedy's side in the campaign. Moreover, he had no personal tie to Kennedy and shared Stevenson's indignation at Kennedy's roughhouse political tactics in the run-up to the nomination.

Stevenson told Ball that after winning the Oregon primary in June 1960, Kennedy came to see him in his Libertyville, Illinois, home. When Stevenson resisted Kennedy's plea for his support at the convention, which Kennedy believed would allow him to clinch the nomination, he "behaved just like his old man," Stevenson said. "Look," Kennedy said, "I have the votes for the nomination and if you don't give me your support, I'll have to shit all over you." Stevenson was outraged by Kennedy's "Irish gutter talk," but he didn't tell him off.

Because he was so closely identified with Stevenson, Ball doubted that Kennedy would ask him to join the State Department. And in fact, Kennedy initially intended to bypass Ball and use the undersecretary's post to satisfy the same political pressures that had moved him to give Republicans a central voice in his foreign affairs appointments: He offered the job to William C. Foster, a liberal Republican. But Stevenson lobbied Fulbright to press Kennedy to choose Ball instead. Fulbright persuaded Kennedy to give Ball the job, warning him that "too many top posts in the three principal departments—State, Defense, and Treasury"—were going to Republicans. It was creating "the impression that the Democratic Party lacked men of stature." But Kennedy brought Foster into the administration anyway as the head of the new Arms Control and Disarmament Agency.

With his national security and State Department teams in place, Kennedy turned his attention to the Inaugural Address—the opportunity to describe his administration's goals in foreign affairs over the next four years to audiences at home and abroad and to the men (there were no female appointees) who would be charged with giving shape to the broad aims of his government. He had been thinking constantly about the speech since his election, but felt compelled first to satisfy public expectations that he name the cabinet officers who would help him meet the national security challenges ahead. Eisenhower thought that Kennedy's concern with choosing advisers made him look weak or like an inexperienced leader who would be too dependent on subordinates. In fact, Kennedy was determined to be his own boss and chart out an inspirational grand design in his inaugural speech.

He hoped that the speech would be a call to unity and action that could help overcome divisions in the country evident in the narrowness of his victory. Galbraith urged him to think hard about how he would achieve liftoff for his administration. "It is evident," Galbraith wrote to Kennedy after he gave his acceptance speech at the Democratic convention, "that in straightforward exposition and argument you are superb. . . . When it comes to oratorical flights and Stevenson-type rhetoric, [however,] you give a reasonable imitation of a bird with a broken wing. You do get off the ground but it's wearing on the audience to keep wondering if you are going to stay up."

Kennedy's address, which he crafted in close collaboration with Sorensen, foretold his almost exclusive emphasis on foreign affairs and his determination to advance international peace and prosperity. "Let every nation know," he declared, "that we shall pay any price, bear any burden, meet any hardship . . . to assure the survival and the success of liberty. . . . To those people in the huts and villages of half the globe struggling to break the bonds of mass misery, we pledge our best efforts to help them help themselves. . . . If a free society cannot help the many who are poor, it cannot save the few who are rich." In Latin America, he promised to convert his words into an Alliance for Progress.

As for the "uncertain balance of terror" that could lead to "mankind's final war," Kennedy urged a new era of negotiation: "Let us never negotiate out of fear," he counseled. "But let us never fear to negotiate. . . . And if a beachhead of cooperation may push back the jungle of suspicion, let both sides join in creating a new endeavor, not a new balance of power, but a new world of law, where the strong are just and the weak secure and the peace preserved." He closed with "a call to bear the burden of a long twilight struggle . . . against the common enemies of man: tyranny, poverty, disease and war itself," ending with the most memorable words of the speech: "And so, my fellow Americans: Ask not what your country can do for you—ask what you can do for your country." He concluded, "Let us begin."

Anyone listening to Kennedy's Inaugural Address heard a president entirely focused on foreign affairs. And given national anxieties about America's apparently weakened international position against a Soviet Union with a seeming advantage in intercontinental ballistic missiles and a Cuban client state threatening to spread communism across Latin America, few were inclined to quarrel with Kennedy's priorities. Indeed, Gallup polls had demonstrated the extent and intensity of public attention to external communist threats and a conviction abroad that America's international standing and power were on the decline. Even Galbraith, who believed that America's strength abroad depended on robust prosperity at home, advised Kennedy to make foreign dangers his first priority: "As to the issues, I would think there are four," Galbraith wrote in August 1959. "The first is to find some durable alternative to the present strategy of deterrence with which we can live in greater safety. The second, is to find some way of assisting the poor lands which takes account not only of their need for capital but also the urgent pressures for political and social advancement. Thirdly, we must find some way of reconciling price stability with full employment and economic growth. Fourth, and finally, we must correct the notable disparity between our comparative private opulence and the poverty of our public services."

Yet, however much foreign affairs was at the center of national concerns, and however little inclined he was to divert attention from dangers abroad, Kennedy could not totally ignore domestic economic and social issues that affected a majority of Americans and made them eager for presidential leadership promoting prosperity and civic peace. Kennedy's election had partly rested on public discontent with Eisenhower's handling of the economy and race relations, and more generally, on a sense of national drift or lost national purpose. Commentators complained of a "bland, vapid, self-satisfied, banal" society that lacked a shared commitment to some grand design. Adlai Stevenson decried an America in which "the bland were leading the bland." He saw a country without "an irresistible vision of . . . exalted purpose and inspiring way of life." Our choice, Kennedy had said in his acceptance speech at the Democratic convention, was "between national greatness and national decline. . . . All mankind waits upon our decision."

Three recessions had roiled Eisenhower's eight-year term, with the last of these downturns still evident in January 1961: Unemployment stood at 7 percent and rising prices added to the woes of some 40–60 million people, between a quarter and a third of the country's population living without adequate nutrition, housing, or medical care. The problem of poverty, chronicled by Michael Harrington, first in *Commentary* magazine in 1958 and then in a 1962 book, *The Other America: Poverty in the United States,* was a national embarrassment for a country urging Third World nations to see capitalism or free enterprise as superior to communist societies, which could not match American living standards.

Kennedy understood that the communist challenge would be only one test of his presidency; he would also need to advance the country's economic well-being, reduce the gap between rich and poor, expand educational opportunities, insulate seniors from the medical and hospital costs afflicting their lives, and defend African Americans against the institutionalized racism that denied them a chance at the American dream. Persuading Congress to free entrepreneurs from confiscatory income taxes of 91 percent, raising educational standards through federal grants to localities, improving the lot of seniors by establishing a federal system of affordable medical insurance, and easing racial tensions with a civil rights act outlawing segregation in all places of public accommodation were challenges demanding the political skills of a leader who could inspire and persuade on a par with a Washington, a Lincoln, and the two Roosevelts. But it would also take an investment of presidential prestige and political capital that Kennedy saw in short supply after his narrow victory. Foreign policy priorities would also leave limited freedom to press the case for domestic reforms.

While he doubted that enacting tax reform, federal aid to education, and Medicare would happen quickly, if at all, during his first term, he saw surmounting the racial divide as a challenge few in 1961 believed any president could achieve in the immediate future. The Supreme Court's 1954 ruling in *Brown v. Board of Education* had pressured school districts across the segregated South to integrate public schools with mixed results. Resistance by Arkansas's Governor Orval Faubus to federal court orders ending racial segregation in Little Rock's Central High School, for example, had forced Eisenhower to federalize the Arkansas National Guard and compel integration by armed troops, which had led to a rise in white schoolchildren attending private schools. Resistance to the nonviolent protests of the Reverend Martin Luther King and his Southern Christian Leadership Conference had erupted in sporadic violence. Having left no doubt during the campaign that he would side with advocates of racial justice by praising peaceful sit-ins at segregated public facilities, as well as promising to integrate public housing "with one stroke of the pen" and to use executive powers "on a bold and large scale" in behalf of civil rights, Kennedy needed to choose a White House staff and appoint cabinet and subcabinet officials who could try to overcome the long odds against his domestic agenda, especially on volatile race relations.

Choosing a White House staff was relatively easy. Kennedy intended to be his own chief of staff. It had the advantage of not antagonizing any of the loyal subordinates who had been with him since his first House campaign in 1946, by elevating one over the others. Five men had been at the center of his political advance to the presidency: Dave Powers, Kenneth O'Donnell, and Larry O'Brien, and more recently, Ted Sorensen and Pierre Salinger. Powers, O'Donnell, and O'Brien were described as the "Irish Mafia": core members of a group that was uncritically loyal to Kennedy and had made his rise possible.

Powers was the first among equals. He had known Kennedy since 1946, when Jack, hearing that Powers had an understanding of the people and issues vital to a candidate's bid for Boston's sprawling Eleventh Congressional District, called on the thirty-four-year-old Air Force veteran of World War II at his three-story walk-up residence to ask his help in the campaign. Powers wondered how "a millionaire's son from Harvard trying to come into an area that is longshoremen, waitresses, truck drivers, and so forth" could possibly win a majority or even a plurality of votes against opponents who could speak the language of the district's Irish and Italian working class. Nonetheless, Powers, a walking encyclopedia of information about the district, saw Kennedy as potentially someone in the right place at the right time: Jack's standing as a war hero and a young man on the make spoke to the aspirations of the district's upwardly mobile blue-collar voters—men and women who took special satisfaction in knowing that one of their own, the Irish Kennedys, had made it so big in America.

Powers's quick wit and easy regard for rogue politicians amused and endeared him to Kennedy. He was at Kennedy's side at every step toward the White House and provided Kennedy with a kind of refuge from the burdens of running for and serving in office. He was a facilitator of Kennedy's considerable affinity for extramarital relations, discreetly arranging the time and places for trysts with a variety of women. Kennedy enjoyed Powers's mischievous sense of humor: During the Shah of Iran's visit to the White House, Powers, bringing the leader into the Oval Office, put his hand on his shoulder and declared, "You're my kind of Shah." To the humorless Soviet first deputy premier Anastas Mikoyan, he said: "Are you the real Mikoyan?"

Initially, Kennedy had to sell himself to Powers, who agreed to attend a talk in January 1946, two days after Kennedy had asked for his help in the campaign. In a speech to Gold Star Mothers who had lost sons in the war, Kennedy not only won over the women with a heartfelt expression of regard for their sacrifice, one his own mother suffered with Joe, Jr.'s death, but also convinced Powers of his viability as a coming political star. "I don't know what this guy's got," Powers recalled. "He's no great orator and he doesn't say much, but they [the mothers] certainly go crazy over him."

Kennedy made Powers a special assistant to the president, in which role he acted as a man Friday watching over the president's personal needs, always with him on trips around the country and abroad. He usually was the first to see Kennedy in the morning and the last to see him at night. He was less a political adviser than a friend with whom Kennedy could relax. They would swim together in the White House pool, where Powers would use a breast-stroke in order to keep up a steady chatter of amusing conversation that Kennedy enjoyed.

Kenneth O'Donnell was Kennedy's principal political adviser or, more to the point, the guardian of the president's political interests. Like Powers, he had an encyclopedic knowledge of the Democratic Party's power brokers, and as the president's special assistant and appointments secretary, he served as the gatekeeper to the Oval Office, welcoming those who could advance Kennedy's standing and barring those whose demands O'Donnell saw as serving no useful purpose. Pierre Salinger, Kennedy's press secretary, recalled that "[e]xcept for members of the official family and state visitors, no one could see the President without first clearing with O'Donnell—and he could say no to a corporation president trying to promote an ambassadorship as readily as he could to a ward politician trying to influence the award of a contract. . . . I found him to be one of the most candid and direct men I had ever met. He would never use five words if one would do, and that word was very often a flat 'no.'" On all political questions affecting Kennedy, O'Donnell's word was final.

O'Donnell's political influence had initially come from his friendship with Bobby; they were roommates at Harvard and teammates on Harvard's football team, where O'Donnell was the starting quarterback and Bobby a favorite pass receiver. The offspring of an Irish family—his father was football coach and athletic director at the College of the Holy Cross, in Worcester, Massachusetts, an Irish Catholic bastion—the twenty-eight-year-old O'Donnell joined Kennedy's Senate campaign in 1952 after service in the Air Force during World War II and a stint as a businessman in a paper company. Between 1957 and 1959, he was an administrative assistant to Bobby Kennedy, who had become a counsel for the Senate Labor Rackets Committee investigating union corruption. In 1960, he became Jack's campaign scheduler, working tirelessly to help him win the White House. His mastery of the men and issues shaping national politics elevated him to the front rank of Kennedy's inner circle.

It wasn't only O'Donnell's friendship with Bobby and political talents that recommended him to the Kennedys; they admired his toughness—his unsentimental determination to win in the political arena as he had on the football field. His wiry physique, sharp facial features, blunt conversation, and "grim, cryptic wit," which gave him a reputation as someone with "the instinct for the jugular," made him more admirable than likable among his White House colleagues. He was usually at the center of "strong and angry disagreements" that erupted among staffers competing for control of policy and the president's approval. Kennedy didn't mind the conflict: "The last thing I want around here is a mutual admiration society," Kennedy told Salinger. "When you people stop arguing, I'll start worrying."

Kennedy's objective was to ensure that every important decision came across his desk. He was determined to hold the reins of authority in his hands; he wanted no hint of a belief that he was too uncertain of himself to exercise the powers of the presidency. He had Franklin Roosevelt's model of divide and rule or encouraging debates among his staff in order to force competing subordinates to come to him for a final decision. But Kennedy, Schlesinger said, "found no pleasure in playing off one subordinate against another." Instead, he valued arguments for what substance they might ultimately give to decisions. His administrative management, however, mainly consisted of staying one step ahead of cabinet and subcabinet officials charged with considering major policy questions. As Schlesinger put it, Kennedy's "determination was to pull issues out of the bureaucratic ruck in time to defend his own right to decision and his own freedom of innovation."

Larry O'Brien, the third member of the Massachusetts Irish Mafia, had an irresistible claim on a White House job as well. O'Brien, like Kennedy, was born in 1917 and served in the military during World War II. He was a politician's politician: less interested in holding elective or appointed offices than in managing campaigns and the give-and-take common to dealings with congressional leaders jealous of their influence. Born and raised in Springfield, in the western part of the state, O'Brien had breathed politics since his earliest years. He was the son of an Irish-born father who had combined success as a tavern keeper and real estate investor with activism in Democratic Party politics. The son's effectiveness as a local organizer had taken him to national conventions and provided an outlet for his resentment toward the state's "WASPs," the Boston Brahmins contemptuous of what they saw as a small-town Irish pol challenging their authority. After law studies at Northeastern University, O'Brien became a political organizer for Foster Furcolo, an Italian American congressional candidate he helped win House terms in 1948 and 1950. Two years as Furcolo's office manager temporarily cured O'Brien of Potomac fever; he returned to Springfield, where Kennedy, aware of his reputation as a keen political organizer, talked him into helping with his successful 1952 Senate campaign. Six years later O'Brien took a leading role in Kennedy's 1958 reelection bid, in which Kennedy won the largest victory margin in Massachusetts's history.

Kennedy readily turned to O'Brien when he mapped out plans for his 1960 presidential run: He was an architect of Kennedy's victories in Wisconsin and West Virginia, worked with Bobby to secure a majority of delegates at the convention, and raced around the country during the fall campaign encouraging state and local Democrats to work hard at putting Kennedy in the White House.

Having earned a choice job in the new administration, O'Brien asked to become Kennedy's point man with Congress, managing the legislative initiatives he expected Kennedy to put forward in 1961. Jack and Bobby, however, were not convinced that O'Brien, for all his skills as a campaign organizer, was the right man for this job. He had no experience leading bills through Congress. Moreover, his affinity for negative thinking, such as an impulse to emphasize impediments to dealings with Congress, made them feel that O'Brien was ill-suited to sell congressmen and senators on bold domestic programs. And even if they were pessimistic about congressional action on major proposals, they did not want it said that they had given up without a fight or had assigned the job of pressuring Congress to someone who had shown little hope of success.

Bobby initially suggested that O'Brien become deputy postmaster general, where he could play the part of party patronage master. But O'Brien wanted no part of what he considered a minor post, and voiced his disappointment by saying he intended to return to Springfield. Since O'Brien's exclusion from a top administrative job would have antagonized many party operatives, who were already complaining about Kennedy's preference for Republican appointees over deserving Democrats, they convinced Kennedy to grant O'Brien's request to manage congressional relations.

Two other considerations influenced Kennedy's decision. The natural alternative to having O'Brien or any other Kennedy political operative deal with Congress was to rely on Vice President Lyndon Johnson. As a twenty-three-year veteran of Capitol Hill with a reputation as the most effective Senate majority leader in

history, Johnson seemed like a natural choice to help enact the
president's legislative agenda. But giving Johnson such authority
would have undermined Kennedy's presidential standing by implying that the younger, less experienced Kennedy needed his more
seasoned vice president to manage Congress. It would have diminished Kennedy's status both at home and abroad and launched his
administration on the wrong note.

Second, Kennedy had little hope of persuading southern conservatives, who chaired the principal congressional committees,
to enact major domestic reforms that threatened to undermine
their region's racial segregation. There was enough to do in foreign
affairs without focusing attention in the press on domestic controversies that Johnson seemed certain to provoke with southern
congressmen and senators. On January 3, when Johnson had tried
to expand his vice presidential control of the Senate by forcing a
vote on his leadership of the sixty-three-member Democratic Senate caucus, he had run into a wall of hostility. "This caucus is not
open to former senators," one opponent of the proposal declared.
The opposition angered and embarrassed Johnson, who told reporters, "I now know the difference between a cactus and a caucus.
In a cactus all the pricks are on the outside." Kennedy believed it
better to let the less experienced O'Brien quietly manage progressive proposals that would have to wait for a second Kennedy term,
if he were fortunate enough to get it.

Pierre Salinger's selection as press secretary was predictable
and uncontroversial. The thirty-five-year-old Salinger had come
into Kennedy's orbit through a career in investigative journalism
that had led him to an appointment as an investigator for Bobby
and Jack's Senate subcommittee probing labor union corruption.
Eagerly signing on to run Kennedy's press operations during the
1960 campaign, Salinger dealt with every facet of Kennedy's public relations from January through the election in November. The
day after his election, Kennedy told Salinger that he wanted him
to stay on as press secretary. It was in fact the assignment Salin-

ger had been thinking about and hoping for since he had started working with Kennedy in 1957.

As spokesman for the president, Salinger could largely stand above the kind of political infighting that emerges in every administration. True, presidential press secretaries carry the burden of measuring every spoken word, which are seen as expressing a president's latest thoughts. But a press spokesman can remain a noncontroversial messenger who does no more than report a president's views. By contrast, White House advisers, faced with choices about what often prove to be insurmountable problems, battle for control of policy. As soon as Salinger wandered onto political grounds, counseling the boss how to manage his public image, he became an advocate with enemies or at least opponents.

Salinger was convinced that Kennedy's instinctual ability to speak spontaneously and effectively about current controversies, as demonstrated in his successful debates with Nixon, could make him the first president to hold live televised press conferences. Salinger believed that the debates turned the election in Kennedy's favor, showing him as "a mature, knowledgeable, attractive man." By contrast, Nixon seemed like "an actor reading a toothpaste commercial." The novelty of live press conferences would expand Kennedy's audience and allow him to get his message directly to an attentive public.

Salinger's proposal provoked a "swift and violent" reaction among "JFK's closest advisers," State Department officials, and White House correspondents. Advisers feared a misstatement that could trigger an international crisis, shaking markets and stimulating talk of war. White House correspondents, who had traditionally enjoyed the prerogative of informing the world about a president's views, were chagrined at losing the privilege of first dibs on a big story. Despite loud complaints about their lost status, Salinger told them they could "take it or leave it." And since it was the president's press conference and not theirs, they had no choice but to take it.

Moreover, the reporters would soon make the adjustment, and Kennedy, who enjoyed the prospect of commanding center stage to a greater extent than any of his predecessors, proved to be masterful at bantering with the press, while also informing the public at home and abroad of where he stood on major issues. It was a chance to not only educate people about his views, but also school subordinate officials in his administration, who had little opportunity for direct contact with the president. His press conferences gave them a better understanding of administration priorities.

Given the importance the new procedure placed on the president's press appearances, he preceded his conferences with intense preparation and breakfast meetings of principal advisers, especially his national security officials, where he would rehearse responses to predicted questions. Some of these answers would convulse his associates with laughter but were usually too undiplomatic to see the light of day. The aftermath, however, was consistently "a superb show, always gay, often exciting, relished by the reporters and by the television audience."

Kennedy's limited focus on domestic problems did not exempt him from choosing appointees who would be stewards of the economy. Although his appointment of C. Douglas Dillon as the secretary of the Treasury further inflamed liberals, Kennedy believed he was a good choice in every way for the job. With the economy in recession and trade imbalances causing a gold drain, there was a premium on finding a Treasury secretary who could generate confidence on Wall Street and in the business community more generally. Kennedy had seen Robert Lovett and Robert McNamara as filling that role, but neither would consider serving at Treasury. Phil Graham, the publisher of the *Washington Post*, and influential columnist Joseph Alsop began promoting Dillon, a prominent Republican, for the job. A former head of his father's Wall Street banking firm, Dillon, Read & Company, Dillon had served as Eisenhower's ambassador to France, undersecretary of state for economic affairs, and then undersecretary, the second-highest job at State.

Although Dillon would meet Kennedy's need for an establishment financier who would quiet Wall Street worries about a spendthrift administration committed to expanding the economy with deficit spending, some of those closest to Kennedy cautioned him against the appointment. Joe Kennedy didn't think Dillon, for all his Wall Street connections, knew enough about finance to merit the job. Schlesinger warned the president-elect against choosing someone who had been a member of the administration responsible for the country's current financial and economic problems, had been a down-the-line Nixon supporter, and seemed likely to favor policies of austerity over deficits promoting economic expansion. Bobby Kennedy warned that after a few months of conflict over administration policy, Dillon might quit with an attack on the White House, which would embarrass them and shake public confidence in their ability to improve the economy.

When Dillon assured Kennedy of his commitment to actions much closer to innovative Democratic solutions than conventional conservative ones, emphasizing his eagerness to promote economic expansion, and promised to go quietly should they come to a policy impasse, Kennedy decided to appoint him.

The fifty-one-year-old, tall, stately Dillon, with his receding hairline and tailored suits, appealed to Kennedy not only as someone with impeccable social credentials but also as an establishment figure with a personal history that belied his public image. He was the grandson of Samuel Lapowski, a Polish Jewish immigrant, whose success in Texas business ventures gave his son Clarence the opportunity for a Harvard education and entrée to the world of high finance. Clarence Dillon made a fortune as the CEO of the investment bank Dillon, Read (after changing his name to Dillon, his mother's maiden name). Clarence's son, C. Douglas, had all the advantages money could buy: elementary education at a private school in New Jersey, with three Rockefellers as school-mates; the storied Groton School in Massachusetts, whose alumni included Franklin Roosevelt, Dean Acheson, J. P. Morgan, Jr., and McGeorge Bundy; and Harvard. The clinching argument in Kennedy's decision to appoint him may have been Dillon's tongue-in-cheek description of Eisenhower cabinet meetings, where high-powered officials sat around batting clichés back and forth. "It was great fun if you didn't have anything to do," Dillon said. Where Dillon would be comfortable with Kennedy, who shared some of his elitist background, he would later find himself horri-fied by Lyndon Johnson, who drove him out of the government by insisting that they hold consultations in bathrooms while Johnson performed a bodily function.

For both substantive and political reasons, Kennedy wanted to balance Dillon's selection as Treasury secretary with more liberal appointees to the Council of Economic Advisers and the Bureau of the Budget. His first choice as council chairman was MIT's Paul Samuelson, a brilliant economist whose 1948 textbook, *Introduction to Economics*, had become the most influential exponent of Keynesian theory, advocating deficit spending as a remedy to economic down-turns. When Samuelson rejected Kennedy's invitation, he turned to Walter Heller, professor of economics at the University of Min-nesota and another prominent Keynesian. Hubert Humphrey had introduced Kennedy to Heller during the 1960 campaign.

Kennedy had grilled Heller about actions that could boost the country's economic growth: Could they achieve a 5 percent expan-sion through a combination of a big tax cut and accelerated de-preciation allowances for businesses? Heller's concise affirmation of these options as the path to sustained development impressed Kennedy, who had little patience for complicated economic expla-nations and wanted a council chairman who could match Dillon's intellect and not be intimidated by him. When he invited Heller to take the council chair he candidly told him, "I need you as a counterweight to Dillon. He will have conservative leanings, and I know you are a liberal."

Before Kennedy selected Dillon and Heller to preside over economic affairs, he had invited David Bell, another Harvard academic, to become the director of the Bureau of the Budget. Because budgetary decisions would face him as soon as he became president, Kennedy wanted a budget director in place as soon as possible. Clark Clifford, Dick Neustadt, and Ken Galbraith suggested Bell; though he was only forty-one, his intellect and background made him worthy of consideration. Bell had served in Truman's budget bureau, worked in Pakistan on economic development, and recently taken a faculty post at Harvard. Kennedy sent his brother-in-law Sargent Shriver to interview him, and Shriver promptly reported that Kennedy would like Bell as a person and see him as well suited to the job. Shriver described Bell as "low-key, well-informed, experienced, un-ideological, sensitive, quick, somewhat ironic, and good-humored." Kennedy did indeed like him—not only because he was everything Shriver described but also because he echoed Kennedy's belief that the budget office should be more than an accounting agency; he urged expanded influence for a budget bureau committed to growing the economy.

A last consideration for Kennedy as he prepared to launch his administration was assigning some aides to develop a civil rights agenda. He was not keen to face up to a domestic issue that seemed likely to provoke conflicts that would distract him from what he believed were more compelling national security challenges. However, he understood that he could not just put the matter aside and have it fester for the next four years, and so he wanted people in place who would persuade liberals that he was not inattentive to reforms they believed should have a high priority. And he hoped that these advocates could find means to satisfy some of the complaints of aggrieved African Americans.

The key figure in assembling personnel and a program for the fight ahead was Harris Wofford. The thirty-four-year-old Wofford was a New Yorker devoted to idealistic goals such as world government and equal rights for all Americans. From 1954 forward, after undergraduate schooling at the University of Chicago and law studies at Yale and Howard University, the black college in Washington, D.C., where whites making a statement in favor of integration were in the minority, Wofford had served on the U.S. Commission on Civil Rights. His connection to Kennedy began when he wrote the senator to praise his thoughtful speeches on anticolonialism. In the spring of 1959, while writing a commission report on civil rights for Congress and the president and serving as a law professor at Notre Dame, Wofford received a personal invitation from Kennedy to join his presidential campaign. Kennedy's compelling appeal to him focused on foreign affairs and included a promise to "break out of the confines of the cold war." Although agreeing only to part-time work for Kennedy, Wofford suggested that Kennedy publish his speeches as a book. *The Strategy of Peace*, which Wofford edited, became a magnet to intellectuals looking for a candidate with an idealistic foreign policy plan.

But a need for campaign help on civil rights made Wofford Kennedy's point man on the issue. "In five minutes tick off the ten things a President ought to do to clean up this goddamn civil rights mess," Kennedy directed him during a car ride to his Senate office in August 1960. It spoke volumes about Kennedy's wish to put civil rights aside that he wanted only a five-minute tutorial on how to manage the greatest domestic challenge of 1960. Wofford's recommendation that a president bypass the Congress, where southern committee chairmen formed insurmountable obstacles to effective legislation, and rely on executive action struck resonant chords with Kennedy. Wofford suggested that he criticize Eisenhower and Nixon for not ending discrimination in federally assisted housing, by saying that here was a problem that could have been resolved "with one stroke of the pen." In collaboration with Shriver, Wofford had persuaded Kennedy to call Coretta Scott

King and help arrange the release of her husband from the Reidsville, Georgia, state prison where he had been sent for having an expired driver's license. In addition, Wofford had helped prepare a statement signed by Democratic senators pledging to carry out their party's platform pledge on civil rights legislation. Wofford also talked Kennedy into sponsoring a national conference on constitutional rights, at which he promised to support legislation and take executive action "on a bold and large scale," including the "moral question" posed by equal rights.

Wofford and other civil rights advocates were disappointed after the election, when Kennedy announced J. Edgar Hoover's reappointment as head of the FBI. Hoover's "agents in the South, all white, had been of almost no help to us," Wofford complained. Hoover's "antipathy to Negroes and the cause of civil rights was well known." In the interregnum between the election and the inauguration, Kennedy "gave only passing attention to civil rights." The economy and foreign dangers were foremost on his mind. Three weeks into the administration, when Chairman John A. Hannah, Michigan State University president and University of Notre Dame president Father Theodore Hesburgh, members of the Civil Rights Commission, complained to Kennedy during a White House meeting that no one had been appointed as a special assistant on civil rights, Kennedy said that he had given that post to Wofford. They replied that Wofford told them that he was working on establishing the Peace Corps. "Oh," Kennedy explained, "that's only temporary."

Within minutes, Wofford received a call to come to the White House. As he waited to see the president, "a solemn-looking man in a dark suit" appeared and asked Wofford to raise his right hand so that he could swear him in. "What for?" Wofford asked. The man didn't know and said the president would tell him after the swearing-in. Ushered into the Oval Office, Wofford was informed that he was now special assistant to the president on civil rights and instructed "to do these things we promised we were going to do." Kennedy explained, "The strategy for 1961 would be 'minimum civil rights legislation, maximum executive action.'"

Despite Kennedy's directive to Wofford, the watchword at the White House on civil rights was caution. Kennedy said more than once: These domestic disputes can wound an administration but unlike international conflicts, they can't kill you. Bobby Kennedy was convinced, however, that controversial rights disputes could create differences that would undermine the president's ability to lead. When Deputy Attorney General Byron White suggested that the White House coordinate all federal action on rights initiatives, Bobby directed that the Justice Department's Civil Rights Division, which had been established in 1957, take the lead. Moreover, Bobby opposed choosing someone with a high profile on civil rights to head the group; better to keep the whole thing as low-key as possible.

Wofford seemed like the natural choice for the post, but Bobby saw him as too "committed . . . emotionally. . . . [W]hat I wanted was a tough lawyer who could look at things objectively and give advice—and handle things properly," he said. Wofford was "a slight madman. I didn't want to have someone in the Civil Rights Division who was dealing not from the fact but was dealing from emotion and who wasn't going to give what was in the best interest of President Kennedy—what he was trying to accomplish for the country—but advice which the particular individual felt was in the interest of a Negro or a group of Negroes or a group of those who were interested in civil rights." Bobby wanted someone who would not automatically agitate protests from southern segregationists complaining that Kennedy was stacking the cards against states' rights advocates and throwing the White House into a distracting political battle.

Kennedy's choice for the post was thirty-eight-year-old Burke Marshall. He was a Yale Law School graduate, working at the Washington, D.C., law firm Covington & Burling, but with no special credentials as a civil rights lawyer; his focus had been on antitrust law for large corporate clients. But people close to Bobby recommended Marshall as a very smart lawyer who could handle a variety of issues. Marshall's interview did not go very well. His conversation with Bobby generated no chemistry between them. Eager for an appointment to the Justice Department, Marshall was terribly nervous, said little, and left Bobby believing that they couldn't work well together. But White and Wofford persuaded Bobby that Marshall would be just the sort of deputy he was looking for—very bright, unemotional about the tough issues facing them, and prepared to apply the law objectively. Marshall proved them right when his personal history and low-key, matter-of-fact responses to questions convinced Senator James Eastland of Mississippi, the chairman of the Judiciary Committee, not to oppose his nomination, though he would vote against him. "I'd vote against Jesus Christ if he was nominated for that position," Eastland told Bobby.

Executive action on civil rights was also a good place to occupy Lyndon Johnson. Kennedy hoped partly to satisfy his appetite for important assignments by telling him that leadership of the new President's Committee on Equal Employment Opportunity (CEEO) could blunt southern concerns: Having a southerner as head of a committee charged with combating discrimination in hiring of federal employees and by private businesses with government contracts seemed like a good way to mute southern fears of an aggressive push by the White House for integration. However, the likelihood that Johnson and Bobby Kennedy would have to cooperate to make CEEO a success put the program in doubt. They would have to overcome an intense dislike of each other that threatened the committee's effectiveness.

The president-elect also asked Johnson to head the National Aeronautics and Space Council, which acted as an advisory board to the National Aeronautics and Space Administration (NASA). NASA was essentially a response to the Soviet's successful launching of the first Sputnik, an earth satellite, which seemed to demonstrate an advantage in rocket technology. As majority leader,

Johnson had led the charge to establish a space agency in response to the Soviet achievement, but he resisted suggestions that it be an arm of the U.S. military; it seemed certain to provoke a battle among the armed services for rich budgetary resources. Instead, Johnson created NASA as a civilian agency principally devoted to scientific exploration. He hoped that the emphasis on greater understanding of the universe would forestall an arms race in space. He also expected NASA to have a significant impact on domestic affairs. The federal monies flowing into industries building spacecraft and the domestic sites that housed space operations were economic and political plums that a seasoned politician could use to his advantage. Kennedy saw Johnson's leadership of the Space Council as an ideal assignment that would provide an outlet for his considerable energy and would serve the administration's political interests.

The person most notably absent from Kennedy's inner circle of advisers was the thirty-one-year-old Jacqueline Kennedy, whose interests were much more in art and literature than in politics. She was never included in discussions of domestic or foreign policies or relied on as a sounding board for how to deal with the administration's daily challenges. Kennedy's idea of a wife's function was revealed in an anecdote Jacqueline related to Schlesinger about the deputy defense secretary Roswell Gilpatric and his wife. At a dinner in which Gilpatric's wife "was saying to Jack . . . 'I say to Ros when he comes home every night, How can they say those things about you? Aren't they all awful?' And he [Jack] said to her, 'My God, you don't say that to your husband when he comes home at night, do you? That's not what you should do. Find one good thing they say, say, Isn't that great? Or bring up something else that will make him happy.' And so, that's how I sensed what he wanted me to be," Jackie said.

Occasionally, however, curiosity about current events got the better of her, and she asked "about something painful. . . . I asked him something and it was at the end of the day. And he said,

'Oh, my God, kid . . . I've had that, you know, on me all day and I just. . . . Don't remind me of that all over again.' And I just felt so criminal. But he could make this conscious effort to turn from worry to relative insouciance." On another occasion when she inquired about a foreign crisis, he said, " 'Don't ask me about those things.' . . . So I decided it was better to live—you get enough by osmosis and reading the papers, and not ask. . . . And I decided that was the best thing to do. Everyone should be trying to help Jack in whatever way they could and that was the way I could do it the best—you know, by being not a distraction—by making it always a climate of affection and comfort and détente when he came home."

She had no desire to imitate Eleanor Roosevelt, who had set the standard for first ladies with high public visibility as an advocate for causes essential to neediest Americans regardless of race, ethnicity, or religion. Mrs. Roosevelt had become a spokesperson for human rights everywhere and after Franklin's death served the world community as America's first ambassador to the United Nations. But while Jacqueline had no intention of reaching for that exalted status, she also did not want to be like Bess Truman or Mamie Eisenhower, her two immediate predecessors, who had stayed in the background as conventional wives attending to household duties. By contrast, Jacqueline became a symbol of good taste and high culture—a first lady who encouraged Americans to see their White House as a monument to the nation's architectural and artistic history. She established the White House Historical Association and oversaw the publication of a historic guide to the White House describing the building's history and treasures. In 1962, she conducted a televised tour of the House that reached into millions of American homes. The Kennedy Center for the Performing Arts, which brought to fruition plans since 1958 for a National Cultural Center, became a permanent legacy of her commitment to American cultural life.

As he launched his administration on January 20, 1961, Kennedy believed that his appointments to cabinet and subcabinet posts had created a "Ministry of Talent." McNamara, Bundy, Rostow, Dillon, Sorensen, Schlesinger, and all the other academics, lawyers, financiers, industrialists, and public servants taking up residence in the White House and executive offices surrounding it were, in the journalist David Halberstam's later use of the term, the "best and the brightest."

Kennedy was innately skeptical about social engineering and human agency to greatly alter either international relations or domestic affairs. But he also believed that an American affinity for grand designs and bold actions meant that the country wanted leaders who would reach for the ideal rather than settle for the ordinary. "I'm an idealist without illusions," he told Schlesinger. His Inaugural Address made clear his ambition for an extraordinary administration of great accomplishments. But he also cautioned, "All this will not be finished in the first one hundred days. Nor will it be finished in the first one thousand days, nor in the life of this Administration, nor even perhaps in our lifetime on this planet."

But with the array of exceptional men he had attracted to Washington, he had high hopes for something more—something better—than what most past presidents had achieved.

"Never Rely on the Experts"

Freezing weather on inauguration day could not dampen Kennedy's evident satisfaction at becoming the youngest and first Catholic president. Despite the twenty-degree temperature and heavy snow the night before, which had threatened to cancel the outdoor ceremony, Kennedy spoke to the thousands before him wearing neither hat nor overcoat, symbolizing his campaign theme of national strength and renewal. During the evening, he made the rounds of the many inaugural balls, dancing and chatting with friends and supporters until 2 A.M., when he slipped away to a private party at columnist Joseph Alsop's Georgetown home. He did not return to the White House until 3:40 in the morning. After less than four hours sleep, he began his day, arriving at the Oval Office at eight minutes before nine.

Kennedy faced the day and the coming challenges of his presidency with confidence. "Did you have any strange dreams the first night you slept in" Lincoln's bedroom, his friend Charlie Bartlett asked him. "No," an amused Kennedy replied, "I just jumped in and hung on."

He believed that the combined power and prestige of the office, joined to his political skills, boded well for a successful administration. Remembering that Woodrow Wilson had launched his administration by breaking long-standing custom and appearing in person before a joint congressional session to urge passage of tariff reform, Kennedy broke new ground five days after his inspiring inaugural speech by holding the first live televised presidential press conference. Couldn't "an inadvertent statement . . . possibly cause some grave consequences?" a reporter asked. Kennedy confidently dismissed the concern, saying the country would have "the advantage of direct communication." Pressed to explain the unusual neglect of domestic affairs in his speech, he answered that the American people are familiar with his national goals, but because his government was new on the world scene, he needed to describe our intentions and hopes to a divided world.

During swearing-in ceremonies for his cabinet and other presidential appointees, he declared himself confident that their service would advance the well-being of peoples everywhere. The Peace Corps and the Alliance for Progress, for example, two executive initiatives launched in the first weeks of his term, confirmed his conviction that he could rely on advisers to help him find the means to compete with communist appeals to the hearts and minds of Third World peoples and make him a successful foreign policy leader.

Kennedy's inspiration for the Peace Corps was partly a response to Nixon's accusation in their last debate, on October 13, that the Democrats were the "war party" that had trapped the United States in Korea. But the idea had been germinating for months. At a rally on the night of the thirteenth at the University of Michigan, Kennedy issued a call for international service by young Americans that, in William James's memorable phrase, could be "the moral equivalent of war." While Kennedy needed no one to tell him that mobilizing American ideals could be an effective response to communist cynicism about the United States as an exploitive imperial power, he looked to the country's history and current ideas for ways to translate this insight into significant programs.

The Peace Corps was grounded in the idealism of missionaries dating from the nineteenth century, who had been acting on convictions that they were God's instruments of enlightenment to non-Christians in Asia, Africa, and the Middle East. In the twentieth century, as U.S. influence expanded around the globe, academics and legislators had been discussing an organization of young volunteers committed to overseas service well before Kennedy asked his brother-in-law Sargent Shriver to tell him how this could be done. Against this backdrop, the Peace Corps, Ted Sorensen said, had "a hundred fathers." And in his response to the president, Shriver pointed to a host of organizations and people, including Democratic congressmen and senators, who had crafted plans for making a corps of volunteers a reality.

Kennedy believed that no one was better suited to head the Peace Corps than his brother-in-law. The forty-five-year-old Shriver was the offspring of a notable Maryland Catholic family dating back to the American Revolution. A graduate of Yale and its law school, Shriver was an idealist who worked with Charles Lindbergh's isolationist group, America First, against involvement in World War II. Shortly before Pearl Harbor, however, he joined the Navy as a lieutenant and won a Purple Heart for wounds suffered during combat at Guadalcanal. After five years in the service, he became an executive at Joe Kennedy's Merchandise Mart in Chicago, where he met Eunice Kennedy, Joe's daughter. They married in 1953 and Sarge, as he was called, became a member of Jack's political team.

As Kennedy understood, Shriver's forte was as a public servant helping the less advantaged at home and abroad. In time, he would become known for his commitment to a war on poverty, Head Start, Volunteers in Service to America (VISTA), the Job Corps, Community Action, Legal Services for the Poor, and, in collaboration with Eunice, the Special Olympics, competitions for handicapped athletes. For the moment, however, it was the directorship of Kennedy's Peace Corps that captured his passion for good deeds. His motto: To do well, do good.

When Kennedy signed an executive order on March 1 setting up the Peace Corps, he viewed the action as coming from a shared conviction with Shriver that it would make a difference not only in helping the less advantaged but also in advancing the national interest. Kennedy hoped the Peace Corps could become a model for how his administration would perform: a collaborative effort of the best minds and most well intentioned to create an innovative program serving both the world and the nation. By 1963, within two years of its founding, the Corps had enrolled 7,300 volunteers serving in forty-four countries.

The Alliance for Progress was a more focused expression of the Peace Corps ideal. With Castro's rise to power in Cuba, Kennedy was eager to counter and, if possible, abort his growing charismatic appeal in the Caribbean and across the Americas. During the presidential campaign, Kennedy had emphasized Eisenhower's failure to address hemispheric dangers from the poverty and misery that gave rise to Castro and made several of the southern republics vulnerable to communism.

Kennedy's inaugural speech partly focused on the region's problems and his eagerness to address them. He announced, "To our sister republics south of our border, we offer a special pledge: to convert our good words into good deeds, in a new alliance for progress, to assist free men and free governments in casting off the chains of poverty. But this peaceful revolution of hope cannot become the prey of hostile powers. Let all our neighbors know that we shall join with them to oppose aggression or subversion anywhere in the Americas. And let every other power know that this hemisphere intends to remain the master of its own house."

The name Alliance for Progress and the promise of a new initiative in dealing with Latin America rested on the suggestions of several advisers, but especially that of Richard Goodwin. As Bobby Kennedy said later, the president credited Goodwin with the felicitous phrase that gave the Alliance its name.

Goodwin was a brilliant twenty-nine-year-old graduate of Harvard Law School. In 1959, after clerking for the storied Supreme Court justice Felix Frankfurter, he became an aide in Kennedy's Senate office. During the campaign, he made himself indispensable to Kennedy as a speechwriter and expert on Latin America.

Goodwin was charged with describing Kennedy's vision of how his administration would improve relations with hemisphere countries. After consulting the *Washington Post*'s expert on the region, who asked a Cuban émigré friend at the Pan American Union to suggest a compelling name for a program of reform, they came up with "Alliance for Progress." Kennedy seized upon it as a worthy successor to FDR's Good Neighbor policy.

When Kennedy announced his commitment to the program in a White House speech before Latin American representatives and congressional leaders on March 13, 1961, he had no illusion that his words would dispel doubts about U.S. motives. He knew that many in the hemisphere dismissed the Alliance as nothing more than anticommunist rhetoric, calling it the "Fidel Castro Plan." But Kennedy hoped that in time it would produce results that could disarm some of the antagonism to the United States. Convinced that Goodwin had impressed himself on Latin American governments as a friend of the region, Kennedy made him deputy assistant secretary of state for inter-American affairs. As important, Kennedy took satisfaction from the belief that his administration had a core of wise advisers who, as with the Peace Corps, were helping him launch productive experiments in developing countries, where communists were aggressively competing for influence.

The need for wise counsel in dealings with Latin America and Cuba in particular had become apparent to him well before his inauguration. When told on January 3 that Eisenhower was breaking relations with Havana, which was aligning itself with the Soviet Union, Kennedy refused to comment privately or publicly on the decision. Nor was he willing to respond to a *New York Times* story on January 10 that the United States was training an anti-Castro force of exiles at a Guatemalan base. Eager to focus on a constructive program for Latin America and uncertain about how to meet Fidel Castro's challenge before hearing from national security advisers, Kennedy preserved his options by saying nothing.

At a pre-inauguration meeting on January 19, Eisenhower pressed Kennedy to adopt an aggressive policy toward Castro. He explained that he had authorized help to the "utmost" of anti-Castro Cubans and recommended an acceleration of support, declaring it essential to oust Castro before he could spread communism across the Caribbean.

Because promises of a new day in U.S. relations with Latin America implied a renewed commitment to self-determination for the southern republics, Kennedy was reluctant to adopt Eisenhower's prescription of forcing Cuba to conform to U.S. designs. But the pressure to do something was intense. CIA director Allen Dulles, who had helped shape Eisenhower's response to Castro, told Kennedy that Castro planned to export communism to other hemisphere countries, including Venezuela and Colombia, where they already had considerable power among the people.

Dulles's knee-jerk anticommunism gave Kennedy second thoughts about keeping him as CIA director rather than choosing someone who might have been less doctrinaire about Castro and his threat to the United States. But national security considerations and domestic politics ruled out any reconsideration of Dulles's tenure. Moreover, Dulles's family and personal history, including a grandfather and older brother who were secretaries of state, and his experience in the State Department, the Office of Strategic Services (OSS), the precursor to the CIA, and in the CIA itself, where he had served as director for seven years, made him an authoritative figure whose views seemed perilous to ignore. A Dulles resignation with leaks about White House opposition to his judgments on Cuba would have created a political maelstrom. Although he masked his doubts about the new young president, Dulles wondered whether Kennedy had the commitment and courage to meet the Soviet challenge. Self-confident that he knew what Kennedy needed to do to meet Moscow's and Havana's threat, Dulles hoped that Kennedy would follow his lead.

In 1961, no one responsible for U.S. national security could imagine dismissing the Soviet menace to the Western Hemisphere.

However strained the comparisons to communist success in winning control of Eastern Europe and China, and however much Joe McCarthy's assault on perfectly innocent people in and out of government had been discredited, Soviet subversion and fear of being labeled soft on Reds constantly shadowed Kennedy and his new White House advisers.

Dean Rusk, the administration's leading subordinate on foreign affairs, also held Dulles in high regard, and spoke about national security against a backdrop in which anticommunist hawks had brought down State Department professionals as having appeased Stalin at Yalta in 1945 and betrayed Chiang Kai-shek's Chinese Nationalists. Rusk advised that ousting Castro represented a sensible defense of U.S. interests in the hemisphere, but he warned that an open use of U.S. power might trigger serious uprisings all over Latin America, which would undermine the credibility of Washington's commitment to an Alliance for Progress. When the military chiefs weighed in with warnings about Castro's strengthening hold on power in Havana and apparent determination to export communism to other Latin American countries, Kennedy accepted the need to act against him. But he doubted the wisdom of an overt U.S. sponsored invasion of the island by Cuban exiles. He saw no alternative leader who commanded Castro's charismatic appeal in Cuba and across Latin America and could set up a representative government.

The great question then for Kennedy was not whether to strike against Castro, but how to mount an assault that brought him down without provoking accusations that the new government in Washington was no more than a traditional defender of selfish U.S. interests at the expense of Latin autonomy. A better question, which none of his advisers posed, was whether Castro represented a genuine threat to national security, and if not, was the administration principally responding to conservative political pressures that could throw Kennedy on the defensive and undermine his freedom to lead? The fact that no such questions were being asked spoke volumes about the mind-set that discouraged discussions of possibly more constructive actions. For all the rhetoric about a fresh approach to old problems, Kennedy and his team were as locked into conventional thinking as their predecessors.

The CIA and military chiefs saw Castro as the advance wave of Soviet control in the hemisphere. In response, they convinced themselves that the exiles could successfully invade Cuba and touch off a civil war that could be seen as an effort to replace Castro's repressive regime with a democratic government more beholden to the Cuban people than to Washington's dictates. But even if they were wrong about the effectiveness of an exile attack, they were convinced that Kennedy would be compelled to take military action if an invasion faltered and his new administration faced an embarrassing and perilous defeat. In short, the CIA and military believed it more important to bring down Castro with direct action, if necessary, whatever the cost to the new administration's image, than to jeopardize U.S. control in the hemisphere.

By contrast, Rusk and subordinates in the State Department advised Kennedy that an invasion could produce very grave political consequences in the U.N. and Latin America. In response, Kennedy asked whether the exiles could "be landed gradually and quietly . . . taking shape as a Cuban force within Cuba, not as an invasion force sent by the Yankees." In short, was it possible to deceive the world about the sources of an anti-Castro attack? Even if they "landed gradually and quietly," who would believe that they arrived in Cuba without U.S. support?

Richard Bissell, the CIA's deputy director of operations and the heir apparent to Dulles, convinced Kennedy, or at least sold him on the fantasy, that it could be done by landing Cuban exiles at the Bay of Pigs, an isolated inlet on Cuba's south coast about a two-hour drive from Havana. But how could an armed force have found the wherewithal—the ships to transport them and the arms and munitions—to fight a pitched battle without the backing of the United States?

A Groton graduate and Yale Ph.D. in economics with a reputation for brilliance as an inventor of the U-2 spy plane, Bissell, with Dulles's full backing, exuded confidence in his plans to oust Castro, which overcame Kennedy's skepticism. Not only did Kennedy, Bobby, Bundy, McNamara, and Rusk mute concerns about the obvious criticism certain to come from complaints about U.S. complicity in making an invasion possible, but they also had to embrace the assumption that a force of fourteen hundred or so exiles could defeat a much larger army of defenders. When former secretary of state Dean Acheson asked Kennedy how many Cubans he could put on the beaches and how many Castro could bring up to oppose them, Acheson responded to Kennedy's response by saying, "It doesn't take Price-Waterhouse to figure out that fifteen hundred aren't as good as twenty-five thousand."

Bissell advised Kennedy that if a battle on the beaches did not produce an immediate outburst of opposition to Castro across the island, the invaders would be able to escape into the nearby Escambray Mountains, where they could become a rallying force for a civil war that would eventually topple Castro. But Bissell neglected to tell Kennedy that to reach the mountains the invaders would have to cross some eighty miles of impassable swampland. Bissell didn't think it would ever come to that; he assumed that Kennedy would feel compelled to use U.S. forces to ensure against such a setback. In fact, the CIA planners did not believe the operation could succeed without direct U.S. military intervention, which they didn't tell Kennedy. Concerned, however, that Kennedy might just stand by his refusal to use American forces, Bissell arranged with American Mafia members, whose profitable gambling and prostitution operations in Cuba had been ended by the new regime, to try to assassinate Castro, transferring funds from the invasion budget to "pay the Mafia types." It was a measure of the shadowy world in which the CIA and Bissell in particular operated that they could consort with shady characters to kill a foreign head of government and use secret budgetary resources to pay for such criminality.

In brief, having been tasked, in the bureaucratic jargon of the day, to take down Castro, Bissell, joined by others in the CIA and U.S. military, would not declare themselves incapable of finding a way; they had extraordinary power at their command and every confidence that they could overturn Castro's government.

The emerging assumption of Kennedy and his White House advisers that they could successfully screen themselves off from domestic and foreign recriminations over a failed attack was more unrealistic than the conviction of U.S. officials that a stumbling attack would compel a White House response. Walt Rostow concluded that military planners as well as Bissell and others at the CIA believed administration passivity in the face of defeat was "inconceivable." As Kennedy later told Dave Powers, the CIA and the military "couldn't believe that a new President like me wouldn't panic and try to save his own face. Well they had me figured all wrong." But not entirely: It was difficult for the CIA-military planners to imagine that a president ready to authorize an operation would not take the next logical step to ensure its success. Their motto could have been, You don't get into a fight unless you intend to go all out to win.

Schlesinger, who had become a part of the conversation as an advocate of the president's domestic and international standing, echoed Rusk's concern that any invasion ascribed to the United States would produce "a wave of massive protest, agitation and sabotage throughout Latin America, Europe, Asia and Africa. . . . Worst of all, this would be your first dramatic foreign policy initiative," he told Kennedy three weeks into his term. "At one stroke, it would dissipate all the extraordinary good will which has been rising toward the new Administration through the world." Schlesinger suggested a possible "black operation" that could lure Castro into "offensive action" that in turn would give Kennedy political cover for striking at his regime. Schlesinger was as intent on protecting Kennedy's standing as on toppling Castro. It was an unconvincing response to Kennedy's hope of finding some way to fool people about an invasion; but in the midst of the Cold War, when almost anything seemed acceptable as an answer to communist expansion, especially in defense of an admired new president whose administration could be put in early jeopardy, even the smartest of advisers succumbed to the lure of international skulduggery. Such an unworkable proposal, however, eroded Schlesinger's credibility with Kennedy and diminished his prospects of remaining a part of the president's inner circle.

Part of Schlesinger's problem was that American military planners unsuccessfully cooked up a similar plan. During the Bay of Pigs attack in April, a U.S. ship with 164 troops aboard dressed in Castro army uniforms were ready to stage an assault on the U.S. naval base at Guantánamo as a pretext for an American military intervention to assure the success of the exiles. The plan had to be scrapped, however, when a small advance element of the force ran into a Cuban patrol. While Kennedy had limited knowledge of the operation—assuring him of plausible deniability—he concluded that the military's eagerness to end Castro's government outran common sense.

During February and March, Kennedy's advisers struggled to come up with a better plan for bringing down Castro. But on March 11, prodded by renewed warnings from the CIA and military chiefs that delay would increase the difficulty of toppling Castro's government, Kennedy declared himself "willing to take the chance going ahead." It was also clear to him that if he dropped the invasion plan, it would mean having the exiles wandering around the United States complaining that Kennedy had lacked the nerve to risk a Cuban attack. It would open him to conservative complaints that he had failed to defend the national security. He feared accusations reminiscent of ones made against him during the campaign—that he was too inexperienced or, worse, that he was naïve about the Soviets and too ambivalent or wishy-washy to deal with the communist challenge.

As Bobby Kennedy said about his brother, "if he hadn't gone

ahead with it, everybody would have said it showed no courage. Eisenhower trained these people [the exiles]; it was Eisenhower's plan; Eisenhower's people all said it would succeed—and we turned it down." Memories of McCarthy's attacks on communist "fellow travelers" or liberal do-gooders too blind to communism's peril and too weak to confront the Sino-Soviet threat influenced Kennedy's decision to strike at Castro. The danger to Kennedy's presidency was more internal than external—Castro, as Senator J. William Fulbright asserted, was "a thorn in the flesh" but "not a dagger in the heart."

And so the safest political ground for Kennedy was to support a coup against Castro, but by means that masked America's role. When the CIA reframed "the landing plan . . . to make it unspectacular and quiet, and plausibly Cuban in its essentials," and Mac Bundy advised that they were close to a workable plan, Kennedy agreed to an invasion in mid-April. He tweaked the proposal by directing that it not be a "dawn landing . . . in order to make this appear as an inside-guerrilla-type operation."

It was an act of self-deception for Kennedy and his advisers to believe that they could effectively disguise the U.S. part in an attack. How they could think that people everywhere would accept the fiction of U.S. passivity is a demonstration that even the smartest men can talk themselves into foolish actions they consider essential for their larger effectiveness. As the philosopher Friedrich Nietzsche put it, "convictions are more dangerous enemies of truth than lies."

All manner of rationalization, however, could not quiet every doubt: Admiral Arleigh Burke cautioned that "the plan was dependent on a general uprising in Cuba, and that the entire operation would fail without such an uprising." It was the Joint Chiefs' way of implicitly pressuring Kennedy to understand that he might have to rely on direct U.S. military intervention to ensure a successful invasion. But Kennedy was determined not to use U.S. forces, which would confirm suspicions that the assault was nothing more than old-fashioned U.S. interventionism and would cripple the Alliance for Progress before it even started. At a meeting with national security advisers on March 29, Kennedy issued instructions telling exile leaders that "U.S. strike forces would not be allowed to participate in or support the invasion in any way . . . and whether they wished on that basis to proceed." When the Cubans said yes, Kennedy gave the final order for the attack.

Undersecretary Chet Bowles was one of the few unpersuaded by the auto-intoxication gripping the White House—the illusion that somehow commentators everywhere would not see a direct U.S. part in the attack and that even if they did, the toppling of a communist dictator like Castro would mute complaints. In a memo to Rusk, which Bowles asked be shown to the president, Bowles cautioned that the risks to U.S. prestige were more than anyone in the administration was willing to acknowledge. He saw the chances of a successful invasion as no more than one in three and the pressure on the president to intervene if the operation faltered as difficult to resist. He told Rusk that it would "jeopardize the favorable position we have steadily developed in most of the non-Communist world . . . by embarking on a major covert adventure with such heavy built-in risks."

Rusk assured Bowles that the operation was being "whittled down into a guerrilla infiltration" and filed away his memorandum. Rusk apparently believed that the attack could be kept so low-key that Kennedy didn't need to hear Bowles's concerns. Or more likely, Rusk was unwilling to press the case against a badly flawed plan Kennedy had decided to follow. It was a demonstration of his reluctance to be little more than a cipher in an administration intent on running foreign policy from the White House. It would do more to diminish Rusk in Kennedy's eyes than to increase Kennedy's regard for him as a smart adviser.

While Kennedy liked Rusk as a person, he came to see him as terribly ineffective in managing the State Department and, more important, as failing to provide helpful advice on Cuba and much else. Jackie Kennedy recalled that he saw Rusk as someone who "could never dare to make a decision. . . . Jack used to come home some nights and say, 'Goddamn it, Bundy and I get more done in one day in the White House than they do in six months in the State Department.' . . . And he used to say that sending an order to Rusk at the State Department was 'like dropping it in the dead letter box.'"

Richard Goodwin also questioned the viability of the invasion, warning that it couldn't succeed without direct U.S. military intervention, which would result in a bloodbath for the Cubans who would fight to save Castro. To "get rid of this irritating young man," as Goodwin recalled it, Bundy, confident that he knew better than the so-called Latin American expert, urged him to go see Rusk. Rusk was as unprepared to give Goodwin a serious hearing as Bundy: Rusk "listened patiently to my monologue, then—I'll never forget it—leaned back in his chair, pressed his fingertips together, hovered for a moment in this pose of thoughtful concentration, and then, slowly, pausing between each phrase: 'You know, Dick, maybe we've been oversold on the fact that we can't say no to this thing.' . . . I was beginning to understand the secret of Rusk's extraordinary staying power—say little, and, above all, go with the flow."

At an April 4 meeting in the State Department between Kennedy and the Joint Chiefs, Senate Foreign Relations Committee chairman Fulbright echoed Bowles's and Goodwin's objections, arguing that taking down Castro would be like swatting a fly with a hammer: A U.S.-sponsored invasion was wildly out of proportion to the threat and would badly compromise America's international standing, he said.

Schlesinger also cautioned that, "no matter how 'Cuban' the equipment and personnel, the US will be held accountable for the operation." On balance, he favored continued quiet anti-Castro actions but opposed an invasion. Against his better judgment, however, he fell into line with Kennedy's command. It is an example of a brilliant critic who sacrificed his independent judgment to the attractions of continuing access to power. Specifically, on April 7, the week before the attack, *New Republic* editor Gilbert Harrison gave Schlesinger an advance look at an article, "Our Men in Miami," describing CIA involvement with Cuba's exiles. Schlesinger believed that publication "would cause great trouble." He struggled over what to do—discourage Harrison from printing it or asking him to put patriotism above the public's right to hear the details of a questionable foreign policy action. He resolved the question by asking Kennedy's judgment.

The president predictably asked that Schlesinger do all he could to stop publication. Schlesinger successfully persuaded Harrison not to go ahead, but it "made me feel rather unhappy," he recalled. Kennedy had no regrets then or later about repressing a story that told the truth about America's role in a reckless operation. When he spoke to the American Newspaper Publishers Association ten days after the Bay of Pigs operation had failed, Kennedy did not commend some in the press for anticipating the administration's miscalculations, but cautioned them to think of the need for "a change in outlook . . . tactics . . . and missions," warning that Moscow was receiving through "our newspapers information they would otherwise have to acquire through theft, bribery or espionage." Although he assured the publishers that he would be scrupulous about press freedom, his Cold War warnings stirred fears of administration censorship.

Kennedy's argument resonated with Schlesinger. Shortly before the Bay of Pigs attack, Bobby Kennedy drew Schlesinger aside at a party to say, "I hear you don't think much of this business," and told him that since the president had made up his mind, he didn't think Schlesinger "should push it any further. Now is the time for everyone to help him all they can." Schlesinger did not quarrel with Bobby's advice. Bobby was not only reflecting his brother's wishes, which would become a constant of their close working relationship in the White House; he was also expressing his fierce determination to force Castro from power. No one in the admin-istration was as loyal to the president as Bobby, nor was anyone more committed to combating the communist threat everywhere.

Adlai Stevenson was yet another skeptic about the plan. But he was kept on the fringes of the operation, receiving on April 8, nine days before the invasion, only an unduly vague briefing by Schlesinger and a CIA official. As the plan moved ahead, Steven-son complained to Schlesinger and an assistant secretary of state for international organization that "he had been given no oppor-tunity to comment on it and believed that it would cause infinite trouble." Anticipating Stevenson's anger at being ignored, Kennedy told Schlesinger that nothing should "be done which might jeop-ardize . . . the integrity and credibility of Adlai Stevenson . . . one of our great national assets." Kennedy assumed that Schlesinger would pass his comments to Stevenson, who would accept that the president was intent on shielding him from sharing in the humili-ation if the operation failed.

But Kennedy had no interest in protecting Stevenson from some public embarrassment. In fact, by leaving him out of the discussion it led to his humiliation. When two planes bombed Castro's forces, Stevenson unwittingly repeated a CIA cover story in a speech be-fore the U.N. General Assembly. He described the raid as con-ducted by defectors from Castro's air force who had taken off from a Cuban airfield. Cuban exiles trained by the CIA actually had flown the planes from Key West, Florida.

When Stevenson learned the truth, he told Rusk and Dulles that he was "greatly disturbed," and asked why he was not "warned and provided pre-prepared material with which to defend us." He was mortified at having described the raid as a "clear case of attacks by defectors inside Cuba. There is gravest risk of another U-2 disas-ter," he warned, referring to Eisenhower's embarrassment at having to acknowledge that a U-2 reconnaissance aircraft was not tracking the weather, as he had publicly declared after it was shot down fif-teen hundred miles inside the Soviet Union, but was in fact a spy plane.

Stevenson understood that the failure to consult him was not an oversight at the White House, but a conscious effort to hold him at arm's length. Kennedy assumed that Stevenson would instinctively oppose the invasion and wanted to isolate him from the decision. Kennedy's assurance to Schlesinger that Stevenson was a great national asset was empty rhetoric: It was Kennedy's way of blunting the annoyance Stevenson was bound to feel at being excluded from the invasion discussion, especially after being told by Kennedy that becoming ambassador would allow him to have an impact on policymaking.

In the final days before the attack, Kennedy pressed to assure plausible deniability, though he had no illusion that the United States could entirely escape accusations of complicity. Yet every effort was made to combat such assertions, especially to defend Kennedy from charges of facilitating the assault: "lies . . . should be told by subordinate officials," and a final decision should be ascribed to "someone whose head can be placed on the block if things go terribly wrong," Schlesinger advised in a memo that may have ingratiated him with the president but does no credit to his historical reputation.

Despite all precautions and efforts to keep the operation as far from the White House as possible, Kennedy feared that he was backing a disaster. The weekend before the invasion, Kennedy joined Jackie and his sister Jean and brother-in-law Steve Smith at Glen Ora, a rented estate in Middleburg, Virginia. Following a long, late afternoon phone conversation with Dean Rusk witnessed by Jackie, which she recalled as "a decisive phone call," she heard him say, "Go ahead." "He looked so depressed when it was over. . . . Jack just sat there on his bed and then he shook his head and just wandered around that room, really looking—in pain almost, and went downstairs, and you just knew he knew what had happened was wrong. . . . It was just an awful thing." Jackie remembers him then as being "really low. So it was an awful weekend."

The operation was a miserable failure. One hundred and twenty-four invaders, including ten airmen, were killed and another 1,202 captured. And despite his determination to bar the U.S. military from a direct role in the invasion, Kennedy could not resist a last-minute appeal to use airpower to support the exiles. Buried in a CIA history of the Bay of Pigs failure, which did not become public until Peter Kornbluh at the George Washington University National Security Archive used a Freedom of Information suit to force its release in 2011, are details about the deaths of four U.S. Navy pilots whom Kennedy allowed to engage in combat as the invasion was collapsing. The White House and CIA directed the pilots to describe themselves as mercenaries if they were shot down and captured, and their valor was recognized only fifteen years after they were killed when the Pentagon honored them in a medal ceremony their families had to keep secret. Even more disturbing in this history is a CIA brief Kennedy never saw predicting a failure without direct U.S. intervention.

Kennedy paid an emotional price for the disaster, suffering considerable anguish about the lost lives and the men confined in Castro's prisons. His repeated refrain: "All my life I've known better than to depend on the experts. How could I have been so stupid, to let them go ahead?" Jackie Kennedy remembered him breaking down and crying in the privacy of their bedroom after the defeat. He "put his head in his hands and sort of wept." He cared about "those poor men who . . . were shot down like dogs or going to die in jail."

Analysts were set to work figuring out what went wrong. Was it the failure to allow exile air strikes from U.S. territory, which Kennedy prohibited after it was clear that initial air attacks came from Florida? Certainly, the CIA and military viewed as the culprit this restraint as well as Kennedy's refusal to allow U.S. forces to ensure a successful outcome after the invasion began. Kennedy saw merit in the argument that America's direct intervention could have changed the outcome. But as he told Dave Powers later and as he tried to make crystal clear before the invasion, the United States would not be the final arbiter in the attack. The invaders were allowed to succeed or fail on the assumption that an anti-Castro uprising would greet their landing or, more likely, would touch off a civil war that eventually would bring down Castro.

But without decisive U.S. military support the invasion was doomed from the start. And even if Kennedy had thrown American might into the attack, Castro enjoyed a measure of popularity that would have made an assault on Cuba a bloody battle with no guarantee of quick success. As the American journalist Henry Raymont, who was a correspondent in Cuba during the invasion, said in an interview in 2000, any diplomat or high school student in Havana could have told Kennedy that there would be no revolt in response to the exiles' attack. He also recalled a meeting with Kennedy in the Oval Office. After Raymont was released from a Havana prison, where Castro had confined him after the invasion, Kennedy asked to speak with him at the White House. He intended to chide the president for not having understood what was so obvious to everyone in Havana about Castro's popularity. But on seeing how distressed Kennedy was at the botched operation, Raymont did not add to his grief by reprimanding him for something that was now transparent.

The failure at the Bay of Pigs lay in allowing the invasion to go forward. And the judgment of most of Kennedy's advisers that it could succeed rested on illusory thinking. CIA and military chiefs assumed that a U.S.-sponsored invasion of so small an island led by a supposedly disliked leader strangling Cuba's economy was pretty much a sure thing. And if it began to stumble, no president could allow it to fail, especially one new to the office whose standing at home and abroad would suffer terribly from so embarrassing a defeat.

But the invasion failed, not only because Castro was more popular than the exiles and their U.S. sponsors understood, but also because it was dictated more by the state of American politics than realistic perceptions of Cuban affairs. The unacknowledged force driving Kennedy and his closest advisers—Bobby Kennedy, a principal advocate of the operation, Bundy, McNamara, Schlesinger, and Rostow—was the concern that calling off an attack would expose Kennedy and his administration to charges of fecklessness in the face of a communist challenge in the Western Hemisphere.

Kennedy did not shy away from this reality: In a conversation Schlesinger had with him on April 7, it was "apparent that he has made his decision and is not likely to reverse it. . . . 'If we have to get rid of these 800 men,' Kennedy said, 'it is much better to dump them in Cuba than in the United States.' I remarked that the political and diplomatic contingency planning was much less advanced than the operational planning. He agreed vigorously."

Kennedy did not want the exiles on the loose in the United States complaining that he had stopped them from taking back their homeland. As Bobby Kennedy said, they feared the repercussions from bringing the Cubans back to the States from Guatemala after the CIA and military chiefs had said that their invasion would succeed. To abandon the plan would have subjected the president to charges of weakness or a lack of courage.

In addition, Kennedy was acknowledging to Schlesinger that too little attention had been paid to the political consequences of a possibly unsuccessful invasion. They should have thought about how they would counter the political outcry should the attack fall short. And so after the invasion failed, they struggled with questions of how to blunt the negative responses at home and abroad. Conservatives Ronald Reagan, the Hollywood actor and TV spokesman for General Electric, Arizona Republican senator Barry Goldwater, and *National Review* editor William F. Buckley, Jr. denounced Kennedy's "do-nothing policy" on Cuba for failing to use American forces to oust Castro. As bad or worse, Moscow took satisfaction from what it saw as the weakness of an inexperienced, young president who failed in his first attempt to defeat a communist regime. By comparison with Soviet actions in Eastern Europe to ensure that their satellite regimes in East Germany and Hungary did not collapse, Kennedy looked like an easy mark for aggressive Soviet moves to drive the Americans out of Berlin.

In a later conversation with Jackie Kennedy, Schlesinger recalled that "[w]e in the White House felt very badly, quite apart from the general horror of the thing, but we felt that we'd served the President badly. . . . All of us felt that we hadn't done the job that the White House staff ought to be doing . . . —we'd been too intimidated by all these great figures and hadn't subjected the project to the kind of critical examination it was our job to do." What Schlesinger neglected to say was that fear of domestic political attacks were every bit as important as bad advice from national security officials.

McNamara and Bundy, Kennedy's most visible national security advisers, tried to give him cover for the failure by suggesting that they resign. McNamara told him, "I was in a room where, with one exception [Fulbright], all of your advisers—including me—recommended you proceed. I am fully prepared to go on TV and say so." Bundy also offered to sacrifice himself: "You know that I wish I had served you better in the Cuban episode, and I hope you know that I admire your own gallantry under fire in that case. If my departure can assist you in any way, I hope you will send me off—and if you choose differently, you will still have this letter for use when you may need it."

Kennedy knew that to accept their offers would make him look too self-serving—throwing loyal aides to the wolves instead of taking responsibility for his flawed decision as commander in chief. Critics would compare him unfavorably to Harry Truman, who said of a president's place in the chain of command, "The buck stops here!" Kennedy was too smart a politician to miss this. Throwing out McNamara and Bundy would have compounded his failure, made it harder to recover his political standing at home and abroad, and created difficulties in finding replacements for the two very talented men he had in place.

Besides, Kennedy understood that McNamara and Bundy were

only following his lead. They were relatively passive observers who assumed that Kennedy knew what he was doing. Moreover, as inexperienced national security advisers, they saw themselves in no position to challenge the likes of Dulles, Bissell, and the supposed experts on an invasion, the military chiefs. Nor should one discount the reluctance of ambitious men like Bundy and McNamara to risk their tenure as powerful officials—jobs they took with the understanding that they had become historical figures. They did not relish thoughts of abandoning their positions before they had a chance to make a positive mark. They had offered their resignations more as a gesture than a realistic possibility. After all, those who deserved to go—and did—were Dulles and Bissell and possibly some of the Joint Chiefs.

Instead of firing anyone, Kennedy did the right and smart thing: He made clear to the press and public that he was the responsible officer of the government, adding, "there's an old saying that victory has a hundred fathers and defeat is an orphan." When questions persisted, he issued a statement: "President Kennedy has stated from the beginning that as President he bears sole responsibility. . . . He has stated it on all occasions and he restates it now. . . . The President is strongly opposed to anyone within or without the administration attempting to shift the responsibility." Privately, he told McNamara, "I am the president. I did not have to do what all of you recommended. I did it. I am responsible, and I will not try to put part of the blame on you, or Eisenhower, or anyone else."

Not only was it smart politics, but it also came from Kennedy's conviction that McNamara and Bundy had been as gullible as he was in accepting assurances of so-called experts that this was a well thought out, entirely doable operation. Kennedy shared McNamara's feeling that they were too little schooled in the ways of the Pentagon and covert operations and had been too deferential to the CIA and the military chiefs. McNamara said that he let himself "become a passive bystander." Kennedy later told Schlesinger

that he made the mistake of thinking that "the military and intelligence people have some secret skill not available to ordinary
mortals." His lesson: "never rely on the experts," or at least take
a skeptical view of their advice and consult with outside observers
who had a more detached view of the policy under consideration.

Kennedy was also genuinely appreciative of the selflessness that
McNamara and Bundy showed in offering to resign. So, instead
of firing them, Kennedy increased their opportunities to consult
with him. He moved Bundy from an office in the Executive Office
Building to one in the basement of the White House West Wing,
where he would have easy access to a Situation Room, a new center
of national security operations. McNamara was also invited to
have a regular presence there.

Kennedy's public response to the failure and his closest advisers was only one side of his reaction. Privately, he was furious
at the CIA's and the military's poor judgment. "Those sons of
bitches with all the fruit salad just sat there nodding, saying it
would work," Kennedy said of the Chiefs. He told Schlesinger,
"Can you imagine being President and leaving behind all those
people there?" He echoed the point to Jackie Kennedy, telling her
repeatedly, "Oh, my God, the bunch of advisers that we inherited!
. . . Can you imagine leaving someone like Lyman Lemnitzer?"
When she repeated this to Schlesinger, she added, "Just a hopeless
bunch of men."

At the same time, a handful of Kennedy's own appointees
greatly irritated him, especially Chet Bowles, who had not only
proved wiser in opposing the invasion but also leaked his dissent
to the press and made clear that the CIA and military had been
the principal advocates of the operation. Kennedy said of the CIA
that he wanted to "splinter" it "into a thousand pieces and scatter
it into the wind." Dulles and Bissell were told that they had to go.
"Under a parliamentary system of government it is I who would be
leaving," Kennedy told Dulles. "But under our system it is you who
must go." Kennedy was especially incensed at Bissell, who Bobby

Kennedy learned had held back a recommendation from a U.S. Army intelligence officer that the Bay of Pigs operation be canceled if air strikes against Castro's forces were not permitted.

The removal of Dulles and Bissell was done slowly, in order not to make direct connections to the Bay of Pigs or to ignore earlier achievements of devoted public servants. As Kennedy told some of his advisers at a breakfast meeting on April 21, he was "concerned that the entire blame for this not be placed on the CIA," and by keeping Dulles in place for the time being, it also "helped keep the Republicans off his back. As long as he was there, they couldn't criticize." As with the decision to allow the operation to go forward, domestic politics shaped Kennedy's handling of the aftermath. Dulles retired gracefully in September 1961, and Bissell left the following year in February, becoming the head of the Institute for Defense Analyses, a Pentagon think tank evaluating weapons systems.

While Kennedy sat on his anger toward Dulles and Bissell and shelved any impulse he may have had to dismantle or reform the agency as too controversial, he had few qualms about striking out at vulnerable liberal critics in the administration. Although Schlesinger had been a team player and gave Kennedy no cause to fire him, his cautionary memos had shown him to be wiser than the president. Kennedy didn't like being one-upped by anyone: When Bundy reminded him that Schlesinger had "opposed the expedition," Kennedy said: "Oh, sure, Arthur wrote me a memorandum that will look pretty good when he gets around to writing his book on my administration. Only he better not publish that memorandum while I'm still alive."

The principal fall guy in the administration was Chet Bowles. At a morning news conference on April 21, a reporter asked if it was true that Kennedy had reached his decision to approve the Cuban invasion against the advice of Rusk and Bowles. It angered Kennedy that the State Department had covered itself by leaking information about opposition from its top officials. Although it would consign Rusk to the fringes of Kennedy's principal advisers, domestic politics again dictated that he not be dismissed or treated so harshly that it became a topic of press discussion. It would deepen doubts about Kennedy's judgment if he had to oust his principal cabinet officer or overtly consigned him to a subordinate role in the administration less than a hundred days into his term.

But Bowles was another matter. He was a less visible representative of the party's liberals, and he closely identified with Stevenson, whom Kennedy had purposely ignored in the run-up to the invasion. To JFK and Bobby, who had been left out of the overt deliberations on the operation but was quietly consulted every step of the way, the liberals were too flabby or irresolute to have risked the hard choice of going forward with the attack. Moreover, they were not to be trusted. Bringing them into the discussion threatened press leaks that could have undermined the operation at the start.

Conflicting memos from Stevenson and Bobby in response to the Bay of Pigs failure deepened Kennedy's antagonism to the liberals. A cable from Stevenson to the president and Rusk on April 19 was a thinly disguised attack on Kennedy's misjudgment on Cuba. The Bay of Pigs disaster was "extremely dangerous to U.S. position throughout the world," he wrote. The Soviets and Castro now had the "moral" high ground. "This is at least partly due to lack of advance planning on how to defend our selves politically," which was code for Stevenson's complaint about not being consulted. "Everyone, of course," Stevenson pointedly told Kennedy, "friend or foe, believes we have engineered this revolution and no amount of denials will change their minds." How could you have been so stupid to believe that it would be otherwise, Stevenson all but told him. "Now we are in for a period of serious political trouble," Stevenson added. He warned against a "prolonged military stalemate in Cuba which we are committed to support." It would produce "grave difficulties in [the] UN" and "would be politically disastrous."

By contrast, Bobby urged his brother to recall that their intervention in Cuba aimed to prevent Castro's efforts to help "Communist agitators in other South American and Central American
countries . . . overthrow their governments. . . . Our long-range
foreign policy objectives in Cuba are tied to survival far more than
what is happening in . . . any other place in the world," Bobby declared with no reference to Berlin or what many saw as the growing
dangers in Africa and Southeast Asia. "Because of [Cuba's] prox-
imity . . . our objective must be at the very least to prevent that island from becoming Mr. Khrushchev's arsenal." Bobby considered
it essential to enlist the support of other Latin American countries
in combating Castro's subversion by whatever it might take. "If it
was reported that . . . Castro's MIGs attacked Guantánamo Bay
and the United States made noises like this was an act of war . . .
would it be possible to get the countries of Central and South
America through OAS [Organization of American States] to take
some action to prohibit the shipment of arms or ammunition from
any outside force into Cuba? . . . Something forceful and determined must be done. Furthermore, serious attention must be given
to this problem immediately and not wait for the situation in Cuba
to revert back to a time of relative peace and calm with the U.S.
having been beaten off with her tail between her legs."

Bobby Kennedy exaggerated the long-term threat posed by Castro. Or at least his analysis had not reflected Fulbright's more ju-
dicious understanding that Castro's Cuba was less of a peril to
U.S. power and influence in the Western Hemisphere than the
White House and most members of the national security establishment believed. The widespread fear that the United States was
falling behind Moscow in a competition for worldwide dominance
spurred unwise reactions, principally a growing military-industrial
complex that Eisenhower had cautioned against in his farewell address and a fear of Castro that continues to bedevil Americans.

Kennedy shared Bobby's concerns and wanted to punish those
who had misled him about the chances for success at the Bay of

Pigs as well as anyone who was undermining his authority by re-vealing that they had correctly anticipated what the invasion would bring. After his April 21 press conference, Kennedy was angry at reporters for pressing him to talk about the Cuban fiasco. "What the hell do they want me to do—give them the roll-call vote?" he told Salinger. "If I'm going to knock some heads together, now isn't the time to do it with everybody looking down the barrel at us." Later that morning "he was still burning." He couldn't under-stand what the newsmen expected him to say: "That we took the beating of our lives? That the CIA and Pentagon are stupid?"

He intended to straighten all this out, and soon. But for the moment the recriminations would have to remain hidden. And because it was too risky politically and to the national security to give direct expression to his hostility to the principal perpetrators of the failure, he attacked the most vulnerable of his perceived tormentors—the liberals, and Bowles in particular.

Harris Wofford got a glimpse of what was coming when "Salin-ger accosted me in a White House corridor. . . . 'That yellow-bellied friend of yours, Chester Bowles, is leaking all over town that he was against it,' Salinger almost shouted. 'We're going to get him!'" Wofford urged him to "get those who got us into this mess. . . . Allen Dulles and Company." But "this made no dent on the President's Press Secretary, who roared down the hall cursing."

The Bowles onslaught came openly from Bobby Kennedy, with the unstated but transparent approval of the president. At a cabi-net meeting on April 20, followed by an Oval Office discussion in-cluding the president, Johnson, McNamara, Bobby Kennedy, and Bowles, and followed by a National Security Council meeting on April 22, Bobby ripped into Bowles with "angry . . . tough, sav-age comments." Speaking for the State Department and himself, Bowles urged caution in the administration's future actions toward Castro. Bobby led the "fire eaters," Bowles called them, in "bru-tally and abruptly" brushing aside his comments.

Dick Goodwin, who was at the NSC meeting, captured the language and feel of the exchange and the president's anger toward a subordinate who had offended him. Bowles's

> tedious, bureaucratic verbiage . . . essentially concluded that . . . nothing could now be done; that Castro's power was secure from anything except an American invasion. (An accurate estimate.) When Bowles finished, Bobby exploded: "That's the most meaningless, worthless thing I've ever heard. You people are so anxious to protect your own asses that you're afraid to do anything. All you want to do is dump the whole thing on the president. We'd be better off if you just quit and left foreign policy to someone else. . . ." As the embarrassing tirade continued, the President sat calmly, outwardly relaxed, only the faint click from metallic pencil cap he was tapping against his almost incandescently white, evenly spaced teeth disrupting his silence—a characteristic revelation that some inner tension was being suppressed. I became suddenly aware—am now certain—that Bobby's harsh polemic reflected the President's own concealed emotions, privately communicated in some earlier, intimate conversation. . . . After Bobby had finished, the group sat silently, stunned by the ferocity of his assault, until the President—without comment on his brother's accusations—named a "task force" to develop a new Cuban policy from which the State Department was pointedly omitted. Shortly thereafter Bowles was fired.

Bobby Kennedy thought his attack on Bowles was entirely justified. Before Cuba, Bobby recalled that Bowles had already "irritated" them no end. He was all "long sentences and big words" that went nowhere and contributed nothing to immediate problems—just pie in the sky. As for Cuba, he talked "to the press too much. . . . He was rather a weeper." After the failure, "He came up in a whiny voice and said that he wanted to make sure that everybody understood that he was against the Bay of Pigs. . . . Everybody rather resented it." And then his recommendations for dealing with Castro were "God awful." Bobby told Bowles that his suggestions were "a disgrace," describing them later as "foolish . . . filled with generalities . . . [and] didn't make sense."

Bowles's open dissent over Cuba persuaded Kennedy to remove him as undersecretary of state and send him overseas. But ousting him provoked a conflict with liberals, who saw it as a loss of influence over foreign policy. In July, when Kennedy tried to send him abroad as a roving ambassador, Bowles resisted the demotion and relied on pressure from liberal allies in and out of the government to force Kennedy to back down. The resistance angered Kennedy, and he privately described his determination to force Bowles to leave. At a press conference on July 19, when reporters asked Kennedy if rumors of Bowles's departure were true, Kennedy's endorsement of him was so equivocal that few doubted his days were numbered. His ouster took until the end of November, when the discussion of Cuba had quieted and Kennedy appeased him and liberals by making Bowles a "special representative and adviser for Asian, African, and Latin American affairs," and naming George Ball, another liberal, as his replacement and Averell Harriman, a prominent FDR diplomat, as assistant secretary of state for Far Eastern affairs.

Kennedy blunted liberal anger with the appearance of continuing influence for Bowles and the elevation of Harriman to a high State Department post. But he could not counter the inevitable doubts planted in the minds of other advisers that open opposition to a presidential policy would diminish their standing at the White House or even lead to their removal from office. Most everyone was convinced, regardless of what Kennedy or Bobby might say, that open or even quiet criticism of a presidential directive would jeopardize an adviser's access to White House deliberations and his own influence. However much Kennedy may have valued genuine debate among Oval Office counselors, Bowles's departure was bound to discourage criticism. In short, subsequent crises did not benefit from Kennedy's overreaction to Bowles's open criticism of the Bay of Pigs failure: Being wiser than the president and Bobby about any high-visibility issue was best left unsaid, certainly in public, but Bowles's ouster made even private dissent less likely from ambitious men excited by the chance to continue in high office.

The principal consequence of the post–Bay of Pigs pressure on White House subordinates to march in lockstep with the president was a sterile approach to Cuba that did far more harm than good. After the Bay of Pigs failure and Bobby Kennedy's warning that they could not afford to ignore the Cuban danger and might have to reconsider armed action, Secretary of Defense McNamara directed the military to "develop a plan for the overthrow of the Castro government by the application of U.S. military force." He cautioned the Chiefs against seeing U.S. military action as probable, but he acknowledged that the defeat had compelled its reconsideration.

Kennedy, however, had no intention of rushing into anything. The Cuban failure had made him more cautious and determined to ensure that any future action would be effective. He was mindful of what Eisenhower had told him after Kennedy had asked his advice about how to avoid another failure like the Bay of Pigs. "I believe there is only one thing to do when you get into this kind of thing," Eisenhower said. "It must be a success." Kennedy assured him "that hereafter, if we get into anything like this, it is going to be a success." Moreover, the "disaster," as many were calling it, intensified whatever doubts he had about listening to advisers, or at least to the men at the CIA, Pentagon, and State Department who had misled him or had passively joined in accepting the bad advice. As Bobby told him, "What comes out of this whole Cuban matter is that a good deal of thought has to go into whether you are going to accept the ideas, advice and even the facts that are presented by your subordinates." Bobby expected the president to ask a lot harder questions of his counselors in the future: "And that is going to be the difference between the President before Cuba and after Cuba. The fact that we have gone through this experience in Cuba has made the President a different man," Bobby asserted. Or at least it had convinced him and Kennedy that they needed to treat advice from experts and even their closest confidants more critically.

Kennedy now wanted a new voice in future discussions of military action. He turned to General Maxwell Taylor, a retired Army commander with impeccable credentials for intelligence and integrity. The fifty-nine-year-old Taylor was a storybook figure—six feet tall, handsome, the model of what a general should look like, in a perfectly pressed uniform with four stars on the epaulettes and twelve rows of battle ribbons beneath the insignia of the 101st Airborne Division, which he commanded when parachuting into France on D-Day in June 1944; he had been the first general to land in France during the invasion. A secret mission to Nazi-occupied Italy the previous year, which Dwight Eisenhower described as the riskiest undertaken by any agent or emissary he dispatched during the war, had made Taylor a prime candidate for command in the post-invasion campaign, which included leadership of the division that won high regard for its courageous stand at Bastogne in Belgium during the Battle of the Bulge.

At the end of World War II, Taylor continued to serve with distinction: first, as superintendent of the U.S. Military Academy between 1945 and 1949, where he established West Point's Code of Honor; then as commander of Western Allied forces in Berlin from 1949 to 1951; followed by combat duty in Korea in 1953 and service as Army chief of staff from 1955 to 1959. An unqualified opponent of President Eisenhower's strategy of massive retaliation, which focused U.S. defense on building a nuclear arsenal, Taylor retired from active duty as a protest against the downgrading of the country's land army. In 1960, he made his opposition public in *The Uncertain Trumpet*, in which he argued that excessive reliance on nuclear weapons, which could not be used in brushfire wars, limited America's ability to meet the variety of threats communist insurgents in Asia, Africa, and Latin America seemed likely to pose. The United States needed an army and a variety of forces that could give it the flexibility to meet every sort of peril. "Flexible response," as Taylor's strategy was called, coincided with Kennedy's criticism of Eisenhower's "brinksmanship" or Maginot Line mentality, which left little room for alternative answers to communist aggression. Described as "articulate, dashing, urbane," Taylor was not only a battle-tested hero, but also a veteran of Pentagon politics, which recommended him to Kennedy as someone who might be more helpful than any of his current military or national security advisers in future debates about how to manage Cold War tensions and conflicts.

In the first weeks of his term, it was not only the Bay of Pigs that had soured Kennedy on the military chiefs; it was also advice they had given him on Laos. Like Cuba, the small landlocked country had seemed like a proving ground of Kennedy's ability to stand up to the communists. But everything told Kennedy that getting drawn into a land war in the Laotian jungles was a losing proposition. At the end of April, as he was reeling from the Cuban defeat, the Joint Chiefs recommended that he blunt a North Vietnamese–sponsored communist offensive in Laos with air strikes and the use of army units ferried into the country's two small airports. Kennedy wanted to know what the Chiefs proposed if the communists bombed the airports after the United States had put a few thousand men on the ground and they were faced with defeat. "You drop a bomb on Hanoi—and you start using atomic weapons!" Lemnitzer replied. In this and other discussions about combating North Vietnam and China or intervening in Southeast Asia, Lemnitzer promised, "If we are given the right to use nuclear weapons, we can guarantee victory." Kennedy dismissed this sort of thinking as absurd: "Since he couldn't think of any further escalation, he would have to promise us victory," Kennedy said.

Kennedy was not indifferent to the fate of Southeast Asia. He had a long-standing concern with how the United States should deal with former colonial regimes. In 1957, after visits to the Middle East and Asia, he had said in a Senate speech, "The most important single force in the world today is neither communism nor capitalism, neither the H-bomb nor the guided missile—it is man's eternal desire to be free and independent." A great test for U.S. foreign policy, he added, was how former colonial nations viewed America's attitude toward imperialism. Current wisdom described the struggle for hearts and minds in the Third World as crucial in shaping the outcome of the Cold War. Nuclear weapons had made an all-out conflict like World War II an anachronism—an event that could only lead to what strategists called MAD, mutual assured destruction. The alternative was the contest of ideologies—communism with its command economy versus democracy and free enterprise. The winning side in that argument seemed likely to flourish in emerging societies, with the likelihood that the less attractive way of life would eventually wilt and disappear.

The most immediate and concrete examples of this competition Kennedy saw were in Latin America, where Castro's Cuba vied to become a model for other Western Hemisphere nations, and Southeast Asia, where the former French colonies of Laos, Cambodia, and Vietnam were battlegrounds between pro-Western and communist forces. Kennedy's support of the Bay of Pigs operation had rested on the hope of turning Cuba away from socialism and toward progressive democracy promoted by the Alliance for Progress.

But the uncertainty of ousting Castro made Kennedy eager to avoid any defeat in Southeast Asia. Laos, however, was not a high priority. With the exception of some of the Joint Chiefs, all his advisers agreed that a military adventure in that landlocked country would be a terrible drain on resources, with little chance of a decisive outcome. Khrushchev's understanding that Moscow as well would not profit from an extended contest in Laos persuaded him to endorse a coalition government that would largely mute Soviet-American tensions over that tiny country, whose only strategic value was as a North Vietnamese supply route to Viet Cong guerrillas in South Vietnam.

The much greater concern than Laos was Vietnam. While hardly anyone in the United States knew much, if anything, about Laos, Vietnam had registered forcefully on Americans attentive to Cold War ups and downs. French defeat at Dien Bien Phu in 1954 had disturbing repercussions in Washington. Was the Viet Minh's victory, a triumph of Vietnamese nationalism, an indication that anticolonial movements across Asia, Africa, and the Middle East were about to be co-opted by the communist drive for world control?

A conference in Geneva, Switzerland, that included the United States, Britain, France, the Soviet Union, and the People's Republic of China divided Vietnam into northern and southern states, with unifying elections slated by 1956. The refusal of the United States and South Vietnam to sign the Geneva accords without assurances that free elections would be supervised by the United Nations left Vietnam divided into opposing regimes. The outbreak of a Viet Cong communist insurgency against the pro-Western South's government led the Eisenhower administration to provide financial and military aid to Saigon. The support became emblematic of the struggle to contain communist expansion. Eisenhower's comparison of South Vietnam to a leading domino that, if toppled, could result in communist control of all Southeast Asia endowed that country with a strategic importance comparable to that of Western Europe. Losing South Vietnam to indigenous Viet Cong guerrillas backed by North Vietnam's communists invoked memories of Neville Chamberlain's appeasement of Hitler at Munich in 1938, which led to World War II and the near defeat of Great Britain. Communist conquest of the South loomed as an apocalyptic event that could translate into a worldwide victory for Soviet Russia.

More than thirty-five years after North Vietnam's conquest of the South, such fears seem ill-considered. But memories of the global struggle against Nazism, fascism, and Japanese militarism that had caused so much suffering shadowed any potential defeat in the new clash with communism for international dominance and discredited suggestions of passivity in response to any acts of perceived aggression. In 1961, Americans were fixated on Churchill's observation about Munich: Chamberlain had a choice between war and dishonor; he chose dishonor and got war. They ignored Churchill's later adage: Oppose the strong and appease the weak.

At the end of January 1961, Kennedy convened a high-level White House meeting to discuss a plan for saving South Vietnam from a communist takeover. He was responding to a report by the counterinsurgency expert Edward Lansdale, who was a deputy assistant secretary of defense for special operations. Following a two-week visit to Southeast Asia, Lansdale declared Vietnam in critical condition. He described the country "as a combat area of the Cold War . . . requiring emergency treatment." Kennedy saw the situation in Vietnam as "the worst one we've got," complained that Eisenhower had never discussed the country with him, despite having committed U.S. resources to preserving Saigon's autonomy after French defeat in 1954, and praised Lansdale's report as giving him "a sense of the danger and urgency of the problem in Vietnam."

Suggestions that helping South Vietnam increase its army from 150,000 to 170,000 men could turn the tide in the conflict, however, left Kennedy unconvinced. He thought that politics and morale rather than military capacity would make the difference in combating the communists. But his advisers weren't so sure. The

French failed to hold Vietnam by strength of arms, but the United States wasn't France. Despite the conviction of some in the State Department and at the Pentagon that expanded military operations in Vietnam would make a difference, Kennedy doubted that this was the key to victory. He believed that South Vietnam's survival depended less on growing its army than on political reforms that could increase the government's popularity. His national security advisers were skeptical about the viability of democratic reforms and believed, at any rate, that military capacity had to come first.

Because political change at best would be slow, Kennedy's advisers said that prompt effective military action to combat the Viet Cong, the communist insurgents, was essential. It needed to rest, however, not on pitched Korean-style battles for which the military had prepared, but on an innovative strategy that met the challenge of guerrilla operations in jungle terrain. In February, remembering the adage that generals always prepare to fight the last war, Kennedy pressed the case for novel counterinsurgency units, telling McNamara and the Joint Chiefs that he expected there to be "more guerrillas and counter-guerrilla activity in Africa and Asia in the near future." McNamara advised the Chiefs that the president wanted these forces and plans to use them to be pursued "with all possible vigor." Rusk cabled the U.S. Embassy in Saigon that the "White House ranks defense Vietnam among highest priorities US foreign policy. Having approved Counterinsurgency Plan, President concerned whether Vietnam can resist Communist pressure during 18–24 month period before Plan takes full effect."

And, Rusk might have added, until political reforms could take hold. No one, however, thought it would happen as long as Elbridge Durbrow, the current ambassador, remained in office. Durbrow was a plainspoken diplomat encouraged to be blunt about the radical reforms needed to save Vietnam, and the Vietnamese, who complained that the Americans had replaced the French as their colonial masters, dismissed Durbrow's hectoring as offensively imperious. During his four years as ambassador beginning in 1957, Durbrow's relations with South Vietnam president Ngo Dinh Diem had deteriorated into mutual contempt. Durbrow viewed Diem as a corrupt dictator, whose indifference to his people's sufferings doomed his regime, and Diem considered Durbrow a neocolonialist trying to reimpose Western control over Vietnam. By 1961, they were barely speaking to one another.

To implement the political side of his strategy, Kennedy appointed Frederick E. Nolting, Jr. as the new ambassador. A highly recommended Foreign Service officer, who had been a U.S. representative to NATO but was unfamiliar with Asia and knew nothing about Vietnam, the fifty-year-old Nolting had a reputation for tact and ability to ingratiate himself with others that made him an attractive candidate for the job. Besides, Kennedy didn't want an Asian expert with fixed ideas about Southeast Asia. Part of Nolting's appeal was that he could bring a fresh perspective to fixing the Vietnam problem. Fearful that Vietnam faced an imminent collapse, Kennedy wanted Nolting to take up his post as quickly as possible. Convinced that the Vietnamese would be more responsive to cooperation between equals—an approach reinforced by Lansdale's recommendation that a pat on the back would be far more effective with the Vietnamese than harsh directives— Kennedy hoped that skillful diplomacy could persuade South Vietnamese president Diem to enact reforms that would outdo the Viet Cong's popular appeal.

It was a mistake. The fifty-year-old Diem, a Catholic who had lived in France and the United States from 1950 to 1954, was unreceptive to Washington's advice, or more to the point, he was an authoritarian and perhaps paranoid personality who resisted American pressure as an extension of Western colonialism. But he was also a shrewd manipulator of Americans who naïvely took his promises at face value.

After Vietnam was partitioned into northern and southern states following the collapse of French rule in 1954, Diem, with the support of leading American Catholics, including New York's

Cardinal Spellman, and the Eisenhower White House, had become prime minister in Saigon. In rigged elections the following year, in which he won 98 percent of the vote, Diem assumed the presidency of the Republic of South Vietnam. A stubborn mandarin who hoped to command U.S. aid with no strings attached, Diem took Lansdale's and Nolting's show of regard as license to do as he pleased. The Kennedy administration's replacement of Durbrow with Nolting encouraged Diem's conviction that he could maintain U.S. support strictly on his terms. From the first, he saw Nolting's eagerness to accommodate him as a godsend.

Yet Nolting's embassy staff did not share his conviction that Diem knew his own best interests. They said of Diem: "He was too weak to rule and too strong to be overthrown. His forces were corrupt, his generals held title on the basis of nepotism and loyalty, his best troops never fought." Despite "mounting terrible pressure," he listened only to "trusted family and sycophants. It was the sign of a dying order."

A National Intelligence Estimate predicted that Diem's control would become increasingly precarious as a consequence of growing communist guerrilla strength and widening discontent with Diem's abuse of power and indifference to public opposition. The intelligence analysts anticipated a repetition of a failed November 1960 coup by noncommunist elements in the coming year. They also saw U.S. prestige as deeply engaged in South Vietnam, and warned that losing the country to the communists would be a severe blow to U.S. security. The White House was less certain that Vietnam's collapse would jeopardize the national safety, but defeat there seemed certain to be a major political setback for Kennedy's fledgling government. No one could forget how politically destructive the loss of China to communism had been for Truman and the Democrats.

Among Kennedy's White House advisers, none pushed harder to beat back the communists in South Vietnam than Walt Rostow. As deputy special assistant to the president for national security affairs and the author of his famous 1960 book on how developing societies could achieve economic growth in a free enterprise system, he was in a position to press Kennedy for action on the conflict in Vietnam. An MIT professor of economic history with unlimited faith in social science engineering, Rostow won Kennedy's support as someone who had identified ways to foster economic and political advance without excessive reliance on military force. His evident intelligence and enthusiasm for rational planning joined with endearing qualities of personal warmth to encourage Kennedy to think that Rostow had a formula for defending Vietnam from a disastrous outcome. But he was leery of Rostow's affinity for excessive enthusiasm about his own ideas. On hearing that Walt was giving a seminar on underdeveloped countries, Kennedy exclaimed: "Jesus Christ! . . . Walt Rostow's got all those people trapped in there, listening to him?" "He really thought Walt Rostow went on and on, and was hard to listen to. . . . But he liked him," Jackie recalled. "He never said anything mean about him."

On April 12, 1961, after Diem had won a second five-year presidential term with 90 percent of the vote, Rostow urged Kennedy to gear up the whole Vietnam operation. He suggested sending Vice President Johnson to Saigon, where he was to underscore U.S. determination to help and would hand-deliver a letter to Diem emphasizing "the urgency you attach to a more effective political and morale setting for his military operations" and the need for a broader base of his government. Rostow also advised Kennedy to expand the number of U.S. Special Forces in the country, but as only a temporary measure that would diminish as Diem introduced the necessary social reforms.

Although Kennedy agreed to increase Special Forces from the 685 Eisenhower had sent to just over a thousand and to write the proposed letter, Kennedy refrained from advising Diem on how to defend his country. A CIA report that the elections were rigged, an NSC warning that it might be impossible to convince Diem that his current behavior could prove fatal, and a Ted Sorensen memo raising doubts about America's ability to save Vietnam unless it moved to save itself, together gave Kennedy pause about becoming too involved in a country possibly beyond rescue. A letter from Galbraith in India gave Kennedy fair warning about the Vietnam problem: Recent developments in South Vietnam suggested "a disintegrating economy which is the cause and consequence of a disintegrating government. . . . American aid, though considerable, is insufficient—perhaps partially as a result of egregious misuse. Maladministration and corruption are general. Underneath is a nauseous social situation in which the landlords and politicians rape the poor with an energy, which they apply to no other purpose." The response always seemed to be to "send a high level mission. This is done partly because no one can think of anything else to do."

For all these doubts, it was clear to Kennedy that he could not abandon Vietnam or effectively refute assertions about its importance to the United States. It might be that no one knew what to do about South Vietnam, but letting the communists in Hanoi seize it did not seem like an acceptable option. On May 3, Deputy Secretary of Defense Roswell Gilpatric sent Kennedy an interdepartmental task force proposal urging consideration of a defensive alliance, including stationing U.S. forces in Vietnam. Speaking for the Joint Chiefs and a larger body of American opinion, which after other communist gains in Europe and Asia saw Vietnam as a line in the sand, Chiefs chairman Lyman Lemnitzer asked: "Does the U.S. intend to take the necessary military action now to defeat the Viet Cong threat or do we intend to quibble for weeks and months over details of general policy, finances, Vietnamese Govt organization, etc., while Vietnam slowly but surely goes down the drain of Communism?"

For all Kennedy's skepticism about involvement in a jungle war that could provoke cries of U.S. imperialism, he also saw Vietnam as a testing ground the United States could not ignore: It fit Khrushchev's description in a January 1961 speech of a war of national liberation, which, if successful, could become a model for other Third World communist insurgencies. Kennedy wanted to discourage the belief that the United States could not defeat these guerrilla movements or that the will and means was lacking to promote democratic freedoms and prosperity, if necessary by military might, but principally by cooperative initiatives of the kind proposed in the Alliance for Progress.

A Johnson trip to Asia from May 9 to May 24 was meant to signal Kennedy's determination to hold the line in the region against communist advance. There is no record of a conversation between Kennedy and Johnson preceding the trip, but based on Johnson's behavior in the seven countries he visited—Laos, South Vietnam, Taiwan, the Philippines, Thailand, India, and Pakistan—it is hard to believe that Kennedy did not give him marching orders to draw all possible attention to his presence in these East-West contested nations or at least did not counsel him against typical high-visibility Johnson pronouncements and actions.

It is clear that Kennedy wanted to give a rudderless, energetic vice president something to fill his time and get him out of Washington, where his frustration at being a fifth wheel was evident to White House observers. As Kennedy told Florida senator George Smathers, an old friend with whom he could speak openly, Johnson was something of an eyesore: He came to cabinet meetings, said nothing, and sat looking forlorn and rejected—a sad child excluded from the circle of most popular teenage boys and girls. When Smathers urged Kennedy to send Johnson abroad to visit countries where he could become the center of attention and "all of the smoke-blowing will be directed at him," Kennedy called it "a damn good idea."

Kennedy didn't need to urge Johnson to draw attention to himself. LBJ had a lifelong affinity for center stage, including outlandish actions that were familiar Washington gossip and a source of much amusement. During his time in the House, visitors to his office described how he would keep a conversation going while he urinated in a sink screened off behind his desk. Or when he was Senate majority leader, how he would engage in recreational sex in what he called his nookie room.

In Vietnam, where a motorcade from the airport turned into a campaign-style event, Johnson repeatedly stopped his limo to shake hands with obedient onlookers directed by the government to give the American a warm reception. Handing out pens, cigarette lighters, and visitors' passes for the U.S. Senate gallery, Johnson told bewildered Vietnamese recipients to come see American democracy at work. In a passionate, arm-waving speech in the center of Saigon, before many who knew no English, he praised Diem as the Winston Churchill of Asia, suggesting that the South Vietnamese president was filling the role of a savior against totalitarian communism just as Churchill had saved Europe from Nazism. A photo op in a pasture with a herd of Texas-bred cattle followed by a press conference in his hotel room, where he unself-consciously changed clothes in front of reporters, were meant to demonstrate that Americans were not haughty imperialists but plain good folks who wanted nothing more for their Vietnamese cousins than the chance to enjoy traditional American freedoms.

Clearly, more was at work here than just a feel-good trip for Johnson. Kenneth Young, the American ambassador in Bangkok, Thailand, saw Johnson's tour as a "timely and gallant enterprise of purpose [that] accomplished the missions originally conceived in Washington. He reached the politicos, the administrators, and the people. Saigon, Manila, Taipei, and Bangkok will never be quite the same again, for a new chapter has opened in US relations with Southeast Asia. The friendship and sincerity of the Vice President and Mrs. Johnson were felt and returned. They came, saw, and won over."

Yet as Johnson told Kennedy in a written report that was promptly leaked to both American and Vietnamese journalists, this was an administration that understood the dangers in Southeast Asia and how to combat them. Johnson warned that "the battle against Communism must be joined in Southeast Asia with strength and determination . . . or the United States, inevitably, must surrender the Pacific and take up our defenses on our own shores." Unless America exercised an "inhibitory influence . . . the vast Pacific becomes a Red Sea." If so stark a description wasn't enough to scare anyone who read it, Johnson added, "The decision in Southeast Asia is here. We must decide whether to help these countries to the best of our ability or throw in the towel in the area and pull back our defenses to San Francisco and a 'Fortress America' concept." No one reading Johnson's report could doubt that the Kennedy administration was determined to save Vietnam. After the Bay of Pigs, Kennedy was declaring that he could not be accused of appeasing the communists.

Yet alongside so stark a description of the red menace, Johnson's recommendations—reflecting Kennedy's caution about unwanted and unproductive commitments—were distinctly restrained. Johnson saw "an obsessive concern with security on the part of many of our mission people." He discounted it by saying that "occasional murders in Rock Creek Park . . . do not mean that the United States is about to fall apart." He stressed that "a mere increase in the level of military aid on our part to Vietnam will not necessarily solve the difficulty. . . . There must be a simultaneous, vigorous and integrated attack on the economic, social and other ills of the Vietnamese peoples. The leadership and initiative in this attack must rest with the Vietnamese leaders." Above all, Johnson saw a need for stronger "democratic institutions in Vietnam." America's "mission people" in Saigon "must by . . . subtle persuasion encourage the Saigon Government from the President down to get close to the people, to mingle with them, to listen for their grievances and to act upon them."

Under prodding from the White House, Johnson offered unqualified advice against a substantial U.S. military effort in South Vietnam's conflict: "Barring an unmistakable and massive invasion of South Vietnam from without," he declared, "we have no intention of employing combat U.S. forces in Viet Nam or using even naval or air-support which is but the first step in that direction. If the Vietnamese government backed by a three-year liberal aid program cannot do this job, then we had better remember the experience of the French who wound up with several hundred thousand men in Vietnam and were still unable to do it. . . . Before we take any such plunge we had better be sure we are prepared to become bogged down chasing irregulars and guerrillas over the rice fields and jungles of Southeast Asia while our principal enemies China and the Soviet Union stand outside the fray and husband their strength."

Would that Johnson had listened to his own counsel beginning in 1965. But when some in the American press criticized him for his flamboyance and public overstatements about Diem during his 1961 trip, Johnson complained that he was only acting "under orders." He told aides, "Hell, they don't even know I took a marked deck out there with me." He prided himself on being a team player, never publicly taking issue with Kennedy and always showing the flag for the administration. In the long run, the lasting effect of the trip would be much more on Johnson than Diem. Whatever his words about recalling French failure and avoiding military involvement, Johnson returned home with a sense of commitment to Vietnam that would only reveal itself in years to come.

Walt Rostow didn't wait until 1965 to ignore Johnson's warnings about overcommitting ourselves to the fight in Vietnam. At the end of May 1961, after Johnson reported his findings, Rostow told Kennedy that Vietnam was endangering world peace and that the United States needed to deflate that crisis. He wasn't ready to recommend direct U.S. military involvement, or he at least understood that Kennedy wasn't yet receptive to any such recommendation. So instead he urged Kennedy to build Diem's strength and encourage the international community to recognize Hanoi's assault on the South. Rostow also advised Kennedy to press Khrushchev to rein in his surrogates in Vietnam, as he seemed agreeable to doing in Laos. Rostow's advice resonated with Kennedy as a way to defuse a volatile situation that could generate demands for U.S. military action. In the meantime, Kennedy told Rusk that he was very anxious to implement the promises that Johnson had made during his recent trip to Vietnam.

Despite Kennedy's directive, Diem complained that the United States was not fulfilling its commitments. Kennedy's aides warned of a collapse without some kind of prompt action. During a mid-June meeting with South Vietnam's visiting secretary of state, Kennedy promised to increase the number of U.S. military advisers in Vietnam but warned against any public discussion of his decision as likely to provoke protests that he was violating commitments made at an international conference on Vietnam in 1954 to limit U.S. forces in Southeast Asia. He also feared that press notices about expanded U.S. support for a collapsing Vietnam would trigger conservative demands in Congress and the press for more decisive action to save Diem's regime.

As it was, White House and Pentagon pressure for a more aggressive response to the Vietnam crisis was more than Kennedy wanted. When the deputy director of a Vietnam task force told Rostow about a plan that worked in British Malaya for sweeping Vietnam clean of communist guerrillas, Rostow "jumped to his feet" and said, "This is the first time I have heard a practical suggestion as to how we should carry out our operations in Viet-Nam." To make any such operation work, Diem wanted U.S. financing for an additional hundred thousand South Vietnamese troops. Kennedy remained skeptical about the effectiveness of such an offensive and was loath to ask Congress for the money—not only because he feared that Vietnam might become an open-ended drain on the U.S. Treasury, but also because it could touch off a public debate about involvement in that country's conflict. On July 3, Kennedy held off Diem by instructing the Pentagon to send a qualified military team to Vietnam to study his request. Still hoping that Diem could find an alternative to a wider civil war,

Kennedy urged him to get on with the economic, political, and social reforms that Kennedy said would nourish the aspirations of the Vietnamese and boost Diem's popularity.

With the Cuban failure still casting a shadow over Kennedy's competence, he was eager to keep Vietnam from erupting into a public dispute that could add to a sense of national drift. By May, only four months into his presidency, he was downcast about his performance and his advisers' ability to help him achieve big things. When asked in an off-the-record discussion if he was "disappointed or frustrated in . . . luring the kind of men you want in your administration," he replied: "It is frustrating. . . . It is hard to get people, and we have had some turndowns on some good men who I think could have been better off in the government." Or at least his administration would have been better served if they had come on board. The immediate task he and Bobby now saw was to find ways to rekindle the excitement and optimism that had been so evident at the start of his term.

○ ○ ○ ○

CHAPTER 5

"Roughest Thing in My Life"

During his first months in office, foreign challenges had tested Kennedy's skills as a political leader, but so had civil rights disputes. On one occasion, when these problems became too exasperating, Bobby said to his brother, "Ah, Jack, let's go start our own country." The levity provided only momentary relief from the unrelenting demands of governing.

America's long history of racial divisions and a Congress dominated by southerners determined to resist pressures for change made civil rights a losing struggle for the most adroit president. Yet after so long a battle for equal justice, civil rights advocates were in no mood to settle for anything less than demonstrable progress. And in Kennedy, they saw someone who impressed them as less than fully committed to their cause. Martin Luther King complained that Kennedy was intent on no more than "token integration." He thought that the president lacked the "moral passion" to fight for equal rights. Expectations of executive action, including pressure on Congress to outlaw segregation, were being disappointed. Kennedy was a "quick talking double dealing" politician, one rights leader said. Bayard Rustin of the Congress of Racial Equality (CORE) believed that Kennedy would only react to pressure. "Anything we got out of Kennedy" would be the product of "political necessity, and not out of the spirit of John Kennedy. He was a reactor." A black official at the Democratic National Committee warned the White House that its timidity on civil rights was allowing the president's political enemies "to charge him with inaction in a very vital area."

Kennedy hoped that the Justice Department might have some suggestions for actions that could advance the cause of equal rights without congressional action. But Burke Marshall in the Civil Rights Division offered little help. He saw distinct limits to what the federal government could do to compel southern state and local officials to promote desegregation and ensure blacks equal access to the ballot box.

An executive order increasing the powers of the Committee on Equal Employment Opportunity (CEEO) became the administration's principal vehicle for meeting black demands. Lyndon Johnson's chairmanship of the committee, however, raised doubts among civil rights leaders that much would come of its charge to compel businesses with government contracts to hire more blacks. As a former Texas senator, Johnson seemed unlikely to press very hard for any kind of integration. Given that 15.5 million people were employed by government-financed businesses, the White House hoped that the CEEO could advance the economic well-being of African Americans. But Johnson, who genuinely wanted to make the committee an effective instrument of black gains, was nonetheless reluctant to pressure southern businessmen or corporations with factories and offices in the South into unwanted social actions. To assure southerners that the federal government would not compel them to act against their accepted norms, Johnson declared "this is not a persecuting committee or prosecuting committee." He hoped that "volunteerism," which was called "Plans for Progress" and relied on business establishments to initiate non-mandated reforms, would bring some results.

Robert Troutman, an Atlanta attorney and friend of the president, who was put in charge of implementing "Plans," echoed Johnson's determination not to threaten or bully anyone. Troutman announced that the CEEO would not be "a policeman with a nightstick chasing down alleged malefactors." It was a formula for no progress: By the summer of 1961 the *New York Times* saw little evidence of increased black employment. It incensed Bobby Kennedy, who served not only as his brother's principal adviser but also as his enforcer—the man everyone had to answer to if they fell short of the president's expectations. And Johnson, whom Bobby already disliked, became the object of Bobby's anger over public criticism of the CEEO's lackluster performance. Johnson's defense of his efforts provoked Bobby to attack him at a committee meeting as insincere and incompetent. Privately, Bobby said that Johnson "lies all the time. . . . In every conversation I have with him, he lies."

The White House came in for additional criticism when it failed to back up a promise to desegregate interstate travel. In May, when an integrated group of CORE members tested the administration's commitment by traveling on interstate buses from Washington, D.C., to New Orleans, the Freedom Riders, as they called themselves, were physically attacked in Alabama. Kennedy pressed Harris Wofford to get his "goddamned friends off those buses." Although the CORE riders gave up the trip, student activists from Nashville, Tennessee, took up the challenge. When the Birmingham, Alabama, police arrested them, Bobby Kennedy stepped in to arrange their release and facilitate their trip to New Orleans. His actions angered southerners but won the administration little credit with rights activists who saw the Kennedy initiative as a belated and token attempt to fulfill the administration's promise. After the riders reached Montgomery, Alabama, a mob attacked them and Bobby had to send federal marshals to save them and King, who had come to speak at a black church in support of the riders.

When riots erupted, Bobby publicly asked rights activists for a "cooling off" period. It provoked ridicule from them. They complained that "Negroes have been cooling off for a hundred years," and would be "in a deep freeze if they cooled any further." The conflict now petered out, but it enraged southern segregationists against the administration and deepened suspicions of civil rights backers that the administration had little resolve to overcome historic wrongs.

To refute civil rights critics and disarm segregationist obstructionism, Bobby Kennedy went to Athens, Georgia, in May to give a speech at the University of Georgia. He felt as if he were entering the lion's den and his hands trembled as he turned pages of his speech. Clearly speaking for the White House, he emphasized the administration's determination to back equal rights for all Americans. To counter segregationist resistance, he explained that they were essential in the struggle against international communism, which continually scored points with people of color by pointing to America's enduring tradition of racism. He warned that acts of violence by segregationists "hurt our country in the eyes of the world." His display of courage by speaking so forthrightly moved the sixteen hundred people in his audience to warmly applaud him. King sent him a telegram of praise, but the good effects were dissipated four days later when the White House announced that it would not press Congress for a major civil rights bill. Roy Wilkins, the head of the National Association for the Advancement of Colored People (NAACP), compared the administration's behavior to giving blacks "a cactus bouquet."

The administration's travails at home and abroad convinced Kennedy that he needed to restate his determination and plans to get the country moving again. On May 25, 1961, almost four months after his State of the Union message, Kennedy took the unusual step of appearing before a joint congressional session to present a "Special Message on Urgent National Needs." It was a transparent attempt to rebuild public confidence in administration policies by giving a second State of the Union speech.

Kennedy began by condemning an unnamed aggressor's assault on freedom around the globe. His spoken presentation omitted a paragraph identifying the culprit as "a closed society without dissent or free information, and long experience in the techniques of violence and subversion." The omission softened the verbal attack on Moscow, which was evident when he said, "[W]e are engaged in a world-wide struggle in which we bear a heavy burden to preserve and promote the ideals that we share with all mankind." Several paragraphs on the administration's attentiveness to expanding the country's capacity to resist nonnuclear aggression and develop its civil defense programs as a guard against the possibility of an accidental war were intended to reassure Americans after the Cuban failure that their security from attack was the administration's highest priority.

The longest part of the speech was devoted to a great new enterprise in space. It was calculated to dispel additional concerns about national security that had been triggered on April 12 by Moscow's success in sending Yuri Gagarin, a Russian cosmonaut, into an orbit around the earth, the world's first manned spaceflight. The combination of this scientific breakthrough and the Bay of Pigs fiasco made Kennedy a little desperate to identify some way of bolstering U.S. spirits in the Cold War.

On hearing the news, Kennedy had instructed Sorensen to assemble a group of scientists who could suggest a U.S. response. The meeting initially added to the gloom that recent setbacks had provoked. The scientists concluded that the Soviets had a considerable advantage in the space race. Specifically, they held an edge on the United States in sending a two-man spacecraft aloft, orbiting a space platform or laboratory, exploring the far reaches of space, and landing an unmanned vehicle on the moon. The one hopeful note was that the United States might be able to beat the Soviets to the moon with a manned spaceship. The challenge was so great that neither the United States nor Moscow had begun to move toward that goal.

On April 20, Kennedy sent Johnson, who headed the administration's space council, an urgent query: "Do we have a chance of beating the Soviets by putting a laboratory in space, or by a trip around the moon, or by a rocket to land on the moon, or by a rocket to go to the moon and back with a man? Is there any other space program, which promises dramatic results, in which we could win? . . . Are we working 24 hours a day on existing programs? If not, why not? . . . I would appreciate a report on this at the earliest possible moment."

Kennedy largely knew what Johnson's answers would be. Everything about the six-foot-two-and-a-half-inch Texan's career bespoke grand designs. As a senator, he worked to become the chamber's greatest majority leader in history, with an unprecedented record of achievement that future leaders could not easily surpass. As the Senate's architect of NASA and an outspoken critic of Eisenhower for not being more aggressive about putting the United States ahead of Russia in a competition for dominance in space, Johnson was more than ready to expand America's program. It appealed to his hopes of becoming a memorable vice president.

Johnson believed that the United States should try to land a man on the moon and that getting there ahead of the Soviets was vital in convincing people everywhere that American institutions and technology were preferable to what the communists had. As he told Kennedy, the Soviets had eclipsed us "in world prestige attained through technological accomplishments in space." But "manned exploration of the moon" would be "an achievement with great propaganda value" and would allow us to win "control over . . . men's minds." Johnson had an apocalyptic view of the competition: It would "determine which system of society and government dominate the future. . . . In the eyes of the world, first in space means first, period; second in space is second in everything." Johnson brushed aside complaints about the proposed costs of a moon mission by saying, "Now, would you rather have us be a second-rate nation or should we spend a little money?" Nor did he reflect on the relatively low Soviet standard of living. In 1961, fear of Soviet might eclipsed all realistic assessments of Russia's economic weakness. Russia was not even a model for China, the other communist giant, which was struggling to emerge from centuries of feudalism. Only after the collapse of Soviet rule in Russia in 1991 did many in the United States concede that the communists never had a chance to catch or exceed America's economic output.

McNamara and Rusk agreed with Johnson's exaggerated view of the benefits from space accomplishments in the competition with communism. Understanding how badly Kennedy needed to distract the country from recent setbacks with a bold initiative, McNamara echoed Johnson's call for an all-out effort: "Major achievements in space contribute to national prestige," he told the vice president. "What the Soviets do and what they are likely to do are therefore matters of great importance from the viewpoint of national prestige." Or, he might have said, favorable comparisons with Moscow were essential in rekindling domestic and international support. Johnson prodded Rusk into telling the Senate Space Committee, "We must respond to their conditions; otherwise we risk a basic misunderstanding on the part of the uncommitted countries, the Soviet Union, and possibly our allies concerning the direction in which power is moving and where long-term advantage lies."

Kennedy was less inclined to believe that beating the Soviets in a space race would determine the outcome of the Cold War. But he shared the conviction that a major victory in space was a way to score political points and win hearts and minds at home and abroad. Moreover, if he were to convince the Congress to spend billions on an unproved scientific and engineering experiment, he needed to overstate the benefits. Buoyed by a successful suborbital flight by the U.S. astronaut Alan Shepard on May 5, Kennedy told the Congress, "If we are to win the battle that is now going on around the world between freedom and tyranny, the dramatic achievements in space which occurred in recent weeks should have made clear to all of us, as did the Sputnik in 1957, the impact of this adventure on the minds of men everywhere, who are attempt-

ing to make a determination of which road they should take. . . .
Now it is time . . . for this nation to take a clearly leading role in
space achievement, which in many ways may hold the key to our
future on earth." Kennedy proposed to land a man on the moon
and return him safely to earth before the end of the decade. Nothing could be more difficult or expensive, but it would be the work
of the entire nation.

Kennedy saw serious risks in gambling so much of his administration's prestige and the nation's money on so uncertain an enterprise. When Shepard met him at the White House after his
successful mission, he thought Kennedy was nervous about his association with a program with such uncertain prospects. Kennedy
joked with Johnson, "Nobody knows that the Vice President is the
Chairman of the Space Council. But if that flight had been a flop,
I guarantee you that everybody would have known that you were
the Chairman." Newton Minow, the head of the Federal Commu-
nications Commission, chimed in, to Kennedy's amusement, "Mr.
President, if the flight had been a flop, the Vice President would
have been the next astronaut." Johnson did not share in the mirth,
understanding that as the principal subordinate responsible for the
moon shot, he would have to take the fall for a failure. Kennedy,
however, knew that, as with the Bay of Pigs, he would be the one
to suffer the greatest consequences. In joking about shifting blame
for any failure to Johnson, he was acknowledging that he was bet-
ting a great deal on a space initiative to improve his public image
as a dynamic leader intent on winning the Cold War.

Kennedy's speech also came against the backdrop of grave concerns about Soviet threats to West Berlin. Since 1945 the city had
been divided into eastern and western zones, with West Berlin 110
miles inside the Soviets' East German area of occupation. In 1948,
Moscow had responded to a western plan to rebuild West Germany by blockading access to West Berlin, which the Allies over-

came through a fifteen-month airlift bringing supplies into the isolated city. The rescue of the city from communist control made West Berlin a symbol of the East-West struggle between freedom and authoritarian rule.

In a January 1961 speech, Khrushchev warned of his intention to sign a peace treaty with the East Germans, who would then control the 110-mile route from West Germany to West Berlin and would insist on integrating the western part of the city into the eastern zone. Leaving it to the East Germans to block access to West Berlin would give Khrushchev some means to avoid a direct confrontation with the United States. But it was a thinly veiled ploy that represented a renewed assault on the independence of a free community at odds with communist Russia. Kennedy had considered addressing the issue in his State of the Union speech at the end of January, but his advisers had persuaded him to mute the threat. Existing conditions in Berlin, however, threatened to desta-bilize East Germany and other communist satellites: The flow of talented migrants escaping the East through Berlin and West Berlin's relative prosperity and freedom formed a painful contrast with the austerity and repression in Eastern Europe's communist regimes.

Kennedy was keen to avoid a crisis with Moscow over Berlin, but the Konrad Adenauer government in Bonn was pressing for private and public reassurances of the new administration's inten-tions to defend West Berlin. On March 10, when Kennedy met at the White House with Ambassador Wilhelm Grewe, he assured Grewe that the United States was determined to defend West Ger-many and West Berlin. But he rejected public pronouncements as seeming to challenge Moscow. Fearful that the coming assault on the Bay of Pigs would raise concerns that he might be recklessly aggressive toward the communists, Kennedy wanted any Soviet-American crisis to be triggered by Moscow, not the United States.

In a meeting a few days later with West Berlin mayor Willy Brandt, Kennedy complained that the West's most difficult post–

World War II legacy was Berlin and that they would just have
to live with the situation for the time being. By meeting with
Brandt, a Social Democrat who was planning to run against Adenauer for the chancellorship, Kennedy was projecting himself into
West Germany's political divide and creating tensions with the
existing government. It carried serious risks: The eighty-five-year-
old Adenauer, the *Alte*, or "Old One," who had become a symbol
of Western resistance to Soviet expansion, enjoyed considerable
standing across Western Europe. But by openly conferring with
the younger, forty-eight-year-old Brandt, who had fled the Nazis,
lived in Norway during World War II, and established himself
in postwar Germany as a progressive proponent of European reconciliation, Kennedy was hoping to get fresh suggestions of how
to resolve or at least mute difficulties over Berlin. He had little
expectation that Adenauer, with whom he was to meet later in the
spring, would offer anything but familiar hard-line anticommunist rhetoric.

Brandt, however, had little to suggest beyond asking Khrushchev
for assurances that if he signed a peace treaty with East Germany,
the German communists would not precipitate a crisis by trying to
seize the city's western zone. Kennedy pressed him for some more
proactive way to resolve the Berlin problem with Moscow and East
Germany. Kennedy asked him what he thought of George Kennan's 1957 proposal to neutralize Central Europe by ending Soviet
dominance of Eastern Europe, demilitarizing Germany, and ending Cold War tensions over this contested ground. Brandt saw no
realistic possibility that the Soviets would relax their grip on their
East European satellites.

Brandt's pessimism about finding a way out of a potential clash
with Moscow over Berlin frustrated Kennedy, who feared the sit-
uation could escalate into a crisis. As Llewellyn Thompson, the
U.S. ambassador to the Soviet Union, told him in a dispatch from
Moscow, without some kind of negotiation and settlement with
Khrushchev beyond telling him that "we would fight rather than

abandon people of West Berlin . . . it could involve real possibility of world war." At a minimum, heightened tensions over Berlin would almost certainly intensify Cold War differences.

Over the next two months, Kennedy cast a broad net in search of some answer to the Berlin question. The State Department, Harvard national security expert Henry Kissinger, former secretary of state Dean Acheson, several U.S. ambassadors, and Germany's Chancellor and foreign minister weighed in. It was a demoralizing exercise. The State Department, for which Kennedy already had little regard, had no good ideas about what to do. The best it could offer was "the possibility of developing and strengthening deterrents other than the pure threat of ultimate thermonuclear war." Kissinger, who struck Kennedy as "pompous and long-winded," urged Kennedy to consider visiting Berlin during a European trip in June. "The Soviets may construe such an action as a provocation," but Kissinger believed that worrying about that would be excessively timid. The Soviets were not going to let the issue rest: Kissinger's answer was to confront them head-on, which was exactly what Kennedy didn't want to do.

Acheson shared Kissinger's call for firmness. He saw Berlin as of the greatest importance and pressed Kennedy to adopt some sort of military response. He suggested that two divisions of ground forces be deployed to ensure continued access to West Berlin. Foy Kohler, the assistant secretary of state for European affairs, echoed the State Department–Acheson–Kissinger view that offering to negotiate wouldn't serve any good purpose. A display of toughness was the best idea. David Bruce and Charles Bohlen, both with decades of experience as U.S. diplomats in Europe, agreed that negotiations with Moscow over Berlin offered no satisfactory outcome. The pessimism of these experts and seasoned diplomats exasperated Kennedy. He diplomatically put them off by saying that he had not yet decided what to do.

Kennedy was not happy with the sterile approach to so volatile a problem. Schlesinger recalled Kennedy sitting "poker-faced, confining himself to questions about the adequacy of existing military plans." Schlesinger later remembered him as more than frustrated by "the apparent impossibility of developing a negotiating position on Berlin." It "left Kennedy with little doubt that the State Department was not yet an instrumentality fully and promptly responsive to presidential purpose." He could have said the same about Acheson and Kissinger, who seemed to have less aversion to a military confrontation with Moscow than Kennedy. By contrast with them, however, Kennedy would have to bear the burden of risking the many lives that could be consumed in a war.

Schlesinger could have told him that nothing had changed since Franklin Roosevelt's presidency. "You should go through the experience of trying to get any changes in the thinking, policy and action of the career diplomats and then you'd know what a real problem was," FDR told a friend. He dismissed the Foreign Service professionals as the "boys in striped pants," "old maids," and "stuffed shirts." Echoing FDR, Kennedy complained about the State Department that "they never have any ideas over there, never come up with anything new." Kennedy toyed with the possibility of having "a secret office of thirty people or so to run foreign policy while maintaining the State Department as a façade in which people might contentedly carry papers from bureau to bureau."

Conversations with Adenauer and German foreign minister Heinrich von Brentano in mid-April yielded little more than platitudes. Kennedy explained that the United States and Britain were still considering what to do about Moscow's threat to Berlin. When he asked Adenauer what he thought might happen in Berlin this summer, "the Chancellor smilingly replied that he was no prophet." But he and Brentano left no doubt that they saw the very future of Germany at stake. "If Berlin fell, . . . it would mean the death sentence for Europe and the Western World."

In 1961, in the immediate aftermath of the Bay of Pigs failure, no issue troubled Kennedy and his advisers more than Berlin. It was the international trouble spot most likely to trigger a Soviet-

American confrontation and a disastrous war. At the beginning of May, when McNamara sketched out military plans to counter a threatened communist takeover of Berlin, he reminded Kennedy that Eisenhower was prepared to fight a full-scale war to defend the city. But McNamara thought that massive retaliation should be preceded by a reliance on conventional forces. Nonetheless, the possibility of a nuclear exchange remained a serious contingency that frightened Kennedy and made him eager to meet with Khrushchev to persuade him to back away from a confrontation over Berlin.

A conversation between Llewellyn Thompson and Khrushchev on May 24 underscored the need for an early summit discussion that could emphasize the dangers to both sides in any Soviet effort to alter the status of West Berlin. Khrushchev declared his intention to sign a peace treaty with East Germany in the fall or winter. But it hardly signaled the end of communist dominance. A Germany under Moscow's control remained a vital part of Soviet national security. A treaty would give the East Germans ostensible control of access to West Berlin. Khrushchev said "he realized this would bring a period of great tension but was convinced would not lead to war." He also declared that "German reunification was impossible and in fact no one really wanted it." When Thompson warned him that the West would respond to any threat to West Berlin with force, Khrushchev waved aside the warning, saying "if we wanted war we would get it but he was convinced only madmen would want war."

Still, unless one side or the other backed away from its existing position, the prospect of a conflict seemed all too real. Yet Thompson did not think that Khrushchev would do anything that would risk a war. The issue between the two sides, Thompson said, had become national prestige, and the need was for a formula that would allow both sides to save face. He suggested that Kennedy explore this in private with Khrushchev. By contrast, the State Department, which prepared a "Talking Points Paper on Berlin and

Germany" as a prelude to meetings with French president Charles de Gaulle in Paris and Khrushchev in Vienna between May 31 and June 4, offered Kennedy no fresh ideas for resolving differences, only suggested preparations for war to prevent further communist expansion in Europe.

The recommendations left Kennedy perplexed. He saw two possibilities: He could tell Khrushchev that negotiations on Berlin were out of the question—the Anglo-French-American access to Berlin was inviolable—or he could say that the future status of Berlin could be discussed, and though nothing would come of these talks, it would paste over differences and delay a crisis in hopes that improved relations with Moscow in time could quiet tensions over Berlin.

On May 30, Kennedy, accompanied by his glamorous wife, who spoke French and had a reputation as a Francophile, flew to Paris for meetings with de Gaulle. A summit in Vienna with Khrushchev and a brief stop in Britain for discussions with Prime Minister Harold Macmillan were to follow. The trip, like the speech to the joint congressional session, aimed to boost Kennedy's standing after the Cuban fiasco and the uncertainties over Vietnam.

At an initial May 31 meeting with de Gaulle at the ornate Elysée Palace, an event marked by high-visibility pomp and ceremony, Kennedy gained instant standing as a world leader. The consultation with the seventy-year-old de Gaulle, who had become president of the Fifth Republic in 1959, served a larger purpose than providing wise counsel on international dilemmas. With Franklin Roosevelt and Stalin deceased, Winston Churchill out of office since 1955 and sidelined by age, at eighty-six, and the seventy-three-year-old Chiang Kai-shek eclipsed by communist control of mainland China since 1949, a meeting on equal standing with de Gaulle, the last storied leader from World War II, "a great captain of the Western World," Kennedy called him, gave the young president heightened prestige and cachet.

De Gaulle's advice on Berlin was largely predictable. Like other close observers of Soviet affairs, de Gaulle saw small likelihood that Khrushchev was prepared to fight a war over Berlin. For almost two years, de Gaulle told Kennedy, Khrushchev has been threatening to move against the city. But if he were going to risk a war, he would have acted already. De Gaulle reported that he had told Khrushchev, "If you want peace, start with general disarmament negotiations." This could lead to changes in "the entire world situation . . . and then we will solve the question of Berlin and the entire German question." De Gaulle warned against any sort of retreat in response to Soviet threats and the importance of making clear to Khrushchev that he would endanger his country's survival if he tried to alter the status quo in Berlin. Kennedy worried that after his failure at the Bay of Pigs, Khrushchev would not believe in his firmness of purpose. De Gaulle urged Kennedy not to mince words with Khrushchev but to make unmistakably clear that any aggression against Berlin would mean an all-out war. Strengthening the Berlin garrisons and airlift capability would put Khrushchev on notice about Western intentions.

Kennedy was not especially impressed by de Gaulle's advice. It was nothing he had not heard before, and as he told an English friend later, de Gaulle's only concern was with the selfish interests of his country. On a question about governing, however, de Gaulle's comment echoed Kennedy's brief experience in office. Charles Bohlen recalled "at one point the President telling me . . . that de Gaulle had said to him, 'you can listen to your advisers before you make up your mind, but once you have made up your mind then do not listen to anyone.' I think the President was somewhat impressed with this as a technique of government and it certainly makes a good deal of sense and is certainly one de Gaulle followed to the extremes."

De Gaulle was a model of what Kennedy aspired to. He shared de Gaulle's view that small men cannot handle great events. Indeed, as the British historian Isaiah Berlin shrewdly concluded from several conversations with Kennedy at White House dinners, the president was "devoted to the idea of great men. There was no doubt that when he talked about Churchill, whom he obviously admired vastly, when he talked about Stalin, he talked about Napoleon . . . about Lenin, and two or three other world leaders, his eyes shone with a particular glitter, and it was quite clear that he thought in terms of great men and what they were able to do, not at all of impersonal forces. [He had] a very, very personalized view of history."

When Kennedy took responsibility for the Bay of Pigs, it was not simply a courageous move to protect subordinates, but also a statement of his conviction that if he were to establish himself as a historical figure, he needed to be seen as at the center of all his administration did—the achievements and the failures. He was always keen to hear the views of others on the great issues of the day. "One of the things which struck me most forcibly," Berlin said, "was that I've never known a man who listened to every single word that was uttered more attentively. His eye protruded slightly, he leaned forward towards one, and one was made to feel nervous and responsible by the fact that obviously every single word registered. . . . He really listened to what one said and answered *that*." Anything he heard from advisers, however, was given the most critical scrutiny, especially after the Cuban failure. Final policy judgments had to be his and his alone. Watching how he decided weighty policy questions, Bohlen concluded that for Kennedy, "the issues and the consequences of mistakes of a serious nature . . . are so great that no man of any character or intelligence will really wholeheartedly accept the views of anybody else."

Kennedy's meeting with de Gaulle made clear that his failure to decide matters for himself about the Bay of Pigs had hurt him. The experience and then conversations with the extraordinary French leader underscored that experts could not be entirely trusted and only he would be in a position to see the broad contours of an issue. He tried to remind himself of this every time he faced a major foreign policy decision.

He saw his dealings with de Gaulle as an opportunity to put

this insight into practice. Before his visit to Paris, he, or Bundy for him, had solicited advice from Nicholas Wahl, a Harvard University political scientist, who had a close relationship with de Gaulle and a rich knowledge of French politics. Although Kennedy valued Wahl's assessment of de Gaulle, as someone who expected deference to his views, he was determined to make his own judgments on how to deal with so imposing a figure. He read de Gaulle's war memoirs as a vehicle for understanding someone he wished to use for his own purposes. He memorized quotes as a way to flatter the older man and make sure that their meeting would resonate as a demonstration of Kennedy's mastery of international relations. It largely worked, since substantive tensions over differences with de Gaulle, especially his determination to build an independent nuclear arsenal against U.S. wishes, did not become an open source of conflict and their exchanges seemed to be those of allies set on defending their nations' security and assuring the peace.

As he prepared to meet with Khrushchev in Vienna, Kennedy sounded out Soviet experts on what to expect from the Kremlin leader. The CIA, the State Department's Soviet division, Kennan, Harriman, Thompson, and Bohlen were the principal advisers. The CIA had prepared a psychological profile of Khrushchev drawn up by twenty American psychiatrists and psychologists. It recounted his rise from humble beginnings, which made him confident that "his vigor, initiative and capacity are equal to his station." It also emphasized his "depression and vulnerability to alcohol" and depicted him as a somewhat erratic character subject to mood swings between elation and depression. His wife described him as "either all the way up or all the way down."

Kennedy had few illusions about the sixty-six-year-old first secretary of the Communist Party. Although born into a Ukrainian peasant family and largely devoid of formal education, Khrushchev was more than a canny peasant survivor of Stalin's ruthless Soviet system, which killed off many of its most ambitious and talented military and political leaders. Kennedy wisely assumed that someone who had risen to the top of the Soviet government through the tumultuous years of revolution and war was a formidable personality who should not be underestimated. However ungainly the burly, overweight Khrushchev might appear, Kennedy understood that it would be a mistake to believe that he faced anyone less impressive than the imperious de Gaulle or any of the high-powered American and British figures he had met in the last twenty-five years.

However, the CIA report boosted Kennedy's confidence that he could handle a difficult character like Khrushchev. The man he saw in those pages reminded him of the powerful American politicians, including his vice president, whom he had managed to deal with successfully during his fifteen years in politics. He worried nonetheless that after the Bay of Pigs Khrushchev would see him as too inexperienced and too indecisive to stand up to someone as assertive as the Soviet leader. "He's not dumb," Kennedy said after reading the briefing books on Khrushchev. "He's smart. He's tough." Yet Kennedy had good reason for concern: When Kennedy failed to bomb or land Marines in Castro's Cuba, Khrushchev said to his son, "I don't understand Kennedy. What's wrong with him? Can he really be that indecisive?"

The State Department was optimistic about the results of a conference. They advised Kennedy that Khrushchev would emphasize peaceful coexistence. He would seek specific agreements and would aim to end the talks on a note of accord. He would also stress the need for a Berlin settlement, a slowing of the arms race, and expanded U.S.-Soviet trade. "Khrushchev might, for effect, strike a note of anger and bluster. . . . But it seems likely that he will generally assume an attitude of reasonable firmness, coupled with a pitch for improved US-Soviet relations. . . . Finally, with the eyes of the world on Vienna, Khrushchev might regard the meeting as an appropriate occasion for some dramatic step intended to demonstrate Soviet progress or peaceful intent." The department's experts predicted the possibility of a new Soviet space shot or a disarmament initiative. In retrospect, the department's analysis became another reason for Kennedy to dismiss his diplomatic experts as less than helpful in assessing international developments and another reason to be skeptical of their advice.

As for Kennedy's "line of approach to Khrushchev," the department, in a draft written by Bohlen, recommended that the president emphasize their shared concern in avoiding nuclear war. Moreover, he would do well to tell Khrushchev that wars of national liberation, which Moscow was sponsoring, constituted a serious threat to world peace. The general impression Kennedy should leave on Khrushchev was of a United States determined to avoid war with the Soviet Union yet also hopeful that Moscow would engage in realistic and responsible actions to preserve the peace.

The department's take on the Vienna meeting struck Kennedy as sterile and predictable. The fundamental issue between the two sides was distrust or the conviction that each intended to subvert and, if possible, destroy the other's political-social system. Professions of peaceful intentions would be of no help in disarming the other's suspicions. Only concrete steps demonstrating a commitment to coexistence could improve relations, and the State Department had nothing to suggest that promised to advance this goal.

Kennedy was more interested in what George Kennan had to say. As the architect of containment, with a long history of astute commentary on Soviet behavior, the fifty-seven-year-old Kennan, who had spent much of his professional life living near or inside the Soviet Union, impressed Kennedy as a detached observer with no political ax to grind and a fearless commentator on Soviet and American policies. Suggestions from Kennan on what to expect from Khrushchev at a summit meeting seemed essential. But Kennan, whom Kennedy had made ambassador to Yugoslavia, where Tito had established himself as a unique independent voice in the communist camp, was loath to offer much advice.

"Fear there is little I can say that would be helpful," Kennan cabled on the eve of the conference. Complaining that he had not been apprised of the reasons for the summit, he doubted the wisdom of holding a conference without clear prospects for useful agreements. Kennedy's interest in scoring political points for himself at home and abroad by meeting with de Gaulle and Khrushchev did not enter into Kennan's calculations. He questioned the need to remind Khrushchev, as Acheson and others had urged, of our determination to resist any overt encroachment challenging U.S. commitments to the U.N. or allies. Kennan thought it self-evident that Khrushchev would not see Kennedy's presidency as representing a significant shift in U.S. determination to defend its interests. He did, however, expect Moscow to exert political pressure by "ruthless exploitation of colonial issue and all-out propaganda attack." Khrushchev should be put on notice that such an assault would undermine any prospects for improved Soviet-American relations and would threaten a nuclear conflict. Such a pronouncement could be helpful to Khrushchev in resisting Chinese pressure to spread communism by all possible means short of war.

Llewellyn Thompson also had doubts about the value of a summit meeting. He thought that Khrushchev would favor a pleasant conference, and he might "make some proposal or take position on some problem which would have effect of improving atmosphere and relations. I find it extremely difficult however to imagine what this could be." He saw slight possibilities that Khrushchev might propose something constructive on China, Vietnam, Laos, Central Europe, arms control, or outer space, but he was hard-pressed to see what these initiatives might be. In brief, he had no high hopes for any sort of breakthrough at the meeting.

Averell Harriman was eager to weigh in as well on how Kennedy should prepare for the summit. Harriman had long experience with Soviet leaders, including Stalin, with whom he had numerous dealings as ambassador to Moscow from 1943 to 1946. He had served Truman as secretary of commerce from 1946 to 1948 and been governor of New York for four years in the fifties.

Kennedy was not keen on having him in his administration. He was notoriously egotistical and seemed certain to demand more presidential attention than Kennedy wanted to give. (In 1975, at a twenty-fifth anniversary Korean War conference at the Truman Library, I watched as Harriman turned a hearing aid on and off during discussions. Why does he do that? I asked someone well acquainted with him. "He only turns it on when he's speaking" was the reply.)

In 1961, Harriman was sixty-nine years old, hard of hearing, and too proud to wear a hearing aid. But he remained someone with unrequited ambitions. He wanted a high post in the Kennedy White House, principally as an adviser on Soviet affairs, which he knew best. As a signal to Kennedy that he deserved appointment as *the* administration's Russian expert, Harriman persuaded Khrushchev to see him. However, two days of conversation in which Khrushchev hoped to send messages to the new U.S. government about Berlin, coexistence, tensions with China, and Soviet might were not enough to convince Kennedy that Harriman should be his man in, or the best one to consult on, Moscow.

Harriman's friends, however, were tenacious in promoting his candidacy for a foreign affairs post. "Are you sure that giving Averell a job wouldn't be just an act of sentiment?" Bobby Kennedy asked Schlesinger. Schlesinger thought not: "Harriman had one or two missions left under his belt," he said. And so Kennedy, after Harriman agreed to wear a hearing aid, appointed him roving ambassador or ambassador at large. Harriman rationalized his peripheral role by remembering how he had managed to expand his influence under FDR, saying, "Oh, you know, all these presidents are the same. You start at the bottom and work your way up."

When he learned that Kennedy was going to Vienna to meet Khrushchev, he saw a chance to demonstrate his value to the administration. He rushed to Paris to see Kennedy, arranging to attend a state dinner with de Gaulle and the president, and to corner Kennedy, to whom he had sent word that he needed to speak to him about Khrushchev. "I hear there is something you want to say to me," Kennedy coolly told him.

Harriman replied, "Go to Vienna. Don't be too serious, have some fun, get to know him a little, don't let him rattle you, he'll try to rattle you and frighten you, but don't pay any attention to that. Turn him aside, gently. And don't try for too much. Remember that he's just as scared as you are, his previous excursion to the Western world in Europe did not go well, he is very aware of his peasant origins, of the contrast between Mrs. Khrushchev and Jackie, and there'll be tension. His style will be to attack and then see if he can get away with it: Laugh about it, don't get into a fight. Rise above it. Have some fun."

Charles Bohlen doubted that meeting Khrushchev would be much fun or very productive of improved relations. But he appreciated that Kennedy needed to find out for himself what Khrushchev and the Soviet leadership were like. Bohlen knew that Kennedy was getting plenty of advice on how to deal with the communists in general and Khrushchev in particular. But he also understood that Kennedy wanted to put what he was being told to a test. So it would be helpful to Kennedy to make Khrushchev's "acquaintance, to get a feel of the type of man he was dealing with and the type of situation he was confronting."

Khrushchev came to Vienna with a relatively weak hand. True, he could boast about Sputnik and Gagarin's spaceflight as well as the shoot-down of America's U-2 over Russia and Eisenhower's embarrassment in having to admit to American spying. He could also threaten to oust Western forces from Berlin, where Soviet and East German troops surrounded them. But he knew that the United States had a significant advantage over the Soviet Union in nuclear missiles, especially submarine missiles that could hit Russia without warning. While Moscow could inflict grievous damage on America's West European allies, it lacked the intercontinental weapons that could strike the United States.

But Khrushchev would concede no weakness and had personal strengths that would serve him in dealings with stronger opponents. Above all, he had survived Stalin's rule and purges. As a member of the Politburo's inner circle, he had spent many nights at Stalin's "frightful . . . interminable, agonizing dinners." Compelled to sample various foods Stalin feared might be poisoned, to drink themselves into a stupor, and to stay awake through the long night's amusements, which included various forms of humiliation Stalin inflicted on associates, the comrades lived in fear of antagonizing or even annoying the dictator. As Khrushchev said later, "when Stalin says dance, a wise man dances." Another member of the inner group whispered to Khrushchev after one of these bacchanalias, "One never knows whether one is going home or to prison."

Khrushchev learned the art of intimidation from Stalin. In 1956, he had told Western diplomats at an embassy reception after Moscow had been embarrassed by the Hungarian uprising and the defeat of its Egyptian ally in the Suez crisis, "Whether you like it or not, history is on our side. We will bury you." In the fall of 1960, when he came to the annual opening session of the United Nations in New York, his boorishness offended and frightened people. During a speech by U.N. secretary-general Dag Hammarskjöld that angered him, he began pounding his fist on a desk until all other communist delegates joined in. A subsequent speech by British prime minister Harold Macmillan, lamenting Khrushchev's cancellation of the Paris summit with Eisenhower the previous spring, provoked shouts of protest: He was compelled to answer the U.S. act of aggression sending a spy plane over Soviet territory, Khrushchev screamed at the PM. When Spanish delegates failed to applaud a speech he gave to the General Assembly, Khrushchev shook his fist, began verbally abusing them, and moved to lunge at one of them when security guards forced his retreat by standing in front of the Spaniards. Most famously, he banged a shoe on the table when a Philippine delegate accused the Soviet Union of colonizing Eastern Europe and depriving it of political and civil rights.

Western diplomats dismissed him as "a crude, ill-educated clown" who was posturing, but no one could ignore the fact that he was the head of a powerful Soviet state armed with nuclear weapons that could devastate much of Europe. However inappropriate his behavior, he had to be taken seriously. Was he another Hitler or Stalin, who might recklessly plunge the world into a cataclysmic war? some wondered.

Khrushchev came to Vienna intent on bullying the young, inexperienced Kennedy, whose temporizing over Cuba made him seem like an easy target. The public response to their respective arrivals in the city angered Khrushchev and made him more combative. Vienna had been chosen as a neutral ground between East and West, but the much larger placard-waving crowd enthusiastically greeting Kennedy at the airport formed a striking contrast to the silent observers of Khrushchev's open car traveling from the train station to the Soviet Embassy. Jackie Kennedy remembered their arrival on a "dark gray day" brightened by "one of the most impressive crowds I've seen" lining the twenty-five-mile route into the city, "weeping and waving handkerchiefs." The shouts to Kennedy of "Give 'em hell, Jack" and "Lift the Iron Curtain" made Austrian preferences clear in the U.S. competition with communism.

The initial meeting at the U.S. Embassy for the opening session also put Khrushchev on the defensive: The taller, younger, more energetic Kennedy towered over the squat first secretary, giving the president the upper hand in the photo op that Kennedy seemed to thoroughly enjoy and Khrushchev smilingly endured at the start of the proceedings. When photographers asked for additional shots of the two shaking hands, Kennedy told the interpreter, "It's all right with me if it's all right with him." Khrushchev grudgingly agreed.

The minute the two sat down to talk, a battle of egos erupted. Khrushchev was determined to put the younger man in his place, especially after the opening sequence. Kennedy was just as eager to assert himself over the Soviet leader, who he assumed saw him as too uncertain to stand up to him. The initial banter reflected each man's bid for an edge. Kennedy recalled meeting Khrushchev in 1959 in

Washington, suggesting that he was no novice coming to his first meeting with a world leader. Khrushchev countered that he remembered meeting him "as a young and promising man in politics." Kennedy replied, he "must have aged since then," implying that he was more seasoned than he looked. Khrushchev answered that the young always want to look older, patronizingly adding that he would be happy to share his years with the president.

The initial formal discussions went badly for Kennedy. His advisers had warned him against letting Khrushchev draw him into ideological arguments. Moreover, after complaining that Eisenhower secretary of state John Foster Dulles had aimed to liquidate communism, but that the West needed to accept that this was impossible, Khrushchev exclaimed that he had no desire to argue about the virtues of their respective systems, saying that he "would not try to convince the President about the advantages of Communism, just as the President should not waste his time to convert him to capitalism." While Kennedy had no illusions about shaking Khrushchev's belief in his Marxist faith, he hoped to show Khrushchev his toughness by warning against any imperial overreach, denouncing the imposition of communism on peoples eager for self-determination. Khrushchev insisted that communist governments reflected popular sentiment and then tried to change the focus of the conversation by saying, "In any event this is not a matter for argument, much less for war." But remembering how well Nixon had done in the famous "kitchen debates" with Khrushchev in Moscow in 1959 and how he had bested Nixon in their first televised debate during the 1960 campaign, Kennedy persisted in arguing the point. The spread of communism was not a historical inevitability, he said.

Never one to back away from an argument, Khrushchev took up the fight. The victory of communism was inevitable, he declared. It was not open to dispute because it rested on "a scientific analysis of social development." Kennedy disputed the argument that there was anything scientific or inevitable about the rise of communist regimes. Khrushchev snidely inquired "whether the United States wanted to build a dam preventing the development of human mind and conscience," which was "not in man's power." He compared such an effort to the Spanish Inquisition. Communism was an idea, which "cannot be chained or burned," he asserted. Kennedy warned against any "miscalculation" that rested on assumptions about predictable political outcomes but instead could lead to war. Khrushchev now lost his temper or at least acted as if he did. Kennedy told an aide: He "went berserk. He started yelling, 'Miscalculation! Miscalculation! Miscalculation! All I ever hear from your people and your news correspondents and your friends in Europe and elsewhere is that damned word, miscalculation! You ought to take that word and bury it in cold storage and never use it again! I'm sick of it!'" He added, "The United States wanted the USSR to sit like a school boy with his hands on his desk."

Khrushchev thought his performance overwhelmed Kennedy. He described the president afterward as "very inexperienced, even immature. Compared to him Eisenhower was a man of intelligence and vision." So weak a president, who seemed unprepared for the rough-and-tumble of Cold War politics, impressed Khrushchev as an easy mark for a diminished U.S. presence in the international competition with communism.

Khrushchev was not much off the mark in his assessment of Kennedy's reaction to his hectoring. Kennedy emerged from the talks dazed. "Is it always like this?" he asked Thompson. *New York Times* columnist James Reston saw Kennedy as "very gloomy." During a conversation at the U.S. Embassy, Kennedy sank onto a couch. He had a "hat over his eyes like a beaten man, and breathed a great sigh." "Pretty rough?" Reston asked. "Roughest thing in my life," Kennedy acknowledged. British prime minister Macmillan saw him after the summit as "completely overwhelmed by [Khrushchev's] ruthlessness and barbarity." Rusk said: "Kennedy was very upset. He wasn't prepared for the brutality of Khrushchev's presentation." Harriman thought Kennedy was "shattered," and Lyndon Johnson, who thought he would have handled Khrushchev much more effectively than Kennedy, said, "Khrushchev scared the poor little fellow dead."

Kennedy complained to Bobby that in his discussions with Khrushchev it was the first time he "had ever really come across somebody with whom he couldn't exchange ideas in a meaningful way." He thought that Khrushchev was "completely unreasonable. . . . It was a shock to him that somebody would be as harsh and definitive as this." When he returned to Washington, Kennedy wanted to send a message to Khrushchev through press leaks that his behavior was unacceptable. He told *Time* columnist Hugh Sidey, "[I] talked about how a nuclear exchange would kill seventy million people in ten minutes and he just looked at me as if to say, 'So, what?'" The criticism of Khrushchev partly expressed Kennedy's outrage at the communist's temerity in being so high-handed and even dismissive of him. But he was also making the case for why they might have to fight a war with Moscow: Khrushchev was a madman who was not open to reason and was forcing the United States into a conflict by threatening its security.

What had particularly agitated Kennedy was Khrushchev's insistence on signing a peace treaty with East Germany that would invalidate all commitments stemming from Germany's surrender at the end of World War II. Khrushchev explained that Germany inflicted terrible devastation on the Soviet Union in the war and he was determined to prevent the reunification of Germany and its revival as a military power that could threaten the Soviet Union with another war. He made no mention of his frustration over the flood of dissidents—roughly a million in 1960—escaping the communist East through West Berlin. Kennedy, recalling all the advice to stand firm against any demands for concessions over Berlin, replied that the United States and its Western allies were in Berlin "not because of someone's sufferance," but because they had fought their way there. "If we accepted the loss of our rights no one would have any confidence in US commitments and pledges." He told Khrushchev that they were talking about fundamental issues of "US national security." Khrushchev contentiously replied that if the president refused to agree to a peace treaty, he would go ahead without him. He dismissed the president's defense of U.S. rights in Berlin, saying that "the US might want to go to Moscow because that too would, of course, improve its position."

Khrushchev insisted that "no force in the world would prevent the USSR from signing a peace treaty" by the end of the year, which he acknowledged would block access to Berlin. He accused the United States of wanting to start a war. "If there is any madman who wants war, he should be put in a straight jacket," Khrushchev said. Kennedy countered that if anyone was threatening to provoke a war, it was Khrushchev, "who wants to do so by seeking a change in the existing situation." The argument along the same lines continued on and off for the rest of the day with no change in position by either of them. Kennedy said he would be leaving Vienna with the view that they were on a collision course over Berlin. "If the US wanted war, that was its problem," Khrushchev responded. "War will take place only if the US imposes it on the USSR." Khrushchev concluded by repeating his determination to sign the peace treaty and end Western rights in Berlin. Eager to leave no question in Khrushchev's mind about U.S. determination to defend its interests, Kennedy said, "It would be a cold winter."

Kennedy stopped in London to see Prime Minister Macmillan on his way back to Washington. Although Macmillan was twenty-three years Kennedy's senior and a Conservative, the two shared an affinity for each other born of a delight in witty exchanges and a conviction that appeasement of aggression would bring another great war. Stopping in London was like being "in the bosom of the family," Kennedy said. Macmillan understood that after his encounter with Khrushchev, whom Kennedy described to Macmillan as "a barbarian," the president needed bucking up. Kennedy found some comfort in Macmillan's support for opposing Khrushchev's threat to Berlin with "all the force at their command."

On returning to Washington on June 6, Kennedy reported to congressional leaders and the public in a nationally televised address on the results of the summit. He made clear to the sixteen Senate and House members that they were facing a crisis over Berlin. He warned against provocative language that would seem to put Khrushchev in a corner, saying that his speech that evening would be restrained and implying that no one in the government should inflame the Kremlin with threats. But he also emphasized the need to convince Khrushchev that reckless action could lead to a nuclear war. In his address from the Oval Office, he said "that the Soviets and ourselves give wholly different meanings to the same words—war, peace, democracy, and popular will. We have wholly different views of right and wrong." He urged Moscow to understand that they owed "it to all mankind to make every possible effort" to avoid a war.

Kennedy now instructed all his diplomatic and national security associates to begin an urgent review of Berlin contingency planning. The feeling that they were facing a crisis that could erupt in war by the end of the year fixed everyone's attention. No one shied away from offering advice, but the division of opinion that opened up between hawks and doves—those convinced that only visible preparations for war would inhibit the Soviets and those persuaded that overt military actions combined with negotiating initiatives were the wiser course—left Kennedy with difficult choices. He was eager to discourage any view of him as appeasing Moscow, but he was equally intent on avoiding some precipitous step that could take both sides closer to war. As never before, he understood what was meant by the "art of diplomacy"—finding a reasonable ground among imperfect alternatives.

Because his White House advisers were less than reassuring on how to get through this crisis, Kennedy was eager to hear from outside voices on what he should do. But other prominent commentators were no more helpful. Mac Bundy brought Kennedy's dilemma into sharp focus when on June 10 he described differing messages from Joe Alsop and Walter Lippmann, two of the country's most respected journalists. Alsop was "for a strong and essentially unyielding position, carried all the way to war if necessary." Lippmann, by contrast, was "for a negotiating solution . . . measures looking toward the genuine neutralization of West Berlin." Alsop, Henry Kissinger, and "most of your advisers would hold that any neutralization of West Berlin would be a form of surrender." Lippmann thought otherwise and predicted that neutralization would bring a net gain in U.S. prestige. Bundy pointed out that despite Macmillan's support for a tough stand, the British would not be so fast to follow Alsop's advice. On the other hand, if Kennedy followed Lippmann's counsel, he would have to face down Adenauer. Whatever Kennedy chose to do, it was essential that he be "in immediate, personal, and continuous command of this enormous question." That Bundy had to remind Kennedy of his need to be central to these deliberations was an indication, after Vienna, of current doubts about his leadership.

Kennedy's first challenge was to bring his military under control. On June 28, at a meeting with the Joint Chiefs, who predictably were urging robust demonstrations of American preparedness in case of a communist interdiction of West Berlin, Kennedy said he valued their counsel but saw them as more than "military men and expected their help in fitting military requirements into the over-all context of any situation, recognizing that the most difficult problem in Government is to combine all assets in a unified, effective pattern." He was making clear that a military solution to the Berlin challenge was not his preference. He was also trying to discourage the Chiefs from stimulating public support for a confrontation with the Soviets or sending messages that might provoke Moscow into more aggressive action. As matters stood, U.S. opinion was already inclined to face down the Soviets: Surveys between June 23 and June 28 showed 82 percent in favor of a continued U.S. military presence in Berlin, despite the view of 59 percent that a nuclear conflict might be in the offing.

Time and *Newsweek* published stories suggesting that Kennedy was reluctant to be as militant as the Pentagon. The leaks to the magazines angered him: "Look at this shit," he told Pierre Salinger. "This shit has got to stop." The same day as his meeting with the Chiefs, Kennedy used a press conference to take the lead on the developing crisis. He refused to take the bait when a reporter asked him what he thought of Vice President Nixon's "dim view of your administration. . . . Never in American history has a man talked so big and acted so little," Nixon said. Instead, Kennedy offered measured comments that could remind Khrushchev of his determination not to abandon Berlin to communist control and "make permanent the partition of Germany. . . . No one can fail to appreciate the gravity of this [Soviet] threat," he added. "There is peace in Germany and in Berlin. If it is disturbed, it will be a direct Soviet responsibility."

On the twenty-eighth, Kennedy also received a report he had asked Dean Acheson to write on Berlin. Because he saw Acheson as "one of the most intelligent and experienced men around," Kennedy believed it useful to have his judgment on how to meet Khrushchev's challenge over Berlin. By no means, however, was he prepared to take Acheson's advice as gospel. After all, judging from earlier policy statements, Acheson was of two minds about Berlin: In 1959 he had said, "To respond to a blockade of Berlin with a nuclear strategic attack would be fatally unwise. To threaten this attack would be even more unwise." But by 1961 he had shifted ground: He not only favored a buildup of U.S. ground forces in the city to confront any possible Soviet attack; he also believed it essential to convince the Soviets that we were ready to use nuclear weapons if they started a war. The issue, Acheson said, was how to restore the credibility of the deterrent. "Nothing could be more dangerous," he wrote, "than to embark upon a course of action of the sort described in this paper in the absence of a decision to accept nuclear war rather than accede to the demands which Khrushchev is now making."

At the same time, however, Acheson was not prepared to go as far as some in the military, who advocated "limited use of nuclear means—that is to drop one bomb somewhere. . . . This I thought was most unwise," Acheson said later. He saw it leading to a larger nuclear exchange. He rejected the suggestion that you could simply stop after a single bomb had demonstrated your willingness to employ these ultimate weapons. "This seems to me irresponsible and not a wise strategy adapted to the problem of Berlin."

In a discussion of Acheson's report at the National Security Council on the day after he gave Kennedy his report and at a later White House meeting that included only Kennedy and Bundy, Kennedy asked Acheson to clarify just when he thought nuclear weapons might have to be used. Bundy recalled that "Acheson's answer was more measured and quiet than usual. He said that he believed the president should himself give that question the most careful and private consideration, well before the time when the choice might present itself, that he should reach his own clear conclusion in advance as to what he would do, and that he should tell no one at all what that conclusion was. The president thanked him for his advice, and the exchange ended."

Acheson's discretion masked his low regard for Kennedy's leadership. He considered Kennedy and his national security team as too young and inexperienced for the challenges they were facing. He wrote a friend, "It seems to me interesting that a group of young men who regard themselves as intellectuals are capable of less coherent thought than we have had since Coolidge. They are pretty good at improvising. . . . But God help us if they are given time to think!" He said later that Kennedy was "out of his depth" and given to "high-school thinking."

Despite outward expressions of deference, Kennedy resisted Acheson's militant advice. Asked later what the president thought of Acheson, Bobby Kennedy replied, "He liked him. No, he didn't like him—that's not correct. He respected him and found him helpful, found him irritating; and he thought his advice was worth listening to, although not accepted. On many occasions, his advice was worthless." Jackie Kennedy thought that Acheson resented Kennedy becoming president at so young an age. It was, what she called, "a jealousy of generations. He couldn't bear to see someone younger . . . come on." The tensions with Acheson ran deeper than his current advice on Berlin or any age gap. Acheson was closely identified with Harry Truman; they shared an antagonism to Joe Kennedy, the "appeaser," and had been unenthusiastic about the son's presidential aspirations. Their hostility to his father and Jack's political ambitions partly colored Bobby Kennedy's assessment of his brother's view of Acheson.

Like so many of his predecessors in the White House, by the summer of 1961 Kennedy had second thoughts about serving as president. No one seemed to have answers to any of the major problems that had descended on him so quickly. The only response to the mounting difficulties he could identify was to appear confident and hope for the best. He saw no miracles on the horizon that would oust Castro or stabilize Vietnam or end segregation or deter Khrushchev from trying to take full control of Berlin. But he continued to believe that rational understanding would deter the Soviets from a mutually destructive war, and that on the domestic front, a majority of southerners, eager to prevent turmoil in their region, would reach some kind of accommodation with the undeniably justified demands for equal treatment of blacks. However much it might be a triumph of faith over fact, Kennedy trusted that he could find ways to ease, if not resolve, each of these dilemmas.

o o o o

Advice and Dissent

Despite all the difficulties of his first six months in office, Kennedy had not lost his sense of humor. He remembered his advice to Bobby when he complained that Jack shouldn't have told the press that he had made him attorney general in order to give him some legal experience: Bobby, you have to understand that to survive in politics you sometimes need to poke fun at yourself and others. So, when the reporter asked him about Nixon's harsh assessment of his foreign policy, Kennedy joked, I sympathize with his problems, adding that he looked forward to 1964, which amused reporters, who thought Kennedy was saying that it was too soon for Nixon to begin his next campaign.

On a lighter note, July began with an eleventh anniversary party at Bobby and Ethel Kennedy's Virginia estate, where wild dancing and singing preceded thirty-year-old younger brother Teddy's leap into the swimming pool fully clothed. "It was all great fun," Schlesinger recorded, "a perfect expression of the rowdier aspects of the New Frontier."

Kennedy's initial problems as president were a prelude to even more troubles in the second half of the year. The testing of nuclear weapons rivaled Cuba as a major concern. Kennedy had come to office against a backdrop of anxiety about nuclear tests and the consequences of a total war. John Hersey's 1946 book, *Hiroshima*, made clear the horrors of a nuclear attack, and explosions in the atmosphere by the United States and Soviet Union had spawned fears of radioactive fallout. The series of U.S. nuclear tests in 1954 over remote islands in the Pacific, which poisoned and eventually killed a Japanese fisherman; the samples of radioactive rain over Chicago in 1955; and evidence four years later of lethal strontium-90 in milk moved the *Saturday Evening Post* to publish a lead article: "Fallout: The Silent Killer." Various cancers and permanent genetic damage were seen as consequences of the Soviet-American race to outdo each other in building nuclear arms.

During his presidency, Eisenhower had grown increasingly worried about the excessive stockpiling of nuclear missiles. He lamented the fact that the military had begun to "talk about megaton explosions as though they are almost nothing." At the end of his term, Ike complained that the U.S. Atomic Energy Commission (AEC) "are trying to get themselves into an incredible position—of having enough to destroy every conceivable target all over the world, plus a three-fold reserve. The patterns of target destruction are fantastic." He worried that "there just might be nothing left of the Northern Hemisphere" after a nuclear war. And this was at a time when Eisenhower's fear was of "radioactive fallout, not of sky-blackening, earth-freezing soot and smoke" scientists later described as a nuclear winter. As a growing group of concerned scientists observed, the only function the excess of nuclear weapons could serve would be to make the rubble bounce.

In 1957–58, the Soviet Union and the United States agreed to a moratorium on atmospheric tests. The first underground nuclear test by U.S. scientists persuaded the Eisenhower White House that it could develop more sophisticated nuclear bombs without aboveground explosions. Then, in March 1958, when the Soviets unilaterally extended their suspension on atmospheric tests, world opinion pressured Washington into a similar decision. A conference of experts in Geneva, Switzerland, during the summer of 1958 concluded that it would be possible permanently to end atmospheric testing by creating detection sites around the world to guard against cheating by any of the nuclear powers. On October 31, a conference convened in Geneva to create an inspection system that would allow the United States, Soviet Union, and Britain to negotiate a test ban treaty. An impasse quickly developed, however, over the conditions governing a control system. The U-2 episode in 1960, followed by the collapse of a U.S.-Soviet summit scheduled for May, left test ban differences unresolved when Kennedy became president.

Despite opposition from his military chiefs and most Defense Department and AEC officials, Kennedy was keen to sign a treaty that would eliminate the atmospheric pollution from testing, reduce Cold War tensions, inhibit proliferation of nuclear weapons, and give the new administration a landmark achievement. Kennedy's appointment of Glenn T. Seaborg, chancellor of the University of California, Berkeley and a Nobel laureate in nuclear chemistry, and Jerome Wiesner, a prominent MIT electrical engineer and outspoken proponent of a test ban, as commission members signaled a shift away from the AEC's insistence on sustained testing. Eisenhower appointees Lewis Strauss and John McCone and commission physicists Ernest Lawrence and Edward Teller warned against undetectable tests that would allow Moscow to eclipse the United States in usable weapons.

In March 1961, Kennedy convened a White House luncheon for the administration's principal defense and arms control officials as well as five senators and three congressmen to discuss Geneva test ban negotiations. The group divided into those who believed an agreement essential for the future peace and those who warned that the Soviets would use the talks to advance their nuclear capacity. The acrimony between the two sides was intense. Kennedy, "with a smile in his voice," as Seaborg described it, thanked everyone for their candid views and said, provoking much laughter, he was "glad to see that there was agreement on the U.S. side." He made clear that he saw much more at stake than just a test ban agreement:

unrestrained increase in nuclear weapons and their spread to other countries. If the negotiations were to fail, it was essential that "the watching world would recognize that we had done our best."

As the talks faltered throughout the spring, Kennedy wrestled with whether to end the moratorium and resume testing. The Joint Chiefs pressed him to ensure the country's safety by a return to testing and described the advantages to the United States. Despite little evidence to support its assertion, the Defense Department warned that "if we don't resume the Soviets may pull ahead." Because the White House feared a hostile world reaction that would undermine U.S. foreign policy objectives, Kennedy decided to hold off on ending the moratorium at least until he discussed a test ban with Khrushchev at the June 1961 Vienna summit.

The exchanges at the conference about nuclear weapons and a test ban could not have been more disappointing. Khrushchev echoed the Soviet position in Geneva, which was dismissive of inspections as a ploy for U.S. espionage. Soviet insistence on barring monitors spoke to the fear of revealing their nuclear inferiority to the United States. Khrushchev insisted that a test ban was of small consequence and that a treaty providing for general and complete disarmament was the best path to improved Soviet-American relations and world peace. As Secretary of State Rusk told his Soviet counterpart, Foreign Minister Andrey Gromyko, and as Kennedy pointed out to Khrushchev, the fact that almost thirty years of disarmament discussions had yielded no results did not discourage Khrushchev from urging additional comprehensive talks. Khrushchev promised that as long as there were discussions, the Soviet Union would not resume testing unless the United States did.

After Vienna, Moscow continued to show little interest in reaching a test ban agreement, but Kennedy refused to lose hope. "It isn't time yet," he told Rusk. "It's too early. They are bent on scaring the world to death before they begin negotiating, and they haven't quite brought the pot to boil. Not enough people are frightened." To signal Moscow and other governments the strength of

Rose and Joe Kennedy: JFK said of his father, "He made it all possible."

President Kennedy and brother Bobby, his closest confidant and most forceful advocate.

LEFT:

Ted Sorensen, the brilliant thirty-three-year-old wordsmith who crafted most of Kennedy's best lines during their eleven-year association.

BELOW:

In March 1962, former president Dwight Eisenhower saw Kennedy as too young and inexperienced to master Cold War challenges.

President Kennedy and Bobby with Vice President Lyndon Johnson and
Secretary of State Dean Rusk. The Kennedys lacked confidence in
their judgments and consigned them to the fringes of the administration.

Historian Arthur Schlesinger, Jr. was
a leading liberal voice in the
White House and the person best
prepared to recount its accomplishments
after Kennedy's death.

McGeorge Bundy, former Harvard dean
and White House national security
adviser, reflected Kennedy's affinity for
the most brilliant men he could find to
staff his administration.

Kennedy discusses the Alliance for Progress with Latin American representatives. The Alliance reflected Kennedy's hopes of winning "hearts and minds" in the contest with communism in the developing world.

Kennedy with Secretary of Defense Robert McNamara, who enjoyed Kennedy's high regard after their successful collaboration in the Cuban Missile Crisis, and General Maxwell Taylor, Kennedy's choice as chairman of the Joint Chiefs of Staff. Taylor was the only one of the military chiefs he fully trusted.

Sorensen in a later meeting with Fidel Castro,
who outlasted all administration efforts to bring him down.

President Kennedy with Allen Dulles and John McCone, the former
and new head of the CIA. Kennedy had grave doubts about their advice, but kept
them on to mute conservative political opposition.

President Kennedy with Adlai Stevenson and United Nations
representatives. Kennedy's failure to consult Stevenson about the
Bay of Pigs invasion embarrassed and angered him.

President Kennedy and the Joint Chiefs. Conflict more than collaboration marked their
exchanges over Cuba, Berlin, Vietnam, and nuclear weapons.

President Kennedy and Charles de
Gaulle during their Paris meeting.
It improved Kennedy's international
standing after the Bay of Pigs debacle.

President Kennedy, Jackie, and the
Khrushchevs at the Vienna
summit, which Kennedy described
as the worst experience of his life.

President Kennedy, Jackie, Lyndon Johnson, and Arthur Schlesinger, Jr.
watching Alan Shepard's return from space—a prelude to
Kennedy's promise to put a man on the moon by the end of the decade.

President Kennedy and the Executive Committee during the Cuban Missile Crisis, the most dangerous moment in the Cold War.

President Kennedy with Averell Harriman, who negotiated a nuclear test ban agreement with Khrushchev in July 1963.

President Kennedy and advisers discussing nuclear arms control, a Kennedy priority, in November 1961.

The Vietnamese leaders Ngo Dinh Diem, Ngo Dinh Nhu, and Madame Nhu. Their resistance to reforms persuaded Kennedy to oust them in November 1963.

President Johnson, Bobby Kennedy, and Jackie leaving the White House for Kennedy's funeral two days after he was assassinated in Dallas, Texas.

U.S. interest in arms control, Kennedy sent Congress a message proposing the establishment of a new "United States Disarmament Agency for World Peace and Security." Khrushchev was not impressed. As Kennedy believed, he seemed intent on bringing the world to the brink of catastrophe before he would negotiate an agreement. On August 30, 1961, Moscow resumed testing in the atmosphere, arguing that U.S. failure to accept Soviet disarmament proposals, an unbending attitude on Berlin, and French nuclear tests encouraged by the United States had compelled them to go forward with the development of "super, powerful bombs" and rockets that could carry them to "any point on the globe." The decision spoke volumes about Moscow's inferiority to the United States in these ultimate weapons.

Khrushchev's deception enraged Kennedy. "Fucked again," he exclaimed when he heard the news. "The bastards. That fucking liar." But he worried that Khrushchev might have been acting on information that under pressure from the Joint Chiefs, Kennedy had agreed to preparations for renewed testing. He had insisted that they be kept hidden so that Khrushchev would not have a justification for ending the moratorium, but he feared that Moscow might have learned of U.S. preparations through spies or leaks from the Chiefs hoping to intimidate Khrushchev. Once the Soviets had tested a weapon, Kennedy saw no recourse but to implement plans for new U.S. explosions. He wished, however, to have only underground tests and to hold off revealing them in order to mine public indignation toward Moscow for polluting the atmosphere. Administration hawks wanted him to announce a resumption of testing at once as a way to cow Moscow, but Stevenson urged Kennedy not to do it, saying the Soviet action was "not a blow but a bonanza." Kennedy agreed, saying that he wanted to get the "maximum propaganda" advantage from the Soviet test.

At the same time, Kennedy wished to make clear to the world that the United States was eager for an end to nuclear testing and agreements that could head off what he feared would be the last

global war. In September 1961, at the opening annual session of
the United Nations, the world body he had described in his inaugural as "our last best hope in an age where the instruments of
war have far outpaced the instruments of peace," he restated his
sense of urgency about the dangers from a nuclear conflict, declaring that "mankind must put an end to war—or war will put an
end to mankind." It was more than a felicitous phrase. He had a
sense of urgency about convincing Moscow and every nation that
the consequences of another global conflict would be humankind's
ultimate defeat.

But the Russians seemed impervious to his appeal or were
convinced that the West was so intent on destroying communism
that only nuclear parity with the United States would ensure its
survival. Kennedy, however, saw Moscow's behavior as not defensive but offensive—a program aimed at victory in the Cold War.
"Of all the Soviet provocations of these two years, it was the resumption of testing that disappointed most," Bundy wrote later—
chiefly because Kennedy now had to commit the United States to
underground testing and lay plans for atmospheric tests as well. By
November 2, the White House knew of thirty-seven aboveground
Soviet explosions, and Seaborg, who had been a consistent voice
for a test ban, warned Kennedy at an NSC meeting "that if the
US tested only underground while the Soviets tested in the atmosphere, we would be in no position to compete with them." Another
scientist that morning echoed Seaborg's conclusion: He feared that
underground tests were less valuable in developing effective nuclear
arsenals. But convinced that the United States was still stronger
than the Soviet Union, Kennedy refused to announce that the
United States would resume testing. The consequence was what
Seaborg described as "a prolonged period of uncertainty regarding
preparations for atmospheric testing. A decision would seem to be
made one day and withdrawn the next. Kennedy wanted to take a
firm stand and be ready; yet he wanted to keep his options open:
he was reluctant to take steps that might bar the way to a test ban."

Kennedy's ambivalence was on display at a December conference in Bermuda with Macmillan. The president described himself as "a great anti-tester" but felt compelled to make preparations for a test series. In private, Kennedy impressed Seaborg as "considerably more in favor of accepting risks and making compromises in order to achieve a test ban than either he or U.S. negotiators ever allowed themselves to be in public." Schlesinger and Bundy reinforced his aversion to testing. Schlesinger believed that "the US has an unmatched opportunity to recover the moral and political leadership of the world" by enlisting the support of "world opinion," which he argued was a match for military power in the contest with the communists. Bundy endorsed Schlesinger's recommendation as "a better argument against testing now than you have yet heard from advisers nearly all of whom personally favor testing." But Kennedy could only delay so long on testing without facing an explosion of political criticism that he was letting the Soviets steal a march on the United States in the arms race.

With little success in turning nuclear arms talks or limits on testing in successful directions, Kennedy hoped that he could make some quiet progress on Cuba. He expected General Taylor to be an organizing force in helping to oust Castro. Initially, Kennedy asked Taylor to come back to Washington to serve at the head of a committee, including Bobby, Allen Dulles, and Admiral Arleigh Burke, to study the Cuban failure. By putting Dulles and Burke on the committee, Kennedy was putting the fox in charge of the henhouse. Kennedy was less interested in assessing blame for the Bay of Pigs failure, however—especially against the CIA or military, which would have produced more domestic acrimony—than in putting off any immediate action about Cuba. His objective in the aftermath of the Cuban catastrophe was to convince hard-line anticommunists in the United States that he was working to bring down Castro, while at the same time keeping the lowest possible profile for his administration in its dealings with Havana. The objective remained to rid Cuba of Castro, but without clear evidence that Washington had arranged his demise.

In August, after attending a meeting of the Inter-American Economic and Social Council held in Punta del Este, Uruguay, Dick Goodwin told Kennedy that they should "pay little attention to Cuba. Do not allow them to appear as the victims of U.S. aggression. Do not create the impression we are obsessed with Castro—an impression that only strengthens Castro's hand in Cuba and encourages anti-American and leftist forces in other countries to rally round the Cuban flag." Goodwin's understanding of what was needed rested on a conversation with Che Guevara, the president of the Cuban National Bank and Castro's close associate, who later became a synonym for leftist Latin American revolutions. Guevara ridiculed the Bay of Pigs failure, saying "he wanted to thank us very much for the invasion—that it had been a great political victory for them—enabled them to consolidate—and transformed them from an aggrieved little country to an equal."

Kennedy agreed with Goodwin, saying he wanted "to play it very quiet with Castro because he didn't want to give Castro the opportunity to blame the United States for his troubles," which Kennedy believed would come from his heavy-handed control of the Cuban economy and secret U.S. operations subverting his government. In September, when Kennedy met with former Brazilian president Juscelino Kubitschek, he explained that he was determined to ignore Cuba, "thus depriving Castro of the publicity on which he flourishes," and avoid complaints that the United States was "a positive threat to the independence, sovereignty and right to self-determination of nations in the Hemisphere." Similarly, when Kennedy met Argentina's President Arturo Frondizi at New York's Carlyle Hotel at the opening of the U.N. General Assembly's 1961 fall session, he said it was important to discourage impressions of "the United States versus Cuba, or of Castro versus Kennedy, because a debate of this kind would only enhance Castro's prestige." It was "important not to leave the impression of the United

States, great imperialist power from the North, attacking poor, brave Cuba, which is the impression Castro wishes to give."

At the same time, however, Kennedy remained determined to undermine Castro's communist regime through all possible clandestine means. He signed on to proposals from Goodwin to quietly intensify economic pressure on Cuba. Economic warfare, including sabotage by anti-Castro paramilitary forces using American-supplied equipment to destroy industrial plants such as refineries, was to be a high priority. Propaganda aiming to convince Cubans and others in Latin America that Castro was sacrificing Cuba's welfare to international communism was another weapon in the anti-Castro campaign. Kennedy told Frondizi that "[i]t was important to take action to discredit the Cuban revolution, identifying it as foreign, alien, and anti-Christian, and not permitting it to be considered as a revolution that was trying to improve the living conditions of the Cuban people." He hoped that others in the hemisphere would understand that Castro aimed to subvert their governments and that the Cuban leader was as much a problem for them as for the United States.

But White House discussions of how to bring Castro down made clear that President Kennedy and his brother were involved in score settling. Castro had bested them in the first clash and they were determined to see to his demise. They assigned responsibility for the project to a new set of advisers. Because McNamara, Bundy, Schlesinger, Sorensen, and Goodwin had fallen short on ousting Castro and his communist government, Kennedy turned to other planners and operatives. On November 3, he and Bobby set up Operation Mongoose, with Bobby at the head of the program. He expected "to stir things up on the island with espionage, sabotage, general disorder, run & operated by Cubans themselves with every group but Batistaites & Communists. Do not know if we will be successful in overthrowing Castro but we have nothing to lose in my estimate," Bobby recorded in some notes he made after initial discussion of the planning.

While Bobby assumed the central role in directing the operations, his principal collaborators were Taylor, Brigadier General Edward G. Lansdale, and William Harvey at the CIA, with a supporting cast of approximately four hundred CIA agents located at agency headquarters in McLean, Virginia, and its Miami station. Bundy was made the chairman of an interagency group known as SGA, Special Group Augmented, which was charged with direction of Mongoose operations, but it was Lansdale and Harvey who were instructed to develop and implement anti-Castro actions. Although McNamara wouldn't say until much later—"We were hysterical about Castro at the time of the Bay of Pigs and thereafter"—his skepticism about investing too much in bringing down Castro had registered sufficiently on the president and Bobby to keep him in the background.

Likewise, Bundy was no enthusiast about making Castro a prime foreign policy consideration. True, he kept on his desk only two boxes, one marked "President's box" and the other "Cuba," but he saw the focus on Castro as a case of overkill. He thought Bobby was trying to repair the damage to his brother's standing from the Bay of Pigs and seemed to view the conflict with Castro as some kind of contest or game: "It was almost as simple as goddammit, we lost the first round, let's win the second." The Alliance for Progress was one counterweight to Castro's threat and Mongoose was the covert side of the campaign.

Whatever ambivalence McNamara and Bundy had about the attorney general as the lead operative in trying to bring down Castro and making this a leading foreign policy goal, they passively went along with the plans. The shadow of Bowles's dissent and displacement hung over the new initiative. They weren't going to challenge something the president and Bobby were so eager to do. But it didn't quiet their doubts about whether the administration could actually affect Cuban affairs. McNamara and Bundy were happy to be largely shut out of the administration's new anti-Castro campaign. They took their counsel from a National Intelligence Estimate. At the same time the White House set Operation

Mongoose in motion, they read an NIE report saying that Castro had "sufficient popular support and repressive capabilities to cope with any internal threat likely to develop within the foreseeable future. . . . The bulk of the population" accepted the regime and "substantial numbers still support it with enthusiasm." Moreover, Castro's "capabilities for repression" were well ahead of "potentialities for active resistance."

Yet the administration or at least the Kennedys and their new team of advisers didn't want to hear about limitations; they wanted to know what could be done. On November 22, Kennedy asked John McCone, the new director of the CIA, who had replaced Dulles, to give him "an immediate plan of action" to overthrow Castro, and Kennedy instructed his principal national security officials to make this a high, if not the highest, priority with a commitment of all available assets. Similarly, Bobby began hectoring Lansdale and Harvey to move aggressively against Castro. "Why can't you get things cooking like 007?" Bobby asked Harvey, whom he and the president hoped might prove to be their James Bond. When Harvey described problems in training CIA agents for infiltration of the island, Bobby sarcastically proposed to take them to Hickory Hill, his home in rural Virginia, where he would train them himself. "And what will you train them in? Baby-sitting?" Harvey snidely asked Kennedy, who had seven children.

Lansdale was the president's and Bobby's great hope for toppling Castro's government through well-disguised skulduggery. There is not a single reference to Lansdale by Kennedy in any of his public comments, but the idea was to keep his involvement secret or at least as low profile as possible. Lansdale had a track record for brilliant secret operations, and if it were common knowledge that he had been tapped to fix his attention on Cuba, it would have revealed Kennedy's obsession with Castro.

The fifty-three-year-old Lansdale was an up-through-the-ranks commissioned officer with a background in advertising. Entering the service in 1943 at the age of thirty-five, he was an OSS operative in World War II. At the close of the war, he served in the Philippines, where he made a name for himself as a counterinsurgency adviser to Ramon Magsaysay, the Philippines' national defense secretary. He helped the Philippine army build its intelligence services, and in collaboration with Magsaysay, he devised a successful strategy for combating the Hukbalahaps, the Philippines' communist guerrillas, who were trying to overturn the pro-American government.

Lansdale's success in the Philippines brought him to Saigon as the CIA station chief, where he hoped to defeat Viet Cong communist guerrillas with the same counterinsurgency strategy: By contrast with advocates of repressive military action like Kennedy's new Green Beret Army units, who declared "When you've got 'em by the balls, their hearts and minds will follow," Lansdale preached a soft policy of wooing the Vietnamese with actions that made the United States appear as "pro-people." It was an approach born of his conviction that American values had universal appeal, and echoed Wendell Willkie's 1943 bestselling book, *One World*, in which Willkie argued that the Russians and Chinese were adopting American ideals. The success of the United States in transforming Germany and Japan from totalitarian societies to representative democracies was evidence to Lansdale that missionary diplomacy was a realistic means for bringing adversaries to our side. As Lansdale said in a later description of his work in the Philippines and Vietnam, "I took my American beliefs with me into these Asian struggles, as Tom Paine would have done."

Two novels in the fifties, Graham Greene's *The Quiet American* (1956) and William J. Lederer and Eugene Burdick's *The Ugly American* (1958), encouraged many to see Lansdale as the principal character in both books: Greene's CIA agent Alden Pyle, whose hopes of finding a "Third Force" in Vietnam between the communists and France's colonial occupiers echoed Lansdale's idea of promoting American democratic values; and *The Ugly American*'s Colonel Edwin B. Hillendale, whose aims to pass along American ideas to Vietnamese peasants resembled Lansdale's strategy for defeating the Viet Cong. Lansdale's appeal to the Kennedys rested on his opposition to working with repressive dictators against communists for the sake of American national security or relying on U.S. forces to overturn leaders tying themselves to Moscow and Beijing. Instead, he urged defeat of left autocratic governments through support for indigenous democrats proposing to serve the people with policies advancing economic development and social justice.

As much as the Kennedys preferred his plans for toppling Castro, Lansdale found himself in competition with the CIA's William Harvey on how to change the Cuban government. The forty-six-year-old Harvey grew up in Indiana, where he earned a law degree and served as an FBI agent from 1940 to 1947. He was an unstable, erratic personality and Hoover had fired him for drunkenness and insubordination. It didn't bar him from entering the newly organized CIA, where he established himself as a master spy by building a tunnel between West and East Berlin that allowed the U.S. military to listen in on Soviet telephone conversations.

In 1961, Harvey became the head of a CIA executive action committee, dubbed "capability," committed to assassinating Castro. Plans to kill Castro had been hatched in August 1960 during Eisenhower's presidency. No scheme was too nutty for the CIA operatives charged with the assignment: The plans included getting Castro to smoke a botulism-filled cigar that could kill him at first puff; poison capsules hidden in a jar of cold cream that a Castro mistress was supposed to put in his drink, but couldn't use when she found that they had melted; and a contract with a former FBI agent with mob contacts who tried to put lethal pills in Castro's drinks or food. After the Bay of Pigs, discussion of assassination plans were suspended; they were reactivated in April 1962, however, with poison pills, assassination teams, and then an exploding seashell at a site where Castro might be skin diving becoming the weapons of choice. When nothing came of these plots, in February 1963 the CIA gave up trying to assassinate Castro.

Harvey's principal assignment in Operation Mongoose was to

promote infiltration by exiles into Cuba, where they were supposed to commit acts of sabotage that would destabilize Castro's government. But nothing he initiated came to any constructive end and Bobby Kennedy and Lansdale saw him as an unreliable loose cannon. He was in fact a paranoid character, notorious for carrying guns, saying, "If you ever know as many secrets as I do, then you'll know why I carry a gun." Although Bobby disliked and distrusted him, Harvey's credentials as a resourceful agent had allowed him to join the Mongoose team. Harvey reciprocated Bobby's antagonism. One of his CIA colleagues said he "hated Bobby Kennedy's guts with a purple passion."

The Joint Chiefs also got into the act. Lyman Lemnitzer endorsed a madcap plan called Operation Northwoods. It proposed terrorist acts blamed on Castro against Cuban exiles in Miami, including assassinations and the possible destruction of a boatload of Cubans escaping the island. Terrorist strikes in other Florida cities were to be considered in order to inflame enough Americans and win support from the world community for an invasion that could bring down Castro's communist regime. Kennedy rejected the plot as too extreme and the memo was too embarrassing to see the light of day until the National Security Archive at George Washington University forced it into the open in 2001. Kennedy may have rejected the plot as excessive, but it was the atmosphere set by Mongoose that gave Lemnitzer license to suggest such outlandish plans.

Unhappily for the president and Bobby, Lansdale had no better luck in deciphering how to topple Castro than did Harvey and the hundreds of other CIA operatives assigned the job. Lansdale's idea was to foment an internal revolution, but Castro's grip on his country was too firm and Lansdale's schemes were no more effective than Harvey's: a chemical assault on the sugar crop that would destabilize the economy; appeals to Castro's government associates to see him as undermining the island's well-being; a bizarre scheme to persuade Cubans that Castro was the Antichrist by exciting expectations of the Second Coming with star shells launched from a U.S. submarine off the Cuban coast—"elimination by illumination," one skeptic called it; and finally, a six-part program with thirty-three steps or tasks that would produce "the Touchdown Play," an internal revolt turning out the communists in October 1962. Cuba had turned into one of those problems about which everyone had an opinion and no one had a solution.

While nuclear talks and Cuba bedeviled Kennedy's hopes for some kind of international progress, worries that Soviet-American differences over Berlin could erupt in war shadowed everything the White House did in the summer of 1961. For Kennedy, the Berlin problem now stood as a kind of two-front war. On one hand, he needed to convince Khrushchev that he could not be pushed around and forced into humiliating concessions on Berlin. On the other, he wished to mute the domestic pressures for military steps toward a confrontation with Moscow. On July 3, *Newsweek* ran a story allegedly coming out of the Pentagon describing preparations for war, including "some demonstration of U.S. intent to employ nuclear weapons," as Acheson had advised. No one could identify the source of the leak, but the White House was a likely candidate: It not only sent a sharp message to Khrushchev but also blunted demands from Dean Acheson and Alsop for stronger leadership in confronting the Soviet threat.

The *Newsweek* article, however, did not appease Acheson. On July 12, at a meeting of an interdepartmental group on Germany and Berlin, he pushed hard for a military buildup. Kennedy must decide "at the earliest possible moment" whether he would implement Acheson's advice or follow Walter Lippmann's conciliatory approach to the Berlin challenge, which Acheson described "as doing it with mirrors." Acheson warned that "there would be a revolt in Congress if it was not given strong leadership soon on the Berlin question." He wanted the president to declare a national emergency, which would then allow the needed strengthening of our military posture. The absence of such a declaration would limit the sort of buildup that could discourage the Russians from intemperate actions. Kennedy held Acheson off by insisting on the need for additional study and discussion before crucial decisions were made. Speaking for the president at a July 19 NSC meeting, McNamara persuaded the group to postpone a national emergency declaration and a call-up of reserve forces until it seemed necessary.

Although in Kennedy's presence Acheson diplomatically went along with the president's decision to defer any action, he doubted the wisdom of further deliberations. In conversations outside Kennedy's earshot, Acheson made no secret of his doubts about Kennedy's understanding of what he needed to do. "Gentlemen, you might as well face it," he told a working group on Berlin. "This nation is without leadership." It was clear to Bundy that Acheson held Kennedy in contempt. He saw him as weak: Kennedy "is not the sort of man that is worth *my* while to be advising," Bundy thought Acheson believed.

On July 25, distressed by the shift of authority from the secretary of state and Foreign Service professionals to the president's national security team, Acheson attacked Kennedy's administration in a public address: He decried the State Department's decline, warning that it was playing havoc with the country's foreign policy. Acheson ended his talk by facetiously declaring that despite his "treasonous" speech he hoped the audience would be willing to hear from him again sometime.

To counter Acheson's assault on his authority and put Moscow and critics at home on notice that he was a resolute leader with a plan to preserve Western rights in Berlin, Kennedy held a press conference on July 19 and gave a nationally televised address from the Oval Office on July 25. At the press briefing, he forcefully declared that any Soviet attempt unilaterally to deny their former allies access to Berlin and deprive the people of West Berlin the freedoms they currently enjoyed would jeopardize the peace.

Asked if he intended to declare a national emergency in order to call up reserve units, Kennedy promised to address these questions in his coming speech. When asked by a reporter whether he agreed with a statement by the Soviet ambassador to the United States that Americans were not prepared to go to war over Berlin, Kennedy replied "that we intend to honor our commitments."

The questions to Kennedy suggested the crisis of confidence in his leadership after the Bay of Pigs failure revealed his reluctance to use American forces to topple Castro. How he intended to defend the Western presence in Berlin without a war and how he would overcome the evident divisions in his administration were the implied questions behind the reporters' questions. It was no secret that Acheson doubted Kennedy's foreign policy competence. It was clear to Washington insiders how on edge Kennedy was about finding his way through these dilemmas.

The questions about his leadership were more than Washington gossip. He was full of anguish about the possibility of a nuclear war, especially since several national security advisers and the Kremlin didn't seem to share his fears. Acheson, Alsop, McCone, and the Joint Chiefs, like Khrushchev, impressed Kennedy as equally oblivious to the costs of such a conflict. After a meeting with the Chiefs, Kennedy remarked that "only fools could cling to the idea of victory in a nuclear war." After Vienna, he thought that convincing Khrushchev was as big a problem: "That son of a bitch won't pay any attention to words," Kennedy said. "He has to see you move"—meaning the United States might have to go to the brink of war before the Soviets would back down and agree to productive talks about Berlin. Schlesinger recalled: "While Kennedy wanted to make this resolve absolutely clear to Moscow, he wanted to make it equally clear that we were not, as he put it to me, 'war-mad.'"

Kennedy saw the speech to the nation as a crucial test of his capacity to prove himself an effective leader. Speaking from the Oval Office crowded with cameras and klieg lights that added to the heat of the July evening, Kennedy strained to keep his poise, recalling how a perspiring Nixon undermined his election chances in September 1960. A larger dose of steroids than he normally re- lied on to control his Addison's disease helped provide the adrena- line boost Kennedy needed to combat the strain of speaking to the hundreds of millions around the world who hoped he could fend off a disastrous war and defend two and a half million Germans from a communist takeover.

The speech aimed to leave no doubt either in Moscow or any- where in the West that the president understood the fullness of the Soviet threat and was determined to meet it head-on. Khrushchev's "grim warnings [in Vienna] about the future of the world . . . his subsequent speeches and threats . . . have all prompted a series of decisions by the Administration," Kennedy said. The Soviet lead- er's intention to end Western rights in Berlin and bring the city under his control would not be permitted. But the danger wasn't simply to the people of that embattled city—it was to free peoples everywhere. The most immediate crisis, however, was in the center of Europe, where the United States intended to stand its ground against aggression. "We cannot and will not permit the Commu- nists to drive us out of Berlin." The United States was ready to talk, "if talk will help. But we must also be ready to resist with force, if force is used upon us." Kennedy then described a defense buildup, which would provide "a wider choice than humiliation or all-out nuclear action." But he had no intention of abandoning "our duty to mankind to seek a peaceful solution. . . . To sum it all up: we seek peace—but we shall not surrender." Kennedy ended with a plea for public understanding and support. "We must look to long days ahead."

Kennedy's speech had its desired effect at home and in Russia. It put Acheson, Alsop, and the Chiefs on notice that he was in command and that while he was determined to avoid a war with Russia, he was also committed to saving Berlin from any act of communist aggression. His speech made clear to domestic critics and advisers that they could not browbeat him into anything he thought too militant or too passive.

Khrushchev initially responded to Kennedy's speech with "rough war-like language," telling John J. McCloy, Kennedy's chief arms control representative, who was in Moscow for talks and was summoned to Khrushchev's Black Sea summer retreat, that Kennedy was declaring war on the Soviet Union. He warned, "If Kennedy started a war, he would be the 'last president of the United States.'" But it was no more than saber-rattling. Khrushchev's military chiefs wanted no part of a confrontation with the West that could lead to an all-out war. "With respect to ICBMs," Marshal Sergei Varentsov, the top commander, said privately, "we still don't have a damn thing. Everything is only on paper, and there is nothing in actual existence."

After so many threats, the pressure on Khrushchev to do something about Berlin was compelling. The exchanges over a showdown had accelerated the flood of immigrants from East Germany to the West. Since the beginning of 1961, more than twenty thousand a month had fled communist control. Khrushchev had initially vetoed suggestions of constructing a barrier between the two Berlins to stem the outflow as too provocative. But indications from Washington, including a statement from Senate Foreign Relations Committee chairman J. William Fulbright, that the United States would not physically oppose Soviet efforts to restrict movement out of the communist zone, persuaded Khrushchev to build a wall. To bar against a U.S. reaction, however, Khrushchev preceded the start of construction on August 13 with a speech decrying "a war psychosis" and reminding everyone that "we have common needs and interests since we have to live on the same planet." He directed the East Germans to build the wall in stages to ensure that the initial construction did not trigger an armed response.

The Wall touched off a new round of debate among Kennedy's national security advisers. Acheson was more adamant than ever about the need for military action. He thought that "the Wall would have come down in a day if Harry Truman had been President." General Lauris Norstad, the head of U.S. forces in Europe, who had no orders to respond, said, "If I had been commander I would have taken a wire and flung a hook over and tied it to a tank and pulled it down." Lemnitzer complained that a passive reaction made everyone in the West appear to be "hopeless, helpless, and harmless." The U.S. diplomatic mission in Berlin reported that Mayor Willy Brandt and newspaper, news service, radio, and television editors felt that accepting the Wall was creating a crisis of confidence that endangered America's position in Europe. The mission also warned that if the Soviets were "able to 'get away' with this fait accompli, other similar actions may be undertaken. . . . Having taken such a big slice of salami and successfully digested it, with no hindrance, they may be expected to snatch further pieces greedily."

Kennedy and his White House advisers were actually relieved by Khrushchev's action. Kennedy saw the Wall as a demonstration of Khrushchev's decision not to force a crisis. "Why would Khrushchev put up a wall if he really intended to seize West Berlin?" Kennedy asked rhetorically. "There wouldn't be any need of a wall if he occupied the whole city. This is his way out of his predicament. It's not a very nice solution, but a wall is a hell of a lot better than a war," Kennedy told Kenny O'Donnell. Bundy reported that there was "unanimity in your immediate staff" for negotiation rather than military action. After all, the Wall was "something they [the Soviets] have always had the power to do; it was something they were bound to do sooner or later." Kennedy's answer was not a military response but the chance to score propaganda points or use this "very good propaganda stick" against the communists for having confined people unhappy living under Soviet control. "This seems to me to show how hollow is the phrase 'free city,'" which Khrushchev had promised a West Berlin under East German rule," he told Rusk. The Wall also showed "how despised is the East German government."

The pressure on Kennedy to act more decisively or give some

indication that he would use force against the Soviets if they over-reached on Berlin persuaded him to satisfy a Brandt request that U.S. troops make a visible show of crossing the Autobahn from West Germany to the city. After the Vienna summit, Kennedy had told O'Donnell, "It seems particularly stupid to risk killing a million Americans over an argument about access rights on an Autobahn. . . . If I'm going to threaten Russia with a nuclear war it will have to be for much bigger and more important reasons than that." But the insistence of the Germans that he provide some demonstration of U.S. opposition to the Wall moved Kennedy to order an armored brigade of fifteen hundred U.S. troops to show the flag by traveling the 110 miles from West Germany to West Berlin. To underscore the importance he put on the gesture, Kennedy directed Lyndon Johnson and General Lucius Clay, the former U.S. military governor of Germany and architect of the 1948 Berlin airlift, to fly to Bonn and then West Berlin to give a message of hope that America was "determined to fulfill all our obligations, all our commitments" and "dare to the end to do our duty."

Johnson was reluctant to take on the assignment. He saw it as more of a symbolic than substantive expression of America's opposition to the Wall and feared that the trip would do more to undermine faith in U.S. commitments than reinforce it. But his presence on the ground, which put him in harm's way if the Soviets challenged the brigade's movement, boosted German convictions that Kennedy meant to hold the line against any communist aggression. After all, it was assumed that Kennedy would have to respond if his vice president came under attack in Berlin. Fortunately, Khrushchev was content to close off the exodus from the East without trying to fulfill threats against West Berlin. And when Johnson, borrowing from U.S. history, spoke passionately about the commitment of "our lives, our fortunes, and our sacred honor" to the preservation of German freedoms and issued a plea to have "faith in your allies," it bolstered German belief in Kennedy's leadership.

But uncertainties remained. And Kennedy, with the support of

Rusk, McNamara, Bundy, Stevenson, and Kennan, wanted to enter into negotiations that could at least reduce the chances of an immediate crisis over Berlin. Kennedy instructed Rusk to "examine all of Khrushchev's statements for pegs on which to hang our position. He has thrown out quite a few assurances and hints here and there and I believe they should be exploited." Kennan reinforced Kennedy's interest in talks: "Unless the West shows *some* disposition to negotiate," he advised, "the hard line is going to be pursued in Moscow not only to the very brink but to the full point of a world catastrophe."

Acheson, the military, and de Gaulle disagreed. They admonished Kennedy against being taken advantage of in any negotiations with Khrushchev. Taylor was invited now to divide his attention between Cuba and Berlin. Kennedy hoped he would be more restrained than the Joint Chiefs about using military power. But Taylor disappointed him, urging a hard line against Khrushchev: He saw "clear evidence that Khrushchev intends using military force, or the threat thereof, to gain his ends in Berlin." He urged Kennedy "to shift into higher gear" on making military preparations to combat Soviet plans. Acheson counseled against appearing too eager to negotiate. Instead, he wanted Kennedy to make clear that he was willing to use nuclear weapons to meet any Soviet aggression. De Gaulle predicted that entering into negotiations would encourage Moscow to increase their pressure on Berlin and "would be considered immediately as a prelude to the abandonment, at least gradually, of Berlin and as a sort of notice of our surrender." It "would be a very grave blow to the Atlantic Alliance."

The clash of opinions demonstrated how imperfect the views of even the most experienced and intelligent observers on current events could be: Taylor, Acheson, and de Gaulle misread Khrushchev's interest in negotiations that would put aside his threats to seize West Berlin. In September, Khrushchev sent a series of messages, including a lengthy letter, encouraging bilateral talks about Berlin that were designed to give him a retreat from his bullying.

The message sent to Kennedy through Pierre Salinger, who received it from a Russian journalist, was "The storm in Berlin is over." When Salinger delivered the message to Kennedy at one in the morning, he declared with no small relief: "There's only one way you can read it. If Khrushchev is ready to listen to our views on Germany, he's not going to recognize the Ulbricht regime in East Germany—not this year, at least—and that's good news."

With the opening of negotiations, Khrushchev announced in a speech on October 17, 1961, at a Communist Party congress that "the western powers were showing some understanding of the situation, and were inclined to seek a solution to the German problem and the issue of West Berlin." And so, "We shall not insist on signing a peace treaty absolutely before December 31, 1961." Schlesinger recorded: "The crisis was suddenly over."

The end of the Berlin crisis was a welcome respite from the constant drumbeat of trouble that had beset Kennedy and his advisers from day one of his presidency. But even with Berlin momentarily put aside as a source of daily tension, Kennedy found no easing of the pressure to save Vietnam. Arguments among his advisers about what to do about Berlin now gave way to sharp and sometimes personally caustic disputes about how to defeat communist aggression against Diem's government in Saigon.

Kennedy's decision in early July to put off expanding U.S. commitments to Vietnam produced an immediate pushback from Rostow, who worried that Kennedy underestimated the need for force in combating the Viet Cong. On July 13, Rostow urged Rusk, who shared his concern about saving Vietnam, to consider three alternatives that he might put before the president: "A sharp increase in the number of Americans in South Vietnam for training and support purposes; a counter-guerrilla operation in the north, possibly using American Air and Naval strength to impose about the same level of damage and inconvenience that the Viet Cong are imposing in the South; and . . . a limited military operation in the North; e.g., capture and holding the port of Haiphong." Except for item one, the proposals were way beyond anything Kennedy was ready to entertain. The stages of economic growth Rostow had written about so persuasively had given way to the stages of military action he saw as the only immediate formula for securing U.S. interests in South Vietnam. Rostow's evangelism about beating back the communist challenge in Vietnam had made him a convert from faith in social engineering to a believer in the use of American power.

Rostow was not alone in believing that Vietnam could be saved by a more assertive U.S. policy. By the middle of July, after only two months in Saigon, Nolting had become a full convert to the can-do school—principally, the work of turning Diem into a liberal reformer ready to take up American ideas. But Diem was anything but a willing pupil. More French than Vietnamese, or at least more schooled in Western ways than traditional Vietnamese culture—a staunch anticommunist Catholic—Diem understood perfectly how to deal with a conventional American diplomat: talk up social and political reform that could outdo the communists in winning the hearts and minds of the country's masses. It was a façade calculated to deceive Nolting. As the journalist David Halberstam, who arrived in Saigon as a reporter for the *New York Times* in 1962, said, "Diem was a Catholic in a Buddhist country, a Central Vietnamese in the South, but most important of all, he was a mandarin, a member of the feudal aristocracy in a country swept by revolution."

By contrast, Nolting saw Diem as the right man in the right place at the right time. He told Rusk that a number of trips around Vietnam with Diem had convinced him that Diem was an honorable man committed to "sound and good" objectives for his country. "He is no dictator": He didn't relish concentrating power in his hands. Nolting thought Diem "would prefer to be a monk rather than a political leader. . . . I think the United States should have no hesitation on moral grounds in backing Diem to the hilt.

Where we think he is wrong, we can bring about amelioration and improvement gradually in proportion to the confidence he has in us. . . . I believe we [are] taking the right track." Few diplomatic assessments in U.S. history were more off target than Nolting's of Diem: He loved power, was contemptuous of American preachments about democracy, and had much less of a hold on his country than Nolting believed.

Others in the Kennedy administration, led by Taylor, Rostow, and Robert Komer, the National Security Council expert on Southeast Asia, shared Nolting's hope of relying on Diem to stem the communist surge in Vietnam, Laos, Cambodia, and Thailand. But they were less optimistic than Nolting and pressed Kennedy to understand that he would have to make some difficult choices. Komer wanted Kennedy to go "all-out in cleaning up South Vietnam." Taylor and Rostow were no less eager to save the country but were less blunt in urging a rescue operation. They urged Kennedy to choose either "to disengage from the area as gracefully as possible," which U.S. domestic politics alone made an unpalatable choice; find "a convenient political pretext and attack with American military force the regional source of aggression in Hanoi; or build as much indigenous military, political and economic strength as we can in the area, in order to contain the thrust from Hanoi while preparing to intervene with U.S. military force if the Chinese Communists come in or the situation otherwise gets out of hand."

John Steeves, a State Department expert on Asia and the chairman of the multi-agency Southeast Asia Task Force, simultaneously warned against the growing dangers of communist aggression in the region: He advised the president to "make the basic decision now to resist this encroachment by appropriate military means. . . . The loss of Southeast Asia to the free world would be highly inimical to our future strategy and interest." Although North Vietnam was not China or Russia threatening to expand its reach in Asia or Europe, it was a communist country about to bring another free people under its control. Most American policymakers believed that if South Vietnam fell into the communist orbit, it would be a blow to free peoples everywhere. Anticommunism was the prevailing mind-set and memories of Hitler made a failure to combat totalitarian aggression anywhere look like a fatal mistake.

Three days after his July 25, 1961, speech addressing the Berlin crisis, Kennedy met with Rusk, Taylor, Bundy, Ball, Rostow, and the task force advisers on Vietnam to discuss the dangers in Southeast Asia. They warned him that the United States was on a treadmill in Vietnam, while Laos remained a matter of grave concern. To protect Vietnam from communist infiltration, they urged Kennedy to consider using U.S., Lao, Thai, and South Vietnamese forces to occupy southern Laos, which Hanoi was using as a supply route to insurgents in South Vietnam. They also described plans for air and naval operations against Haiphong or Hanoi, North Vietnam's principal port and capital.

Kennedy thought their suggestions absurd. But he politely said that he doubted the "realism and accuracy in such military planning." As for Laos, "optimistic estimates were invariably proved false." He did not think that any operation could save southern Laos, "and he emphasized the reluctance of the American people and of many distinguished military leaders to see any direct involvement of U.S. troops in that part of the world." When some of the advisers predicted that "with a proper plan . . . the results would be very different from anything that happened before," Kennedy put them off by pointing out that "General de Gaulle, out of painful French experience, had spoken with feeling [to him] of the difficulty of fighting in this part of the world."

Kennedy made no mention of a letter from Galbraith that supported his reluctance to join the fighting in Southeast Asia: "These jungle regimes, where the writ of government runs only as far as the airport," Galbraith warned, "are going to be a hideous problem for us in the months ahead. . . . The rulers do not control or particularly influence their own people; and they neither have nor warrant their people's support." As for relying on Southeast Asian forces, Galbraith thought that "the entire Laos nation is clearly inferior to a battalion of conscientious objectors from World War I." He counseled that losing Laos would not be the disaster some were describing. "We must not allow ourselves or the country to imagine that gains or losses in these incoherent lands are the same as gains or losses in the organized world."

Most of Kennedy's White House advisers disagreed with Galbraith's assessment. They said that "it would be most helpful in planning if it could be understood that the President would at some future time have a willingness to decide to intervene if the situation seemed to him to require it." Kennedy refused to commit himself to anything. He was especially resistant to sending Americans to block Hanoi's supply route through Laos. He thought "that nothing could be worse than an unsuccessful intervention in this area." He was willing, however, to soften his rejection of this advice by having studies done of how to deal with the region and sending a high-level team to check the facts on the ground. For all his skepticism about sending military forces into far-off places, where they would come up against skillful guerrilla fighters and former colonial peoples suspicious of another Western power compromising their autonomy, Kennedy could not entirely dismiss hawkish demands for military intervention in Southeast Asia. He would shortly say in a speech at the U.N.: "The very simple question confronting the world community is whether measures can be devised to protect the small and the weak from" communist attackers threatening their independence. "For if they are successful in Laos and South Vietnam, the gates will be opened wide."

But opened wide to what: the defeat of the West, of freedom? Hyperbole had become the accepted wisdom about communist dangers. For all Kennedy's reluctance to rely on military action in a region of questionable importance to long-term U.S. security, he gave voice to the undertone of fear reflected in his rhetoric about protecting "the small and the weak" and ultimately the nuclear-armed United States from insurgents in Laos and Vietnam.

Mindful that Kennedy was under pressure to expand U.S. involvement in Southeast Asia, Galbraith offered another counter-argument. "South Vietnam is exceedingly bad," he reported in a July letter. "I hope, incidentally, that your information from there is good and I have an uneasy feeling that what comes in regular channels is very bad. Unless I am mistaken Diem has alienated his people to a far greater extent than we allow ourselves to know. This is our old mistake. We take the ruler's word and that of our own people who have become committed to him. . . . I fear that we have one more government which, in present form, no one will support."

Galbraith thought that Kennedy and the United States would be best served if Vietnam were left to work out its own problems. But the collective wisdom was against giving up on the country and for pressing ahead in search of solutions. And because Kennedy said he was prepared to hear policy proposals, on July 20 Rostow reported that he and Taylor had come up with questions that if answered wisely could turn failure into success. Eager to keep the initiative on meeting what they considered a crisis, they sent Kennedy a memo in line with his position. They echoed his reluctance to rely on force of arms but emphasized that military action was not being ruled out: "You would wish to see every avenue of diplomacy exhausted before we accept the necessity for . . . fighting" in South Vietnam. They also understood that he would prefer using economic assistance as fully as possible, having indigenous forces do the fighting, and that "should we have to fight, we should use air and sea power to the maximum and engage minimum U.S. forces."

In an August meeting with Rusk in Paris, French foreign minister Maurice Couve de Murville, drawing on the French experience, cautioned against excessive optimism on what any westerner could achieve in Vietnam: "The real problem is always the same," Couve de Murville said. "The difficulty is to change the present government, which is a strong government, into a popular government. . . . We had all more or less failed in our efforts."

The continuing reports from Vietnam echoed Couve de Murville's doubts. The journalist Theodore White, whose 1946 book, *Thunder Out of China*, had foreseen the collapse of Chiang Kai-shek's Nationalists, visited Vietnam in August 1961. He reported that "no American wanted to drive outside of Saigon even during the day without military convoy." At the beginning of September, the Viet Cong gave fresh evidence of how uncertain Diem's future was by launching their largest attack to date. On October 11, White wrote to warn Kennedy against sending troops to South Vietnam, where they would "be useless—or worse. The presence of white American troops will feed the race hatred of the Viet-Namese. This South Viet-Nam is a real bastard to solve—either we have to let the younger military officers knock off Diem in a coup and take our chances on a military regime . . . or else we have to give it up. To commit troops there is unwise—for the problem is political and doctrinal."

With the Berlin Wall having just gone up, Kennedy wanted no part of a crisis in Vietnam or Laos. Moreover, he tried to rein in the Rostow-Taylor talk of air and naval strikes against Hanoi by emphasizing that world public opinion would see any U.S. military action against North Vietnam as an act of aggression. But Rostow tried to convince him that striking at North Vietnam ultimately would be seen as comparable to the Truman Doctrine: "Your decision here is not easy," he told Kennedy. "It involves making an uncertain commitment in cold blood. It is not unlike Truman's commitment on Greece and Turkey in March 1947; for, in truth, Southeast Asia is in as uncertain shape as Southeast Europe at that time." It was a false analogy: Southeast Asia was not Europe, which millions of Americans were much more ready to save from communism by investing hundreds of millions of dollars, as Truman had requested. For Rostow the threat to Southeast Asia was another crucial moment in the Cold War, and he believed that world opinion would rally behind a bold policy of expanded containment. From a post–Cold War perspective, however, it is clearly an all-too-familiar misreading of history: The defense of Southeast Asia had nowhere near the importance of the eastern Mediterranean for the United States and its European allies.

But no one could deny the precariousness of the Saigon government. And no one in the White House believed Diem could survive without some kind of U.S. intervention. With the Berlin issue finally quieting down, the focus on Vietnam continued to grow. In a memo to Kennedy on October 5, Rostow warned that unless they acted soon the United States would face a "slow but total defeat" in Vietnam. At a White House meeting with his full national security team on October 11, Kennedy came under intense pressure to respond proactively to the crisis. Rostow predicted a possible catastrophe: "The gut issue . . . is this: We are deeply committed in Viet-Nam," or at least Rostow was. "If the situation deteriorates, we will have to go in; the situation is, in fact, actively deteriorating; if we go in now, the costs—human and otherwise—are likely to be less than if we wait." With the advantage of hindsight, it is clear that a large commitment then would only have moved forward the disaster that was to befall the United States.

Whatever the timing of U.S. involvement, no one endorsed passivity. The Joint Chiefs, the State and Defense departments, and the Southeast Asian Task Force called for some kind of decisive action, whether through the Southeast Asia Treaty Organization (SEATO), the U.N., an appeal to Moscow for a neutrality agreement, a direct use of American military power, or a plan for quickly turning the South Vietnamese army into a more effective fighting force. The cry was for an effective response. "It *is* now or never if we are to arrest the gains being made by the Viet Cong," Deputy Assistant Secretary of Defense William Bundy, McGeorge's brother, insisted. He later recalled the mood of all the planners "that we had to act fast and hard if we were to act at all." Trying to anticipate every contingency following a deployment of

U.S. forces, the Joint Chiefs imagined a Chinese intervention that would compel consideration of "whether to attack selected targets in north China with conventional weapons and whether to initiate use of nuclear weapons." If they were forced into a war with China, the Chiefs wanted no half measures.

Kennedy wisely refused to be rushed into anything. The failure to vet the Bay of Pigs invasion sufficiently, the Galbraith-White warnings, and the Chiefs' talk of nuclear bombs on China made Kennedy determined to gather as much information as possible before he took an irretrievable step, or at least to hold off doing anything by studying the situation. He directed Taylor, Rostow, and Lansdale, as well as two leaders of the Vietnam task force, to make a fact-finding trip to Vietnam, where they were to assess the need for dispatching U.S. forces and the alternative of relying on additional economic aid and military advisers. It was one way to ease the pressure for quick action. In the meantime, Kennedy avoided any open discussion of sending U.S. troops. An aide leaked a story to the *New York Times* that American military chiefs were opposed to deploying forces to Southeast Asia, which of course was untrue, and that they wished to rely on local troops guided by U.S. advisers. The leak speaks volumes about Kennedy's continuing fear of involvement in an Asian land war and the degree to which he felt compelled to counter domestic political demands for military intervention in South Vietnam.

In choosing the team he selected, Kennedy could hardly expect any of them to recommend less than a forceful U.S. effort to save Vietnam. Every one of them had already made clear that they favored the most aggressive possible support for Diem in combating the communist insurgents. Moreover, the presence on the trip of the Washington columnist Joe Alsop seemed to make it essential that Kennedy show resolve to meet the communist threat by sending the most tough-minded members of his administration to assess the risks. Alsop was a hawk of hawks: "Is there any real foundation for all the talk about the Kennedy administration 'lack of firmness?'" he asked in a column. "On the way to troubled Vietnam where the administration's firmness is once again being tested, the forgoing question looms very large indeed."

Like Johnson before them, the White House investigative team went to Saigon with, in Johnson's words, a "stacked deck." Kennedy had already made clear to them what the limits of his commitments would be. "We would like to throw in resources rather than people if we can," Rusk said privately after the meeting. Taylor told Lemnitzer that he was instructed to "give most discreet consideration to introduction of U.S. forces if he deems such action absolutely essential." Because they were known hawks eager to meet the communist threat head-on everywhere, proposals they might make—short of sending U.S. troops, which was favored by the Joint Chiefs, congressional conservatives, and Alsop—could not be attacked as the reluctance of administration liberals favoring negotiations over militancy.

On October 16, the day before he left for the sixteen-day trip to Vietnam, Rostow had already made up his mind on what to do. He urged Taylor to instruct the commander of America's Pacific forces to prepare a plan for "systematic harassment by U.S. naval and air power of North Viet-Nam." The United States could not rely on any prompt and radical improvement in Diem's forces. Whatever Rostow and Taylor might conclude at the close of their mission, "especially those relating to U.S. forces," Kennedy cautioned them against discussing it prematurely or "outside your immediate party in terms that would indicate your own final judgment. . . . Rumors of your conclusions could obviously be damaging."

Despite Kennedy's directive, as the mission concluded on November 2 speculation was rife that he would be pressured into sending U.S. combat troops to Vietnam. A Nolting report that the South Vietnamese were virtually unanimous in their desire for U.S. participation in the fighting became public, and rumors that the Joint Chiefs were eager to enter the war moved Senate Majority Leader Mike Mansfield to caution the president against being drawn into "a quicksand" war. It could be considered a revival of colonialism by millions of Southeast Asians, the Montana Democrat predicted. Additional military and economic aid was one thing, but combat troops were an entirely different matter. Mansfield could only imagine four ruinous outcomes: an ultimate retreat in failure, a drawn-out, indecisive conflict like in Korea, "a major war with China while Russia stands aside," or "a total world conflict."

Mansfield's memo was part of a torturous debate that now exploded about Vietnam. Kenneth T. Young, the U.S. ambassador in Thailand, and a leading expert on Southeast Asia, had a view diametrically opposed to Mansfield's. Young saw nothing but disaster from losing Vietnam. "Denial of Southeast Asia to Viet Cong, Chinese or Russian control," he asserted, "is indispensable for United States interests and purposes in the whole world. . . . Southeast Asia is the critical bottleneck stopping Sino-Soviet territorial and ideological expansion—territorial in Asia, ideological in the whole world." Defeat in Vietnam would mean losing all of Southeast Asia, with the United States "forced off the mainland of Asia." It was the view of a regional advocate blind to other considerations, but it resonated with most of the country's foreign policy experts, who shuddered at the prospect of a renewed attack on the State Department and others in the administration for losing yet another country to communist aggression.

Taylor and everyone on the mission came away from Vietnam with a heightened sense of alarm. Taylor described a country suffering from "a collapse of national morale." The answer he and his colleagues saw was "vigorous American action . . . to buy time for Vietnam" to save itself. But they warned that time was running out, and that Kennedy had to act quickly. He needed to endorse a shift in American policy "from advice to limited partnership." The alternative was nothing less than disaster: Vietnam's collapse would bring global wars of national liberation, raise universal doubts about U.S. resolve to resist communist expansion, and provoke a domestic debate about the administration's competence and wisdom in defending the national interest.

But what specifically could be done to save Vietnam? The Taylor group recommended a multi-tiered partnership—economic, military, and political, showing the Vietnamese how to finance and fight the war and how to bring a majority of their countrymen to the government's side. The addition of six to eight or possibly ten thousand U.S. military advisers, who would counsel the Vietnamese on strategy and tactics, was essential. The mission also considered promoting a coup to replace Diem with a military dictatorship, but they rejected the proposal as too risky and based on a hasty dismissal of the possibility that U.S. advisers could compel necessary political reforms to save Diem's regime.

After seeing President Kennedy on November 4, Taylor said that the president's initial reaction to his recommendations was to raise "many questions. He is instinctively against introduction of U.S. forces." McNamara and Rusk shared the belief that saving Vietnam was essential and initially supported sending U.S. troops immediately, with the option to add reinforcements later if necessary. McNamara, Rusk, and the Chiefs worried that a limited force would "get [us] increasingly mired down in an inconclusive struggle." Eventually we would have to send six to eight divisions of about 220,000 men. They also favored a warning to Hanoi that it was risking U.S. retaliation if it did not halt its assault on South Vietnam.

Pushback against sending combat troops and a large-scale involvement came immediately from Undersecretary of State George Ball. Ball saw Vietnam as a "serious" problem but believed it was "hopeless" as a nation the United States could defend against a communist takeover without a massive military commitment that could last for years without much success. He worried that we would replicate the French experience: French friends had taught him that "there was something about Vietnam that seduced the toughest military minds into fantasy." During a White House meeting with Kennedy on some international economic matters,

Ball pressed him not to accept the recommendations of the Rostow mission, telling him that committing "American forces to South Vietnam would be a tragic error. Once that process started . . . there would be no end to it. Within five years we will have three hundred thousand men in the paddies and jungles and never find them again. That was the French experience. Vietnam is the worst possible terrain both from a physical and political point of view. To my surprise, the President seemed quite unwilling to discuss the matter, responding with an overtone of asperity: 'George, you're just crazier than hell. That just isn't going to happen.'"

But not because Kennedy believed that a limited U.S. force could secure Vietnam; rather, as he told Schlesinger after reading the Taylor-Rostow report, he did not "like the proposal of a direct American military commitment. 'They want a force of American troops,'" he said. "'They say it's necessary in order to restore confidence and maintain morale. But it will be just like Berlin. The troops will march in; the bands will play; the crowd will cheer; and in four days everyone will have forgotten. Then we will be told we have to send in more troops. It's like taking a drink. The effect wears off, and you have to take another.' The war in Vietnam, he added, could be won only so long as it was *their* war. If it were converted into a white man's war, we would lose as the French had lost a decade earlier."

McNamara recalled that no sooner had he endorsed the Taylor-Rostow recommendations than he had second thoughts. And the more he considered the matter, the more doubtful he became. Rusk and his advisers at the State Department came to the same conclusion. They composed a memo to Kennedy "advising against sending combat forces in the way Max and Walt had recommended. While acknowledging that such forces might be necessary someday, we pointed out that we were facing a dilemma: 'If there is a strong South Vietnamese effort, may not be needed; if there is not such an effort, U.S. forces could not accomplish their mission in the midst of an apathetic or hostile population.'"

The memo gave Kennedy support for his resistance to sending

troops. During a White House meeting on November 11, 1961, Kennedy put eight questions before eleven advisers, including Mc-Namara, Rusk, Bundy, Taylor, Rostow, Lemnitzer, and Bobby Kennedy. It was the forty-third anniversary of the end of World War I and the horrors of that conflict were never far from Kennedy's mind. The list was a window into Kennedy's thinking: Could the Taylor-Rostow program be effective without including the introduction of a U.S. troop task force? He wanted to know how they could turn down a request by Diem for U.S. troops without antagonizing him. What circumstances might compel a reconsideration of not sending troops? Should they go public with their decision to save South Vietnam or keep it secret? Would help to Diem depend on his implementation of requested reforms? The rest of Kennedy's memo asked about plans for realizing the team's recommendations.

At the meeting, Kennedy raised additional concerns about the Taylor-Rostow proposals. He warned that Congress would be less than supportive: The chairman of the Armed Services Committee, Georgia senator Richard Russell, and other senators were opposed. Kennedy said, "Troops are a last resort. Should be SEATO forces." He expected a decision to send U.S. forces to create a domestic problem. He wanted to keep any commitments to rescue Diem and South Vietnam as quiet as possible. Reflecting Kennedy's wishes, Bobby described the president's public response to the Taylor mission as an emphatic denial of any major military commitments: "We are not sending combat troops," Bobby declared. Armed force might be necessary, but the United States and the South Vietnamese needed to rely on SEATO as much as possible. William Bundy came away from the meeting convinced that Kennedy clearly opposed the dispatch of "organized forces" as "a step so grave that it should be avoided if . . . humanly possible." Nor was Kennedy prepared to support a "categorical commitment to prevent the loss of South Vietnam."

A fierce argument now erupted over what to do. Harriman

weighed in with a political-diplomatic proposal that would forestall heavy military U.S. commitments. He urged discussions with Moscow. He also favored additional U.S. military aid in conjunction with SEATO to avert a Vietnamese collapse while pressing Diem to get on with internal reforms. At the same time, Rusk instructed Nolting not to return to Washington for consultations, in order to be available for delicate negotiations with Diem.

Rostow was outraged at the resistance to bringing U.S. military power to bear against the communists. He thought discussions with Moscow a terrible idea, and sensing Kennedy's worries about the political fallout from losing Vietnam, he told him: "I submit that it would be unwise and contrary to the lessons of the past and current experience to negotiate with the Communists before we have moved to buy time in Viet-Nam." Discussions would provoke "a major crisis of nerve in Viet-Nam and throughout Southeast Asia. . . . There will be real panic and disarray." Rostow was apocalyptic about the consequences of inaction: "The whole world is asking . . . what will the U.S. do . . . ?" The outcome of indecisive U.S. action would be nothing less than the fall of Southeast Asia and a larger war. Schlesinger privately attacked Rostow as a "Chester Bowles with machine guns."

Rostow wasn't the only one to raise warnings about appeasement. Stuart Symington, who was Air Force secretary under Truman and at this time a leading Senate Democrat on national security, cautioned that U.S. prestige was on a global decline and required demonstrations of firmness to shift the balance in the Cold War. "Whether it be in Saigon, or Berlin, or some other place," he told Kennedy on November 10, "I do not believe this nation can afford to bend further." A policy of "whatever is necessary" was essential to save Vietnam and all of Southeast Asia.

In so heated an environment, it was difficult to chart a reasonable course. But Bundy tried to find a middle ground among the competing opinions. With so many advisers voicing strong judgments on Vietnam, Bundy was reluctant to say anything. But during a midday break at the White House swimming pool, where Kennedy would retreat from the pressures of decision-making, he pressed Bundy to add his voice to the mix. Bundy did not think a loss of Vietnam would resonate all that much globally. But he believed that "a victory . . . would produce great effects all over the world." And so he recommended that Kennedy agree to send one division when necessary. The troops didn't need to go now, but such a commitment would signal U.S. determination to save Vietnam. In the meantime, Kennedy should replace Nolting as the chief U.S. representative in Saigon with a military man who would make "a much clearer statement that Diem must take U.S. military counsel on a wholly new basis." It was the sort of response that gave something to both those urging action and those counseling caution.

Bundy's advice resonated with Kennedy. Unconvinced that losing Vietnam would be so catastrophic, but unwilling to risk the public outcry that would follow such a collapse, Kennedy responded ambiguously to the pressure for a coherent policy. At a November 15 NSC meeting, "he expressed the fear of becoming involved simultaneously on two fronts on opposite sides of the world. He questioned the wisdom of involvement in Viet Nam since the basis thereof is not completely clear." U.S. involvement in the conflict seemed likely to provoke "sharp domestic partisan criticism as well as strong objections from other nations. . . . He could even make a rather strong case against intervening in an area 10,000 miles away against 16,000 guerrillas [fighting] a native army of 200,000, where millions have been sent for years with no success." Sending U.S. troops to Vietnam would mean struggling against "phantom-like" guerrillas.

Taylor challenged the president by saying he was optimistic that the United States could work its will in Vietnam if it took clear-cut actions to defeat the communist guerrillas. McNamara cautioned that this could lead to the need for U.S. troops, planes, and other resources. Kennedy asked McNamara if he favored U.S.

action. When he said yes, Kennedy asked for his reasoning. Before he could answer, Lemnitzer stepped in with a reply, offering the familiar argument "that Communist conquest would deal a severe blow to freedom and extend Communism to a great portion of the world." Kennedy wanted to know how he could justify action in Vietnam while ignoring Cuba. Lemnitzer had a ready answer: Even after the Bay of Pigs, the Joint Chiefs supported going into Cuba.

Kennedy refused to sign on to anything until he had a chance to discuss his options with the vice president. It was a ploy to delay making any decisions: Kennedy was not in the habit of discussing anything of importance with Johnson, a fact that had left LBJ frustrated and angry. At the same time, Kennedy instructed Rusk and McNamara to consider Harriman's proposals and asked whether they thought he should write to Khrushchev about Vietnam, explaining "how dangerous we thought the situation was."

Yet for all his skepticism, Kennedy could not resist the pressure for a demonstration of U.S. determination to save Vietnam. He asked McNamara and Rusk to consider Bundy's proposal to have a four-star general command U.S. operations in Saigon. He also agreed to have the Defense Department plan to send combat forces to Vietnam. Plans, of course, were not the same as action, but they certainly increased the possibility of active military participation, especially after Kennedy ordered the Chiefs to send additional advisers to Vietnam to help with military operations. Nolting was to make an immediate approach to Diem to propose a great increase in U.S.-Vietnamese cooperation, but only if Diem would promise a total mobilization of his own resources. Kennedy wanted a letter drafted in the State Department and signed by Diem stating this commitment.

McNamara later asserted that the pressure on Diem to pledge domestic reforms and all-out mobilization had the ironic effect of drawing the United States into irreversible commitments. The letter described an international partnership for the benefit of the Vietnamese people and a "mutual determination to defend the frontiers of the Free World against Communist aggression. . . . Together we have laid the material foundations of a new and modern Viet-Nam," Diem was asked to say. "Together we have checked the thrust of Communist tyranny in Southeast Asia. . . . If we lose this war, our people will be swallowed by the Communist Bloc." Diem pledged to mobilize all his country's resources. But because Vietnam lacked the wherewithal to meet the onslaught, "we must have further assistance from the United States."

At the same time, Kennedy wrote Khrushchev that the United States viewed the threat to Vietnam "with the utmost gravity. . . . Our support for the government of President Ngo Dinh Diem we regard as a serious obligation." Rusk then told a press conference that the communist attack on Vietnam worried all free nations and represented a threat to the peace. As a result, the United States was increasing its commitment to supplying and training of Vietnamese forces. The full meaning of the U.S. commitment was not lost on the *Washington Evening Star*, which reported that the White House was pressing Diem "to broaden participation in his Government and has offered him every aid short of combat troops if he does."

In retrospect, McNamara saw these developments as the beginning of America's substantial and long-term commitment to save Vietnam. In his later recollections about the war, he anguished over the failure of the Kennedy administration to ask five basic questions before becoming deeply involved: "Was it true that the fall of South Vietnam would trigger the fall of all Southeast Asia? Would that constitute a grave threat to the west's security? What kind of war—conventional or guerrilla—might develop? Could we win it with U.S. troops fighting alongside the South Vietnamese? Should we not know the answers to all these questions before deciding whether to commit troops?" Recalling the terrible consequences of America's involvement in the conflict, McNamara found it "beyond understanding, incredible, that we did not force ourselves to confront such issues head-on."

The mistakes he saw were innocence, overconfidence, ignorance about the region, inexperience in dealing with crises, "other pressing international matters [that] clamored for our attention during that first year," and perhaps most important, problems "for which there were no ready, or good, answers." All of it generated a tendency "to stick their heads in the sand."

Writing thirty-five years after the 1961 events, McNamara forgot or overlooked the fact that questions about Vietnam's importance in heading off communist domination of Southeast Asia and its impact on long-term U.S. national security were very much in the forefront of discussions about Kennedy's response to the crisis in Saigon. And while it is certainly true that no one could confidently predict the outcome of increased U.S. involvement, and that anticipating a constructive result from American intervention was not without plausibility, the most compelling reason for Kennedy's decision to expand U.S. commitments in Vietnam was not a conviction that we might lose the Cold War if that country came under communist control. As Kennedy told *New York Times* columnist Arthur Krock in October 1961, "United States troops should not be involved on the Asian mainland." Truman's decision in the 1940s not to send U.S. forces to fight the communists in the Chinese civil war and the stalemate in Korea powerfully resonated with Kennedy.

Kennedy was more concerned about the political ramifications of "losing" Vietnam. He told Galbraith: "There are limits to the number of defeats I can defend in one twelve-month period. I've had the Bay of Pigs, and pulling out of Laos, and I can't accept a third." In short, a communist takeover of Saigon would raise concerns abroad and at home: The Soviets and Chinese might see him as irresolute or weak and might become more aggressive about ousting the West from Berlin and/or try to subvert other weak governments in Asia; and conservatives or militant anticommunists in the United States, borrowing from Joseph McCarthy in the early fifties, would launch another "Who lost China?" or "Who lost Vietnam?" campaign.

In November 1961, Kennedy hoped to muddle through on Vietnam: Send more military advisers, increase the financial and material support of the Saigon regime, and press Diem into effective reforms that improved his popular standing. And maybe, just maybe, it would fend off a communist victory and keep Diem's government afloat.

In the meantime, Kennedy believed it essential to keep questions about U.S. military involvement in Vietnam as low-key as possible. George C. McGhee, the State Department counselor and chairman of the Policy Planning Staff, warned against "prolonged involvement of American soldiers in . . . indecisive anti-guerrilla operations." Recalling the collapse of Truman's public standing, he worried that "we would be back in the atmosphere of Korea 1950–1953—only more so." He predicted that a faltering conflict in Vietnam would agitate the public and stimulate demands for more forceful measures to prevent another Korean stalemate. Pressures for escalation could propel us into an "all-out struggle with Peiping." As increased U.S. involvement became a reality, a principal administration objective became guarding against press leaks about U.S. operations in Vietnam. Rusk cabled the embassy in Saigon: "Do not give other than routine cooperation to correspondents on current military activities in Vietnam. No comment at all on classified activities." Theodore White's warning—"a real bastard to solve"—was more evident than ever.

As 1961 came to an end, Kennedy understood what John Steinbeck meant when he said, "We give the President . . . more pressure than a man can bear." Kennedy described himself as "always on the edge of irritability." The strains on him were so evident that a reporter asked, "I wonder if you could tell us if you had to do it over again, you would work for the presidency." Reporters asked Bobby, "Do you think your brother can handle the presidency without harming his health?" Bobby assured them that the pressures on him were no more than what he had dealt with during the presidential campaign. He admitted, however, that "the responsibilities are so great and weigh so heavily on him that it is bound" to affect him. Kennedy was learning what Jefferson meant when he said that the presidency is a splendid misery.

o o o o

"The Greatest Adventure of Our Century"

At the start of 1962, Kennedy felt pressured to speak force-fully about the country's domestic challenges. But his heart wasn't in it. Although he devoted the first half of his State of the Union address to the economy, civil rights, health, and education, Schlesinger complained that "the domestic section . . . had been reduced to a laundry list." And though Kennedy agreed to add a paragraph Schlesinger wrote giving the program "a philosophical coherence" that related it to the New Frontier, Kennedy remained grudging about having to appease liberals. "What more do the liberals want me to do that is politically possible?" he asked.

In an era before television prompters, Kennedy read the speech haltingly, turning pages and looking down at his text rather than keeping continuous eye contact with his congressional audience. His remarks evoked only occasional applause and less overt en-thusiasm than his first State of the Union a year before. While that speech also began with a discussion of the country's economic travails, it, like the Inaugural Address and the second State of the Union speech, principally focused on foreign affairs.

What they were trying to do at home "gives meaning to our efforts abroad," Kennedy said. "The successes and the setbacks of the past year remain on our agenda of unfinished business. . . . Yet our basic goal remains the same: a peaceful world community of free and independent states." He did not see that goal within reach "today or tomorrow. We may not reach it in our own lifetime. But the quest is the greatest adventure of our century." America "had been granted the role of being the great defender of freedom in its hour of maximum danger." While hyperbole is not uncommon in presidential annual messages, Kennedy had good reason to worry about threats to the peace.

But he could not escape questions about domestic change, especially about civil rights. In January 1962, when a reporter pressed Kennedy on his administration's civil rights record, he defensively asserted that his White House had "made more progress in the field of civil rights on a whole variety of fronts than were made in the last 8 years." Kennedy could point to the fact that the majority of the government's contractors had agreed to plans for progress, with compliance now mandatory rather than voluntary, efforts to expand integration in the armed services, seven lawsuits against southern states to compel school integration, seventy-five suits to force southern counties into facilitating black voting, and the nomination of Thurgood Marshall to serve on the U.S. Circuit Court of Appeals—only the second black to be appointed to that federal court. At the same time, however, a failure to issue a promised order to integrate federally supported public housing, five southern racists appointed to federal judgeships, and a refusal to put a comprehensive civil rights bill before Congress gave resonance to liberal complaints about excessive White House caution in advancing equal rights. As Roy Wilkins told Kennedy, he had not "gained anything [in 1961] by refusing to put a civil rights bill before" Congress. His hope that he could garner southern support for a tax cut, federal aid to education, and medical insurance for seniors by not pressing for desegregation was unfounded.

Kennedy did not dispute the failure to do big things at home. But for the time being, he felt that foreign dangers still had to command his primary attention. And Latin America, where he continued to see communism as an aggressive competitor for regional control, posed a grave potential setback for the United States in the Cold War. He was determined to counter the threat with the Alliance for Progress. True, it was just getting started in improving the southern republics, but Kennedy described the hemisphere as alive with "the quickening of hope" and the Latin republics as committed to "a new and strenuous effort of self-help and self-reform." Yet "the one unchangeable certainty is that nothing is certain or unchangeable."

In Kennedy's view, the greatest danger to Western Hemisphere freedom remained Castro's hopes of exporting his revolutionary fervor. At a National Security Council meeting on January 18, 1962, Kennedy wanted to make sure that Castro would be isolated at a coming meeting at Punta del Este. But even if they could blunt his influence at the conference, Kennedy expected Castro still to be a very large problem. While he believed that some way would eventually have to be found to deal with the Cuban dilemma, he saw nothing that could be done at once. The next day, when Bobby Kennedy met with CIA and military officials, Bobby explained that the administration had been lying low since the failure at the Bay of Pigs. But because Cuba was so rapidly becoming a communist police state, the ousting of Castro was "the top priority in the United States Government—all else is secondary—no time, money, effort, or manpower is to be spared." There could be no misunderstanding on the responsibility of the country's defense agencies "to carry out this job." Bobby quoted the president as telling him that "the final chapter on Cuba has not been written," and Bobby added, with a pugnacity that reflected his combativeness, "it's going to be done and will be done."

On January 20, Lansdale gave marching orders to members of a Caribbean Survey Group. Invoking Bobby's directive, he said that "it is untenable to say that the United States is unable to achieve its vital national security and foreign policy goal re Cuba. . . . We have all the men, money, material, and spiritual assets of this most powerful nation on earth." Every member of the group was instructed to meet and if possible exceed deadlines in reaching the goal of turning Cuba away from communism. In February, Lansdale set a timetable stretching from March to October 1962, when Castro was to be overthrown and a new government put in place. In the meantime, the Joint Chiefs were to make contingency plans for U.S. military intervention in Cuba. In March, however, Kennedy put a damper on plans for direct military action. Still concerned that U.S. military intervention would undermine the Alliance for Progress, Kennedy discouraged all talk of air or ground attacks on the island. At a meeting with Bobby, McGeorge Bundy, McCone, Gilpatric, Taylor, and Lemnitzer, he foresaw no immediate circumstances that would justify military steps.

Yet the administration's determination to bring down Castro remained as evident as it had been during the Bay of Pigs fiasco. Khrushchev, who saw Castro's government in Cuba as a valuable encroachment into America's sphere of control and a blow to U.S. prestige, believed that renewed Soviet threats to Berlin could keep Kennedy from attacking the island. He had no doubt, however, that the Americans, short of an invasion, would continue to do everything possible to subvert Castro's regime.

Khrushchev also worried that the U.S. planned a first nuclear strike against Russia. When Georgi Bolshakov, the ostensible Soviet Embassy press officer in Washington, asked Bobby Kennedy about the influence of war hawks in the government, Bobby explained that there were some in the Pentagon as well as other opinion makers who were eager to attack the Soviet Union. He had in mind E. M. Dealey, the conservative publisher of the *Dallas Morning News*, who made no secret of his belief that Kennedy headed an administration of "weak sisters." During a White House luncheon, he told the president that the country needed "a man on horseback. Many people in Texas and the Southwest think that you are riding Caroline's tricycle." Kennedy angrily replied that "wars are easier to talk about than they are to fight."

Yet he could not discount war talk as long as Khrushchev engaged in provocations that made people in the West think that he was intent on destroying anticommunist opponents. During the first half of 1962, Khrushchev renewed his threats to sign a peace treaty with East Germany, which would once more endanger U.S. access to West Berlin. Kennedy could not understand Khrushchev's belligerence. He described him as unstable and irresponsible, which were frightening traits in someone who could trigger a nuclear war. But Khrushchev saw his own behavior as calculated pushback. During a visit to Moscow by Secretary of the Interior Stewart Udall, Khrushchev made clear that he would not be intimidated. If "any lunatics in your country want war," he said, "Western Europe will hold them back. War in this day and age means no Paris and no France, all in the space of an hour."

Khrushchev's aggressiveness made Kennedy all the more eager to end a U.S.-Soviet deadlock over limits on nuclear testing and an arms control agreement that would reduce the possibility of a nuclear war. As he wrote Prime Minister Macmillan on January 13, 1962, "we must do all we can to turn the nuclear spiral downward, and to save mankind from the increasing threat of events of surpassing horror."

But Seaborg saw "the realities of American politics" as more compelling in shaping Kennedy's actions, particularly the constraints put on him by winning Senate approval of any test ban agreement he might negotiate in the future. And as Rusk, who had been well disposed to a test ban, now advised, the Soviets "did not seem to be suffering greatly from the public indignation which had greeted its tests last fall." Rusk also thought that it would be a mistake to reach a point where anyone thought that we had fallen behind in the arms race. In April 1962, after the Soviets stubbornly continued their opposition to inspections of any kind in their country, Kennedy concluded that prospects for a test ban treaty were all but gone and that he had no choice but to order atmospheric tests in the Pacific over Britain's Christmas Island. He remained skeptical about the wisdom of testing and did not want them on U.S. territory, where a mushroom cloud could frighten Americans sensitive to the dangers from nuclear pollution. Nonetheless, the political pressure demanding a response to the Soviet challenge, coupled with advice from responsible scientists warning of national security perils, made a decision to test in the atmosphere irresistible.

Yet Kennedy was unwilling to let matters rest there. He agreed to continue negotiations in Geneva and urged Khrushchev to join him in reaching for "real progress toward disarmament and not to engage in sterile exchanges of propaganda." Kennedy also told his advisers that he "wanted the world to know that we were prepared to walk the last mile to obtain" a treaty. At a meeting with Gromyko, Rusk said that he did not see disarmament as a hopeless problem, and recalled French foreign minister Aristide Briand's precept that "disarmament should be such as would leave no one a dupe or a victim."

Kennedy assumed that meaningful negotiations were unlikely until both sides completed their current round of tests toward the end of 1962, and then they could decide whether to make another offer limited to tests in the atmosphere or work for a comprehensive test ban that included underground explosions. A limited ban had the advantage of being something the Soviets could accept without feeling that they had backtracked from earlier positions. Stevenson thought it a fine idea that could prevent "a non-stop series of competitive nuclear tests in the atmosphere." Pressure from the Joint Chiefs to exclude a comprehensive agreement that could not ensure "full verification" through "unhampered verification" largely persuaded the White House to propose a limited ban.

It also had the advantage of simpler, less intrusive verification. Unlike underground tests, which seemed to require monitoring stations on Soviet and U.S. territory, atmospheric tests of any significant size were impossible to hide. In addition, the unintended release of a Defense Department seismic study concluding that detection facilities in the Soviet Union might be unnecessary to track underground explosions undermined Pentagon insistence on such stations as an essential element of a comprehensive treaty. When Arthur Dean, the U.S. representative to the Geneva disarmament talks, acknowledged that the verification stations might be superfluous, it made convincing Moscow of their need impossible.

The inability of Kennedy's diplomatic and military advisers to find the means to negotiate a nuclear test ban with the Soviets had frustrated and demoralized him. The blunder in releasing information that heightened the difficulties of finding some common ground between Soviet opposition to monitors and Pentagon insistence on them left Kennedy feeling angry at advisers who were not only falling short in identifying means to overcome differences but also now adding to them. Kennedy complained that "the U.S. had worked itself into a deplorable situation by releasing the . . . data on enhanced detection possibilities." Bundy told a member of Dean's delegation that "the president was very upset. He liked to have things done well and the idea that we had made a proposition and now we were saying something else—he had a rather adverse reaction to that, to put it mildly."

In a conversation with Rusk and Bundy, Kennedy was scathing: Of the Foreign Service types or professional diplomats like Dean, he said, "I just see an awful lot of fellows . . . who don't seem to have cojones. . . . The Defense Department looks as if that's all they got. They haven't any brains. . . . And I know that you get all this sort of virility over at the Pentagon and you get a lot of Arleigh Burkes: admirable, nice figure, without any brains."

The disappointments over Castro's unshaken control in Cuba and Moscow's refusal to come to terms on a test ban agreement were compounded by Kennedy's continuing struggles with health prob-

lems and a stroke that left the seventy-three-year-old Joe Kennedy
barely able to speak or walk. Joe's illness depressed Kennedy. The
sight of someone as active and vital as his father being so disabled
heightened his sense of vulnerability to his own medical difficulties and his premonition that he would not have a long life.

The resurfacing of the civil rights struggle added to Kennedy's
worries. In March 1962, civil rights advocates complained that
the White House had not laid the proper groundwork for a major
rights law. "Negroes are not convinced that the Administration is
really on their side, " Kennedy was told. He hadn't made clear that
this was a "moral issue." Moreover, unlike Bobby and the president, who felt that pressing Congress to do something about civil
rights would jeopardize the rest of their legislative program, rights
supporters predicted that if he lost the fight for a civil rights law,
"the President's whole program will go down the drain." Kennedy
was urged to take the issue to the public in a nationally televised
Oval Office address.

The Civil Rights Commission, headed by Father Theodore
Hesburgh, weighed in with demands for bolder action. Bobby and
Hesburgh clashed over the commission's decision to hold hearings
in Louisiana and Mississippi to underscore abuses of blacks by
police and local authorities. Fearful that the commission's presence
in the South would provoke violence, Bobby urged them to spend
more time looking into violations of black rights in the North. But
seeing the commission as a "burr under the saddle of the administration," Hesburgh would not give ground. After the hearing,
which passed off without riots, the commission's report described
the terror tactics that Mississippi officials used to inhibit black
voting. "You're making my life difficult," Kennedy told two com-
missioners. They were not very sympathetic. They thought that
Kennedy was too insensitive to the miserable conditions under
which so many southern blacks lived.

Kennedy took comfort in knowing that the public was on his
side. A January 1962 Gallup Poll showed a 77 percent approval

rating, while 62 percent said they had a highly favorable view of the president. Moreover, the public sided with him on what it saw as the most important problems facing the country: 63 percent said it was war and peace and only 6 percent mentioned racial tensions. 32 percent of Americans thought that Kennedy was pushing racial integration too fast, 35 percent thought his actions were "about right," and only 11 percent said he was not moving fast enough. His press conferences were a huge hit—91 percent viewed them favorably. When Gallup asked voters in a trial heat who they preferred between Kennedy and Nixon, Kennedy won decisively—65 percent to 35 percent.

In the spring of 1962, part of Kennedy's upbeat ratings resulted from his effectiveness in winning a battle against corporation executives. On April 10, the nation's steel companies announced a 3.5 percent price increase, which threatened to trigger greater inflation and an economic downturn. Kennedy was furious at what he described as a "double-cross" by industry chiefs, who had promised to cooperate with labor unions and the White House in holding down prices and avoiding a recession. At a stormy Oval Office meeting, Kennedy told aides that steel executives "fucked me. They fucked us and we've got to try to fuck them." They had "made a fool of him." He quoted his father as having told him that "businessmen were all pricks. . . . God, I hate the bastards," he said. "They kicked us right in the balls." He not only saw the steel chiefs' action as a personal blow to his political standing and the country's well-being; he was also irritated at having to shift his focus from more important foreign policy issues to something he considered an unnecessary fight. They had reached a settlement that had served the national interest, and now he had to spend time, energy, and political capital on a needless struggle. The episode also put his impatience with domestic affairs on display. He was hardly indifferent to the country's economic condition or its fundamental tie to his political standing, but foreign dangers, especially the fear of Soviet or American missteps that might bring them to the brink of war, remained his greatest concern.

He now took his case to the public, holding a press conference in which he declared the price increase an "irresponsible defiance of the public interest" and a "ruthless disregard of public responsibilities." Invoking the sacrifices being made by reservists risking their lives in Vietnam, he denounced "a tiny handful of steel executives whose pursuit of private power and profit exceed their sense of public responsibility" and demonstrates "utter contempt for the interests of 185 million Americans." At the same time, Bobby Kennedy, at the president's behest, unleashed the FBI to investigate and intimidate the steel executives, while the IRS threatened to audit their tax returns. After the companies relented and announced a rollback in their price increases, Kennedy, with feigned horror, joked at a dinner party that his brother would never have investigated steel executives' tax returns or tapped their phones, which is exactly what he or, more precisely, his subordinates did. As far as the Kennedys were concerned, they were simply meeting fire with fire.

The struggle to find answers to the conflict in Vietnam was less satisfying. On January 3, 1962, when Kennedy met with the Joint Chiefs, General Paul Harkins, the U.S. commander in Vietnam, Johnson, and McNamara at his winter retreat in Palm Beach, Florida, he instructed that "no publicity would be given, at least for the time being, to General Harkins' new mission." Kennedy made clear that he did not want the United States more greatly involved in Southeast Asia than we already were. He wanted the number of U.S. military personnel in Vietnam to be kept as quiet as possible. Moreover, any discussion of what U.S. troops were doing there "should emphasize their role as advisers and deny that they were in any way engaged in combat."

Three days later, when McNamara lifted a Kennedy-imposed embargo on information about the meeting, Lemnitzer informed Admiral Harry Felt and General Lionel McGarr about the change in the command structure in Vietnam and the importance of keeping this secret. He said nothing, however, about Kennedy's determination to limit and mute the U.S. role in the fighting. He and the other Chiefs were determined to challenge Kennedy's injunction against the introduction of ground forces. They shared Nolting's belief, which he articulated before the Senate Foreign Relations Committee on January 12, "that the situation in Vietnam was primarily the result of Chinese Communist expansionism" and that the United States could not fail to meet this challenge—even at the risk of deploying combat forces. Nolting's image of red China winning control across Asia echoed the fears of millions of Americans.

On January 13, the Chiefs sent McNamara a memo for consideration by the president. They urged Kennedy to reconsider his ban on direct U.S. participation in the fighting, arguing that it was essential to prevent the loss of Vietnam and that American troops were the best and perhaps only way to ensure this. Unwilling to challenge Kennedy's clear opposition to ground forces, McNamara refused to endorse the Chiefs' views but waffled on their recommendation by saying that the present U.S. program in Vietnam should be tried first. However, he would not rule out in the future joining in the Chiefs' call for U.S. fighting men to take on the Viet Cong.

But Kennedy remained determined to limit U.S. involvement in the conflict. He knew, of course, that attacks by U.S. aircraft and increasing the number of advisers, who would accompany South Vietnamese troops on search-and-destroy missions, would put Americans in harm's way, with inevitable casualties. But this was to be kept out of the news, with private and public declarations that American advisers were not actively fighting the war. At a press conference on January 15, although one U.S. adviser had already been killed and U.S. aircraft were providing cover for Vietnamese forces, Kennedy emphatically denied that American military personnel were involved in combat.

As the American role in the fighting grew, Kennedy and his advisers became more aggressive about hiding the truth. To acknowledge that the United States was becoming the principal combatant in the conflict with the Viet Cong would make Diem's dependence on a Western power transparent and would strengthen the insurgency's appeal as an anticolonial defender of Vietnamese independence. It would also provoke unwanted pressures in the United States for a victory over the communists that had eluded Kennedy in Cuba, Berlin, and Laos, and the Democrats in Korea. Nolting urged that all press briefings on Vietnam "should give full credit to the GVN [government of Vietnam] and not make it look as though the U.S. were running the war in SVN, making the plans, or pulling all the strings."

Hiding America's role in the conflict was a terrible error. Franklin Roosevelt had understood that the country could not fight World War II without full public backing. The lesson was lost on Harry Truman, whose decision to cross the 38th Parallel and fight a wider war in Korea without building a national consensus for the coming sacrifices had destroyed his political support, with his approval rating falling to 23 percent. An undeclared and secret war in Vietnam risked Kennedy's credibility and ability to lead the nation.

Yet nothing gave the lie to denials of U.S. control over the fighting in Vietnam more than a seventeen-page paper prepared at the request of the president and Taylor. On February 3, Roger Hilsman, the director of the State Department's Bureau of Intelligence and Research, who had visited Vietnam for two weeks in January, outlined "A Strategic Concept for South Vietnam." The plan described in precise detail the matériel and numbers of Vietnamese and American forces needed to fight the war and how they were to be used.

As a Joint Operations Center with limited Vietnamese participation began controlling air strikes and the U.S. Military Assistance Command began directing ground operations, U.S. newsmen in Saigon complained about an embassy blackout or exclusion from helicopter missions; they also objected to the "clamming up by U.S. officials" to prevent them from writing "bad stories" about "press censorship." *U.S. News & World Report* described a "Curtain of Secrecy" that had descended on Saigon, "a U.S. Embassy effort to confuse and disguise the situation." The repression had the opposite effect than intended: It attracted a large number of experienced, responsible American journalists to Vietnam to cover what they saw as the participation of U.S. military men in South Vietnam's war and the South Vietnamese government's attempt to conceal it from the public.

On February 14, the *New York Times* made the issue a subject for national discussion when it ran an editorial asserting that Washington was hiding America's growing military involvement in Vietnam, and predicting that this could lead the country into a major conflict. In his widely read column, James Reston asserted that the United States was already "involved in an undeclared war in South Vietnam."

At a press conference later that day, a reporter asked Kennedy to respond to an allegation in a Republican National Committee publication that he had "been less than candid with the American people as to how deeply we are involved in Vietnam." Kennedy was prepared for the question. He launched into a lengthy explanation of American commitments to South Vietnam's independence dating from 1950, noting a military training mission and economic assistance during the Eisenhower presidency: He described a long history of working to prevent Vietnam from falling under communist control. As the insurgency had become more aggressive, the United States had responded in kind. When a reporter asked Kennedy if he was telling the American people the full story of our involvement, he acknowledged that assistance had been increased but denied sending "combat troops in the generally understood sense of the word." The president added that he was being as frank as he could be without jeopardizing U.S. security needs.

But the reporters in Vietnam didn't think so. They complained that embassy regulations were making it impossible for them to do their job. They were being refused permission to travel to areas where they could report on what U.S. military men were doing. A press officer in the State Department thought the censorship would have the exact wrong effect. He warned that it was creating a hostile press, which was writing unfavorable stories criticizing Diem's regime and describing deepening U.S. involvement in directing the war against the Viet Cong. The department's public affairs expert predicted that a more flexible policy would make the reporters more cooperative and easier to manage. He was on the mark, but his recommendation was not seen as important enough to reach the White House.

Kennedy and his national security advisers refused to bend on allowing reporters in Vietnam to accompany U.S. advisers serving with Vietnamese units. William Shannon at the *New York Post* complained that "Kennedy devotes such a considerable portion of his attention to leaking news, planting rumors, and playing off one reporter against another, that it sometimes seems that his dream job is not being Chief Executive of the nation but Managing Editor of a hypothetical newspaper."

Press officers in the State and Defense departments and in the United States Information Agency (USIA) thought that press problems in Saigon were becoming serious and recommended that a skilled public relations expert be sent to Vietnam, where he could try to foster greater cooperation with the newsmen. Kennedy grudgingly agreed to give Nolting authority to allow a limited number of journalists to monitor some U.S. air operations, but only if the reporters were willing to emphasize the U.S. support function and South Vietnam's primary role in fighting the war. Convinced that speculative stories were doing more harm than any reporting of the facts, the White House agreed to a policy of "maximum feasible cooperation, guidance and appeal to good faith of correspondents."

At all times, however, the embassy needed to reinforce the idea that "this is not a US war" and that participation was only in training and advising the local forces. Managing the news remained a central part of the agenda: Criticism of Diem was to be discouraged as undermining U.S. aims in the conflict, and "correspondents should not be taken on missions whose nature such that undesirable dispatches would be highly probable." As in World War II, when the press accepted broad and effective censorship, self-restraint now was just as important. But this was not World War II, and reporters wondered whether expanded involvement in the conflict served national security interests; at the very least they believed that an open discussion of America's growing part in the fighting was essential before any additional future commitments were made.

In February, as the U.S. effort to save Vietnam expanded, Rusk and Lansdale gave voice to the administration's determination to combat the insurgency. Rusk reassured Nolting, who feared that divided authority between himself and Harkins might undermine the effort to save Diem and defeat the communists, that he and the president were fully committed, promising to help all they could. Lansdale met with the editors of *Life* magazine to convince them to get behind the war effort. Instead of "treating Vietnam as some strange and quaint place, full of peculiar little people, out at the end of nowhere where our good 'American boys' don't belong anyhow," he urged, the editors should present the Vietnamese "as real people not too unlike us, people fighting against tyranny today." The editors needed to tell the American people that U.S. advisers in Vietnam are "good guys" and that every American should feel it when any of them suffered a casualty rather than wondering what they were doing in harm's way. No one should think of American support as anything less than a defense of U.S. national security. Lansdale's intimidating message: Criticism is unpatriotic.

Nothing demonstrated U.S. intentions in Vietnam more clearly than a statement to reporters by Bobby Kennedy during a refueling stop in Saigon on a return journey from Asia. Asked about America's role in the conflict, Kennedy did not hesitate to avow the administration's determination to defeat the communist insurgency. "We are going to win in Vietnam. We will remain here until we do win," he declared, as defiant of the reporters as he was toward the Viet Cong. When Wayne Morse, a skeptical Oregon senator, asked the White House whether the attorney general's statements represented administration policy, the State Department, speaking for the White House, replied that Kennedy's remarks had not been cleared by the White House but did reflect the government's outlook. Although Bobby's comments could be attributed to his characteristic combativeness in his brother's behalf, especially against the backdrop of the administration's recent setbacks, no one could doubt that the president and his advisers were all speaking with one voice on Vietnam.

On February 27, when dissident South Vietnamese pilots unsuccessfully bombed the presidential palace in an attempt to kill Diem, Nolting and the White House sprang to Diem's defense. The State Department approved a request to have U.S. helicopters temporarily provide close support of operations by the Army of the Republic of Vietnam (ARVN) while Saigon investigated its air force. In a press conference on March 1, after the press in the United States and abroad asserted that the conflict in Vietnam was rapidly becoming a U.S. war, Rusk reassured reporters that the United States would not send combat troops. He emphasized, however, that the United States remained determined to assist the South Vietnamese until the threat to their autonomy ended. Tensions between the embassy in Saigon and U.S. reporters increased when the latter described the attack as evidence of Diem's basic unpopularity. Nolting, by contrast, dismissed the incident as of no great consequence and predicted that the South Vietnam government was minimizing the effects.

As all too often with embassy staffs in a friendly country, Nolting identified with the existing regime. His reporting was the captive of a favorable bias or an inclination to put the best possible face on unwanted events. With the administration in Washington eager for good news or evidence that, unlike in Cuba and Berlin, it had found the right formula for success, Nolting and his military counterparts in Saigon talked themselves into believing that Diem could rally his country against the Hanoi-backed Viet Cong. The American journalists watching developments with greater detachment saw a distinctly less rosy outcome to Diem's governance and the civil war.

While Kennedy battled the steel companies, Secretary of Defense McNamara, the U.S. military, and the embassy in Saigon were more than ready to take over the management of America's expanding role in Vietnam. As McNamara recalled later, "I increasingly made Vietnam my personal responsibility. That was only right: it was the one place where Americans were in a shooting war, albeit as advisers. I felt a very heavy responsibility for it, and I got involved as deeply as I could and be effective. That is what ultimately led people to call Vietnam McNamara's war." Meetings in Hawaii at Pearl Harbor with the U.S. military commander in the Pacific, the principal officers in Vietnam, and civilian national security officials in Washington and Saigon became a monthly exercise.

Regular visits to Vietnam following the ten-hour trips to Hawaii were part of the routine that brought McNamara in repeated contact with Diem and his military and civilian advisers. It gave Diem the opportunity to persuade McNamara and his aides that he was intent on making Vietnam free and democratic. He impressed them as someone who had absorbed Western values during his studies at a Catholic seminary in New Jersey in the early 1950s. Seeing that Diem was under the influence of his brother Ngo Dinh Nhu and his wife, the "bright, forceful, beautiful" Madame Nhu, who was "also diabolical and scheming—a true sorceress,"

McNamara had some misgivings about the reliability of the South Vietnamese government as a partner in the conflict. But the need to get on with the war and the belief that there was no good alternative to Diem and the Nhus disarmed McNamara's doubts.

In the first half of 1962, the Pentagon took over the complete development of an overall military strategy for South Vietnam, mapping out resources needed from the United States, organizing Vietnamese forces, and how they should be used to defeat the Viet Cong in the quickest and least costly way. Rusk told the embassy that the administration in Washington saw the program as workable and worthy of all-out support. A closed hearing before the Senate Foreign Relations Committee made clear to Oregon's Senator Wayne Morse that U.S. military personnel were transporting South Vietnamese troops into combat, engaging in firefights with the North Vietnamese, patrolling the sea approaches to South Vietnam, and dropping propaganda leaflets over guerrilla-held areas.

By late March, Rusk had pressed to ensure that Hanoi could not resupply the Viet Cong by airdrops; he urged that hostile aircraft over South Vietnam be shot down. Kennedy wanted an assessment of progress in the conflict by April 1, with contingency planning in case current efforts to defend South Vietnam were falling short. McNamara instructed the Pentagon to plan the introduction of U.S. ground troops if the interior of South Vietnam were in danger of collapse.

Kennedy had a sense of urgency about removing the threat to South Vietnam's autonomy and ending the national discussion about a wider U.S. role in an Asian war. During February and March, the *New York Times* had carried numerous front-page stories about America's expanding role in the conflict, the prospect of a prolonged fight with increasing U.S. casualties, and Soviet warnings that American actions were jeopardizing world peace. With reporters revealing that U.S. forces were actively engaged in combat operations both in the air and on the ground, the embassy and the White House became more eager to mute the talk of America's engagement. Kennedy approved a directive for U.S. fighter planes to intercept and destroy communist resupply aircraft over South Vietnam, but he wanted to assure that "public handling will be simply that Communist plane crashed, thus attempting avoid problem of degree to which Americans engaged in active hostilities in SVN." The cover-up of U.S. actions in Vietnam was a clear administration aim.

The message to the field went out over George Ball's signature— even the most critical of Kennedy's advisers on the expanding conflict followed the party line. "We're heading hell-bent into a mess, and there's not a Goddamn thing I can do about it," Ball recalled telling his chief of staff. Kennedy resisted a large, high-visibility military commitment, but a shadow war was another matter, and Ball, however great his doubts, followed the president's lead.

On April 4, Harriman, who had become the assistant secretary of state for Far Eastern affairs and chairman of the Southeast Asia Task Force, cabled Nolting about growing concern over the flow of news stories about U.S. direction of South Vietnamese forces and American participation in the fighting. The stories could only lead people to believe that the conflict was becoming more of a U.S. than a Vietnamese war. It was essential that U.S. involvement be seen as strictly advisory. Harriman complained that U.S. military actions were too conspicuous and needed to be conducted under a greater cloak of secrecy. It gave the communists a propaganda advantage that helped their war effort—not to mention the danger of provoking a firestorm of criticism in the United States.

Kennedy's eagerness for a solution to his Vietnam problem made him receptive to a proposal from Ken Galbraith for escaping what Galbraith saw as a losing effort. Our involvement, he had told Kennedy in a March 2 letter, is increasingly like that of the French, a "colonial military force" that stirs resentments. The Russians were delighted at the prospect of an America spending "our billions in distant jungles where it does us no good and them no harm. Incidentally, who is the man in your administration who decides what countries are strategic? . . . What is so important about this real estate in the space age? What strength do we gain from alliance with an incompetent government and a people who are so largely indifferent to their own salvation?"

Galbraith understood perfectly well that the man setting policy on Vietnam was the president. But cautious about challenging Kennedy directly and eager to move him in a different direction, Galbraith laid the blame for "the political poison" shaping policy on the military and State Department. Knowing Kennedy's "distaste for diagnosis without remedy," he proposed four rules for change in dealing with Vietnam: Keep up the commitment against deploying U.S. forces, ensure that U.S. civilians and not military men manage policy in Saigon, stay alert to the possibility of any kind of a political settlement, and be open to any alternative to Diem, which was bound to be an improvement. Galbraith urged Kennedy to understand that "politics is not the art of the possible. It consists in choosing between the disastrous and the unpalatable. I wonder if those who talk of a ten-year war really know what they are saying in terms of American attitudes. We are not as forgiving as the French." Was he fearful that the United States might unleash unprecedented atomic attacks on the North Vietnamese if it found itself in the sort of stalemate it had in Korea?

Skeptical of his military and State Department advisers, Kennedy invited Galbraith, during an early April visit to the United States, to spend an evening with him at Glen Ora, his four-hundred-acre estate in Middleburg, Virginia. Encouraged by Kennedy to provide a more formal policy statement that he could use in prodding his advisers to consider an alternative to the current reliance on military and economic support of Diem, Galbraith sent a memo the next day that repeated the points in his March letter. He added the suggestion that the United States initiate the search for a political solution by approaching the International Commission for Supervision and Control in Vietnam, an agency with Canadian, Indian, and Polish representatives that was set up in 1954 to monitor the Geneva accords ending French occupation of Southeast Asia and was responsible for the election of a government for all of Vietnam. Galbraith also suggested possible talks with the Russians and Indians about an end to the Viet Cong insurgency in exchange for phased U.S. withdrawal and a goal of unifying Vietnam under a noncommunist, progressively democratic government.

The same day Galbraith sent his memo to Kennedy, Chet Bowles, back from a fact-finding tour of Africa, Asia, and the Middle East, sent the president a fifty-four-page memorandum that echoed Galbraith's advice on Vietnam. Where Galbraith punctuated his letter with witticisms and a bit of self-deprecation that amused Kennedy, Bowles was his usual intense, ponderous self. His analysis was every bit as keen as Galbraith's: He saw our involvement in Vietnam as producing "neither total victory nor total defeat but rather the development of an uneasy fluid stalemate with the Viet Cong unable to crack the U.S.-supported government forces, yet still able to maintain an effective opposition." He thought it provided the opportunity to reach for a permanent negotiated political settlement. He also thought it could open the way to the neutralization of all Southeast Asia.

Bowles did not discount the importance of a U.S. military presence in the region to discourage Chinese adventurism. But the length of Bowles's report instantly reminded Kennedy of Dean Acheson's view of him as "a garrulous windbag and an ineffectual do-gooder," which had been confirmed for the president and Bobby by his response to the Bay of Pigs failure and had triggered his demotion from undersecretary to roving ambassador. A fifty-four-page sermon on how to set matters right in a contested Vietnam by someone consigned to the fringes of the administration did not command Kennedy's attention. He ignored Bowles's request for a meeting to discuss his proposals.

By contrast, on April 6, the day after Galbraith wrote him, Kennedy discussed his suggestions for neutralizing Vietnam with Harriman. Harriman agreed with Galbraith's advice to minimize the U.S. military role in the conflict. He showed Kennedy his April 4 cable to Nolting emphasizing the need to keep U.S. military activities in Vietnam as quiet as possible and saying that we had no intention of fighting Vietnam's war. Harriman, however, rejected the proposal to seek a neutral solution in Vietnam and the suggestion that the U.S. dump Diem: While "Diem was a losing horse in the long run . . . there was nobody to replace him." Despite Harriman's response, Kennedy told him to forward Galbraith's memo to McNamara and to instruct Galbraith to discuss a mutual withdrawal of North Vietnamese and American forces from South Vietnam with the Indians.

Kennedy didn't trust that Harriman would follow his directive. He saw Harriman as intent on making Vietnamese policy without regard for his wishes. He sensed that Harriman thought he knew better than the president how to deal with this crisis. Kennedy continued to worry that Harriman and other hawks in the State Department and Pentagon were more inclined to fight in Vietnam than he was. He wanted to be sure that Galbraith would be instructed to seize upon any opportunity that might allow the United States to reduce its involvement in Vietnam. Kennedy wasn't ready to sign on to negotiations, but he didn't want to exclude that possibility; at least, not yet.

After reviewing Galbraith's argument for a political solution, McNamara and the military chiefs emphatically dissented. They argued that the policies in place had not been given a fair trial and that a negotiated settlement in Vietnam would have disastrous effects on U.S. relations with allies everywhere. Galbraith's proposals were "tantamount to abandoning South Vietnam to the Communists" and losing Southeast Asia. The country was "a testing ground of U.S. resolution in Asia. . . . The Department of Defense cannot concur in the policy advanced by Ambassador Galbraith," the Chiefs told Kennedy, "but believe strongly that present policy toward South Vietnam should be pursued vigorously to a successful conclusion."

When Bowles followed in June and July with additional memos

urging consideration of neutralizing Vietnam, Rusk told him, "You realize, of course, you're spouting the Communist line." The State Department's Far Eastern Bureau called Bowles's initiatives "unrealistic, impractical and premature." Bowles warned that a likely deterioration in South Vietnam would force a choice between increased troop commitments and an embarrassing withdrawal, but his prediction was dismissed as unworthy of further discussion. It was a measure of how intense the argument over Vietnam had become that Rusk would accuse Bowles of communist sympathies. Vietnam had taken on meaning in the Cold War that greatly exceeded its importance for U.S. national security. But any officeholder who might have pointed out Vietnam's inflated importance would have exposed himself to attacks not only on his judgment but also on his commitment to defeating the communists.

On May I, Kennedy had already put aside consideration of neutralizing Vietnam and indicated as much to advisers who doubted the wisdom of trying to negotiate a settlement with Hanoi. During a White House discussion with his national security officials about the merits of Galbraith's suggestion for negotiating a neutralized coalition government for South Vietnam, Harriman and Roger Hilsman "vigorously opposed this recommendation and the President decided against it." After the "loss" of China to Mao Zedong's communists in 1949 and Truman's failure to oust Kim Il Sung's communist regime in North Korea, the domestic political consequences, more than the national security perils, made "losing" South Vietnam through a political arrangement too risky for Kennedy to accept or openly favor in the first half of 1962.

At the same time, however, he remained quietly receptive to hearing about any possible interest by Hanoi in discussing a settlement. On May 16, Rusk told Galbraith that Kennedy was interested in a conversation Galbraith had had in New Delhi with the Indian representative to the international control commission on Vietnam. The Indian thought that there might be a chance for negotiations in the future and Kennedy wanted Galbraith to continue informal discussions about the likelihood of talks on ending the conflict in Vietnam and ending U.S. participation in an unwanted war.

Administration resistance to a negotiated settlement partly rested on the conviction that, despite relying on the unpopular Diem, U.S. sponsored anticommunist military and political actions were showing positive signs—or at least that's what American officials in Saigon were telling McNamara. At a March hearing before the House Appropriations Committee, which was considering the annual foreign aid budget, McNamara voiced unqualified optimism about U.S. policy in Vietnam. In fact, he saw a definite endpoint without the need for U.S. ground forces. Vietnam's troops, aided by U.S. advisers, were "terminating subversion, covert aggression and combat operations."

When Secretary of the Army Elvis Stahr returned from a visit to Vietnam in April, he reported that the Vietnamese were "greatly encouraged by our policy toward them and by our strong support. Slowly but surely they are working out the techniques of counterinsurgency and of civic action." The South Vietnamese army was becoming an effective fighting force. Harkins and Nolting were working well together, and since Stahr had served for two years in Asia during World War II and recently spent much time focused on the problems of the Far East, he was able to make authoritative judgments on conditions in Vietnam, or so he and others were ready to believe. At the same time, a U.S. Embassy official who visited four South Vietnamese provinces, where he spoke with various officials and local residents, saw the great likelihood of a substantial improvement of security in a year or two. Things were "not too rosy" at the moment, but "we are moving in the right direction." Sterling Cottrell, the director of a Defense Department task force on Vietnam, returned from meetings in Saigon convinced that "we have found the right formula."

A four-day visit to South Vietnam by McNamara from May

8 to 11 gave added support to rising hopes. But McNamara's inspection tour was not an excursion by a commonly tough-minded
skeptic; rather, it was a ceremonial glimpse at the war front as portrayed to him by General Harkins, who was intent on persuading
him that they could defeat the Viet Cong. When Harkins looked
at a briefing map showing more areas under communist control
than he wished McNamara to see, he directed a junior officer to
doctor the map, which then gave the secretary a much brighter picture of the fighting. As eager for a victory as Harkins, McNamara
uncritically accepted what he was shown and signed off on a Defense Department report describing an atmosphere of restrained
optimism in every area he visited. In Pentagon-speak, McNamara
declared that "victory is clearly attainable through the mechanisms
that are now in motion." The whole operation from McNamara
down was at best an exercise in auto-intoxication and at worst a use
of unmitigated deception. If the facts did not support a rosy war
scenario, Harkins was determined to make it appear that way and
McNamara was all too ready to embrace good news.

Journalists who trailed McNamara on his tour were puzzled
by his seeming acceptance of all that the military was telling
him. When he echoed the feel-good reports of his briefers to reporters during an informal session in Nolting's home, one asked
McNamara if he might have a different view if he extended his
stay. "Absolutely not," he said. He thought his optimism would be
strengthened. Neil Sheehan, a young United Press International
reporter who had only been in Saigon for two weeks, had heard
enough from his more seasoned colleagues to confront McNamara
privately about his conclusions. As he was about to get in his car,
Sheehan pressed him to say how he could be so optimistic after so
brief a visit to the front lines. McNamara abruptly replied: "Every
quantitative measurement we have shows that we're winning this
war."

Did he believe it? Probably. It was an expression of his un-
bounded confidence in social science engineering. And so what-

ever the current realities, he was confident that they could find the means to defeat a smaller, less well equipped enemy. Moreover, after sixteen months in which he had been repeatedly at cross-purposes with his military subordinates, the defense secretary was keen to support their plans for winning a conflict that might become a model for fighting other guerrilla wars. Besides, he and Kennedy didn't want to "lose" Vietnam, and there seemed to be no current alternative to what the military proposed to do.

McNamara's snub of Sheehan reflected the White House and Pentagon view that the journalists were an impediment to winning in Vietnam. The reporters, however, had no desire to undermine the Kennedy administration's reach for victory in Vietnam. But they had a different take on what they saw in Saigon and the provinces and believed that they were not only doing their job by reporting what they learned but also serving U.S. and Vietnamese interests.

Consequently, they could not accept McNamara's upbeat view of the conflict. They remained critical of Diem and continued to publish accounts questioning prospects for success in the war. July 25 and July 29 stories in the *New York Times* reported that some American embassy staff and military advisers thought that the war was not going well and that McNamara's optimism was unwarranted. Administration spokesmen responded with fresh pronouncements on the importance of saving Vietnam from the communists and the likelihood that it could be done without U.S. combat forces—only advisers and matériel.

Homer Bigart, the *New York Times* correspondent in Saigon, who in 1962 had a reputation as a tough-minded seeker of truth, thought that U.S. actions in Vietnam were ineffective and began saying so in his dispatches. A two-time Pulitzer Prize winner for reporting during World War II and in Korea, the fifty-four-year-old Bigart angered Diem, Nolting, and American military chiefs, who considered him a subversive force. In the middle of May, after Bigart published a critical story about the South Vietnamese government's shaky hold on the provinces, Lemnitzer told a meeting of a Special Counterinsurgency Group that Bigart's report ignored the fact that the defense of various provinces was going well. Always alert to any published story that might embarrass the administration, Bobby Kennedy urged Lemnitzer to send the president a note "pointing out Bigart's inaccuracy."

Despite his reservations about the administration's growing involvement in Vietnam, George Ball also defended its policy. In speeches in Chicago and Detroit, he announced that U.S. national security demanded a proactive policy in Southeast Asia, where the communists were aggressively trying to seize control of South Vietnam and dominate the region. Although conceding that it would take years, he asserted that we would definitely win by relying on the South Vietnamese. "We are not running the war," he asserted. Harriman as well weighed in with a *New York Times Magazine* article explaining "What We Are Doing in Southeast Asia." Bundy and Kennedy, who remained keen to negotiate a settlement in Vietnam, worried that the Ball speeches might reduce chances of persuading Hanoi to talk peace.

Because serious negotiations seemed like a distant reality, Kennedy and his advisers focused on hopes of aiding South Vietnam to find the wherewithal to defeat the Viet Cong. No one discounted the difficulties, but wishful thinking blotted out harsh truths. U.S. military and embassy officials put a positive face on any glimmer of hope. At the end of May, the embassy reported that a Strategic Hamlets program, which aimed to defend some sixteen hundred villages across the provinces from the Viet Cong, who compelled villagers to join their forces and supply foodstuffs, showed "considerable momentum behind [a] promising idea." Despite troubling "weaknesses in the GVN administration, . . . US counsel and advice are becoming increasingly acceptable and should produce further dividends," the embassy told Washington.

The positive reports became a spur to calls for more action and encouraged additional expressions of optimism. Rostow once again urged a more robust military campaign against Hanoi, including bombing raids against North Vietnam's transportation and power grids and the mining of Haiphong Harbor. In June, Nolting reported that, in spite of some setbacks, he saw considerable improvement. On June 18, picking up on the hopeful signs coming from Vietnam, Hilsman told Harriman that deterioration in political and military conditions in Vietnam had been arrested, with "heartening progress" in the effectiveness of South Vietnam's fighting forces. Chances of success in the war were "good" if Saigon made continuing progress in its current strategy. In the cliché of the day, there was growing light at the end of the tunnel.

o o o o

"If We Listen to Them, None of Us Will Be Alive"

After eighteen months of interactions with his counselors, Kennedy had diminished confidence in most of the men advising him on policy. With the exception of Bobby, who was principally a sounding board and instrument for testing out ideas on others, he thought it best to rely less on his associates and more on himself for the hard decisions he seemed to be confronting all the time.

Neither Rusk nor McNamara nor Bundy nor Rostow nor Taylor had impressed him as all that masterful about any of the big issues they had faced on Cuba, Berlin, or Vietnam. As for Sorensen and Schlesinger, they had been impressively helpful in composing speeches and preparing him for press conferences, but they were as much at sea as everyone else about how to solve his foreign policy dilemmas.

And Schlesinger in particular had become something of a liability. The conservative press was describing him as promoting socialist ideas. In June 1962, Schlesinger confided to his journal: "I have a feeling that JFK is a little edgy about all this and may even be beginning to wonder whether I am not more of a liability than a working asset." At the end of the month, when some high jinks at a Bobby Kennedy party, with Arthur shoving Ethel Kennedy into a pool, and accusations that he was unpatriotic and antireligious and had violated government policy by accepting payments for magazine articles became front-page news, Schlesinger was mortified and justifiably convinced that "whatever limited effectiveness I may have had will be diminished." Kennedy advised him not to "worry about it. . . . All they are doing is shooting at me through you. Their whole line is to pin everything on the professors—you, Heller, Rostow." Nonetheless, it rightly persuaded Schlesinger that Kennedy would need to hold him at arm's length.

In the summer of 1962, no group of advisers was less helpful than a Vietnam task force describing great progress in the conflict. They saw slow but clear forward movement. On a visit to the United States, South Vietnam's Foreign Minister Vu Van Mau echoed their optimism. But Lansdale and McNamara wanted a sharper picture of what was happening on the ground. When Lansdale designed a questionnaire that U.S. officers could use to gather data, McNamara told Lansdale: "An excellent set of questions Ed—it is this kind of info I need & am not receiving."

Kennedy was reluctant to take the happy talk at face value. In July, during negotiations in Geneva about concluding a neutrality agreement on Laos, an intermediary asked Harriman if he was willing to meet privately with North Vietnamese foreign minister Ung Van Kiem. When Kennedy gave Harriman the go-ahead, James Barrington, the British undersecretary for Burmese affairs, arranged a meeting in his hotel room. Harriman and Ung talked past each other, exchanging accusations of blame for the fighting. "We got absolutely nowhere," Harriman's deputy recalled. "We hit a stone wall." It was enough to encourage feelings that the only route out of Vietnam was through more U.S. aid and counsel that gave Saigon the wherewithal to beat back the insurgents.

At the end of July, McNamara returned to Hawaii for another meeting with Harkins, Nolting, and other embassy and military officials stationed in Saigon. Given the dead end in possible negotiations with Hanoi, the impulse to see a military solution to the Vietnam problem was greater than ever. Moreover, Harkins and Nolting could not have been more upbeat about prospects for a successful outcome. Harkins told McNamara that Diem was winning the war. McNamara wanted to know "how long a period before the VC could be eliminated as a disturbing force." Harkins was unprepared for the question and clueless as to an answer, but McNamara's aide recalls him "jumping up in his chair," collecting himself, and answering: a year after South Vietnamese forces were "fully operational and really pressing the VC in all areas."

McNamara, however, believed it best to take a conservative view and assume that it would be three years before they could declare victory. It was all guesswork, resting more on hope than anything concrete—an astonishing bit of flimflam from someone who so prided himself on statistical analysis. Their problem during this time, McNamara believed, would be maintaining public support. U.S. losses in the fighting would raise questions about the wisdom of being in Vietnam. McNamara wanted plans for reducing U.S. involvement in the conflict that he could make public at the same time that the United States expanded its operations.

To delay public discomfort with the fighting, he directed his public affairs officer to begin getting "good material in the press." Nothing was as perceptive on McNamara's part about the growing U.S. role in Vietnam as his understanding that public dissent would become a major problem in fighting the war. But none of the national security advisers wanted to ask why, if the communists are actually losing, weren't they willing to salvage something by talking? Instead, the objective was to issue upbeat reports and hope that the good news would eventually reflect reality.

In August, the State Department told Nolting that the White House and the whole government were committed to winning in Vietnam. But the president wanted hard data on why the embassy in Saigon thought the war was going so well. Michael Forrestal, the National Security Council's expert on Vietnam, told Bundy that they wouldn't really know if the war was progressing until the end of November, when the rainy season ended and military activity increased. He expected to know then whether sending more advisers and facilitating Strategic Hamlets were showing any results. He thought that American casualties would increase and "we will be in for real trouble" unless the public believed that they were winning the war.

In September, Kennedy sent Taylor to Vietnam for four days to get a clearer picture of what was happening, and more important, to counsel the president on how to turn the war more quickly in a positive direction. Taylor's trip was essentially an acknowledgment by Kennedy that he had turned Vietnam policy over to McNamara and the military.

For the moment, Kennedy had another priority: a clash with Mississippi governor Ross Barnett over James Meredith, a black Air Force veteran trying to end segregation at the University of Mississippi. "I won't let that boy get to Ole Miss," Barnett defiantly told Bobby Kennedy, who, as attorney general, was the administration's point man in the conflict. "I would rather spend the rest of my life in the penitentiary," Barnett declared, than enforce Meredith's court-ordered enrollment. After private conversations with Barnett, in which the governor tacitly agreed to let Meredith enter the university, Kennedy gave a national speech from the Oval Office noting Barnett's acceptance of the rule of law and assuring Americans that the conflict was about to be peacefully resolved. But as detached as ever from domestic crosscurrents, Kennedy misread Barnett's willingness to accept federal direction. Barnett saw political advantage in disregarding the president's public declaration of an agreement and withdrew state troopers from the campus, leaving five hundred federal marshals at the mercy of a segregationist mob trying to bar Meredith from going on campus. It forced Kennedy to send in regular U.S. Army troops, who arrived too late to prevent two deaths and hundreds of injuries, including twenty-seven U.S. marshals wounded by gunfire. New doubts surfaced about the president's competence, with some journalists saying that Barnett had played Kennedy for a fool. "I haven't had such an interesting time since the Bay of Pigs," Kennedy moaned. Bobby told his brother, "We are going to have a hell of a problem about why we didn't handle the situation better." Their response to the conflict looked to him like "one of the big botches."

Aside from conversations with Bobby, who had been conferring with associates at the Justice Department, Kennedy had made no effort to convene a group of administration troubleshooters, an executive committee counseling him on how to resolve this crisis with the least destructive consequences for the country and the White House. But unlike the Bay of Pigs, where consultations with experts had been a constant in the run-up to the invasion, Kennedy had refused to see Mississippi or any other domestic problem as worthy of similar crisis management.

While Kennedy temporarily fixed his attention on a domestic crisis, Taylor filled the vacuum on Vietnam. He used his trip to Saigon to confirm administration hopes that the United States could help Diem defeat the communists. Everything he heard from Nhu, Diem, embassy officials, and the military advisers on the ground boosted his confidence in what they were doing. Nhu enthusiastically recounted the accomplishments of the Strategic Hamlets program and predicted more gains in the future. Taylor responded that "this situation resembled that which exists during any war. There is a period during which an impasse exists, and then, suddenly, a sudden surge to victory." Taylor congratulated Nhu on Vietnam's rapid progress. It allowed Nhu to describe South Vietnam's success in the war as a model for other Third World countries threatened by communists and eager for democratic development.

Diem was equally upbeat. He disputed skeptical press accounts, saying that with U.S. help the Vietnamese would achieve the ultimate victory. William Trueheart, the U.S. minister in Saigon, echoed the Nhu-Diem predictions: He described Saigon's military progress as "little short of sensational. . . . The Strategic Hamlet program had transformed the countryside." Nhu's and Diem's hopefulness was meant to keep the U.S. aid flowing, while Trueheart's enthusiasm issued from an advocate's optimism rather than the observations of a detached analyst assessing the facts.

On his return from Asia, Taylor told the president that although he had had only a short stay in Vietnam, he had seen many people during travels around the country. His conversation with eight junior U.S. officers advising South Vietnamese units was most encouraging. He wished the president could have heard from them directly. He advised Kennedy not to read press reports from Saigon as accurate measures of what was happening. "You have to be on the ground to sense a lift in the national morale," he said. And yet, he acknowledged, they needed more information to be confident that things were going as well as Diem and Nhu said. And although there could be doubts about significant progress in the war, Taylor encouraged Kennedy to believe that they were winning. Like some true believer, Taylor, the tough-minded soldier, thought that they could will their way to victory. It was a testimony to how little advocates, even military men mindful of battlefield uncertainties, could be trusted to make objective assessments of policies they favored. And it raised questions with Kennedy about Taylor's reliability as an adviser.

No one close to Kennedy cared to hear dissenting opinions about progress in the war in the press or from embassy officials or military advisers observing the combat. If they had accepted the possibility, even the likelihood, that the communists were winning, they would have had a different picture from the one Nhu, Diem, and Taylor painted. And with so much politically and emotionally already invested in Vietnam, administration policymakers didn't want to accept that they were on the wrong track.

Journalists reported that friction between U.S. advisers and South Vietnamese military commanders was imperiling the war effort and that assertions about the impact of South Vietnamese offensives on the Viet Cong were questionable. Communist losses and their diminished willingness to fight were exaggerated. David Halberstam, who had become the *New York Times* correspondent in Saigon in June, described "a frustrating hunt for elusive foes."

Joe Mendenhall, the political counselor in the embassy in Saigon, had a very different take on conditions than what Taylor heard and reported. Mendenhall saw the Viet Cong as strong and resourceful, despite recent government attacks. They controlled much of the countryside, with government authority largely confined to the cities and towns. He saw a future of "gradual deterioration." So "why are we losing?" he asked. The fundamental cause was Diem's ineffective governance and his unpopularity with the masses. Mendenhall doubted that the South Vietnamese could win the war with the existing government in power. The only solution he saw was a U.S.-backed coup. But Mendenhall's doubts did not reach Kennedy's ears. He was not high enough in the chain of authority to have his reports land on Kennedy's desk.

Instead, it was the good news or the hopeful assessments that filtered through. Mike Forrestal told Kennedy on September 18 that the Saigon embassy's bullish review of political and military developments gave hope that they had found the means to succeed in combating the insurgency. Forrestal was the son of James Forrestal, the first secretary of defense, and a Harvard law graduate who had served as Harriman's naval attaché at the U.S. Embassy in Moscow and later practiced international law. His association with Harriman and ties to Bundy had brought him into Kennedy's National Security Council, where he focused on Asian and Vietnamese affairs in particular. He was a leading voice for finding the right answers to the challenges in Vietnam.

His principal worry was not the weakness of the Diem government or the administration's inability to work its will in the conflict, but rather the bad publicity generated by the journalists in Saigon. The problem seemed to be that "the newspapers and news magazines have not sent top drawer people to the area." The attitude of Diem's government toward the press was part of the difficulty. (On September 4, Diem had expelled *Newsweek* reporter François Sully from Vietnam for criticizing him as corrupt and alienated from his people.) Forrestal urged Kennedy to use discussions with U.S. editors and publishers to discourage negative reports about Vietnam.

Taylor echoed Forrestal's complaints. In a September 20 memo to the president, he described the Saigon press corps as "uninformed and often belligerently adverse to the programs of the U.S. and SVN Governments." They needed "the support of publishers in obtaining responsible reporting." Kennedy, who had his own tensions with the Washington press, especially the *New York Herald Tribune*, a Republican paper that had consistently attacked him, sympathized with their complaints. When he met with a high South Vietnamese government official on September 25, Kennedy urged him "not to be too concerned by press reports. . . . Inaccurate press reporting . . . occurs every day in Washington."

Whatever the remaining challenges in South Vietnam, reining in the messengers' bad news was not going to save the country from a communist takeover. But everyone at the White House from Kennedy down believed that a pliable press more sympathetic to Diem could make a difference in helping him win his war. It was only one part of the growing illusions about Washington's ability to save Vietnam.

At the center of the administration's increasing investment in Vietnam was the rationalization that it mattered to America's security. At the start of October, Kennedy asked the State Department to draft a paper updating progress in the war and explaining U.S. involvement in the conflict. The department summarized the recent gains in the fighting and described the war as important in convincing all our allies that we stood by our commitments. In addition, "a victory for us would prove that . . . underdeveloped nations can defeat 'wars of liberation' with our help [and] strike a telling blow to the mystique of the 'wave of the future.'" If such a victory could be won without the involvement of American ground troops or, so to speak, on the cheap with advisers and material support, Kennedy was an enthusiastic supporter. No one at the State Department considered the likelihood that America's prime European allies might view a growing U.S. involvement in Vietnam as a distraction from more important commitments. Moreover, no one seemed to think that winning in Vietnam was tied to saving Southeast Asia from communism—or at least they said nothing about this.

While the administration struggled to solve problems in Vietnam, Cuba remained a minefield of uncertainty and bad advice. With information flooding in by late August 1962 about a Soviet military buildup on the island, Bobby Kennedy urged Rusk, McNamara, Bundy, and the Joint Chiefs to consider what new "aggressive steps" could be taken, including "provoking an attack against Guantánamo which would permit us to retaliate." McNamara favored heightened sabotage and guerrilla warfare, and the Chiefs urged Castro's elimination, which could be done "without precipitating general war." They suggested that "manufactured . . . acts of sabotage at Guantánamo . . . faked assassination attempts against Cuban exiles and terrorist bombings in Florida and Washington, D.C." could trigger U.S. intervention. But Bundy, speaking for the president, cautioned against action that could provoke a Berlin blockade or a Soviet strike against U.S. missile sites in Turkey and Italy.

John McCone, the acting CIA director, was as eager as the Chiefs to identify ways to overthrow Castro. With Allen Dulles about to retire in November 1961, the fifty-nine-year-old McCone had become effective head of the agency. His selection by Kennedy bothered liberals, who saw him as a conservative hawk. His background only reinforced this view. The silver-haired, bespectacled McCone was the offspring of a wealthy California family: Educated at Berkeley in mechanical engineering, he had worked in the family iron foundry business, Bechtel-McCone. During World War II, the corporation had made $44 million in shipbuilding, which a General Accounting Office official described as wartime profiteering. "At no time in the history of American business," he said, "had so few men made so much money with so little risk and all at the expense of taxpayers."

An outspoken Republican with ties to the Eisenhower administration, McCone became head of the Atomic Energy Commission in 1958. He had been an early and forceful proponent of nuclear weapons. In 1956, after Adlai Stevenson pledged support for a nuclear test ban as a presidential candidate, and scientists at the California Institute of Technology endorsed his proposal, McCone, a trustee of the institute, attacked them as taken in by the Russians. McCone's principal advocate for the CIA directorship was Bobby Kennedy, who saw him not only as an ally in urging all-out action against Castro, but also as a firewall against Republican criticism of the administration's failed Cuban policy.

The minute McCone saw evidence of the Soviet buildup in Cuba he was convinced that they intended to turn the island into a missile base. At an August 21 meeting with other national security officials, he described the Russian shipments to Cuba as equipment either to guard against a future air assault or for missile sites. After the meeting, McCone privately told Bobby Kennedy that "[i]f Cuba succeeds, we can expect most of Latin America to fall." Although the president considered McCone's prediction hyperbolic, he accepted his advice urging an analysis of "the probable military, political and psychological impact of the establishment in Cuba of either surface-to-air missiles or surface-to-surface missiles which could reach the U.S."

No other top administration official besides McCone voiced similar concerns. Schlesinger disputed the conclusion that the Soviet decision to make a major investment in Cuba signaled a readiness to challenge the United States head-on. "Any military construction will probably be defensive in function; a launching pad directed against the U.S. would be too blatant a provocation," Schlesinger told Bundy.

Roger Hilsman, a West Point graduate who had served in the OSS during World War II, behind the lines in Burma and China, and became the director of the State Department's Bureau of Intelligence and Research, shared Schlesinger's view of the Soviet buildup. As someone who had passed the Kennedy test of physical courage and knowledge of counterinsurgency through personal experience, Hilsman enjoyed standing with the president as tough-minded. Kennedy especially enjoyed Hilsman's nerve at a briefing when he spent ten minutes correcting General Lemnitzer about Laos. Kennedy sat smiling as Hilsman spoke and Lemnitzer did a slow boil. And so Hilsman's view that the Russian matériel and military personnel arriving in Cuba were meant to help Castro defend himself against another U.S.-sponsored invasion carried weight at the White House. Hilsman did not discount Moscow's direct military involvement in Cuba but rejected suggestions that it grew out of a risky plan to turn Cuba into an offensive base like those the United States had encircling Russia.

Having been so badly burned by the CIA's miscalculations about the Bay of Pigs, Kennedy was inclined to agree with Schlesinger and Hilsman and see McCone's warnings as unconvincing and likely to stimulate pressure to invade Cuba, which step he resisted as certain to undermine the Alliance for Progress. Kennedy wished simply to shelve Cuba as an issue in the developing fall congressional campaign. At the end of August, when the CIA showed him photos of surface-to-air missile sites on the island, Kennedy saw them as defensive installations and acted to repress press leaks that could stimulate Republican demands for an attack. He instructed Marshall Carter, who was temporarily standing in as CIA director while McCone was away, to "limit access to the information. . . . 'The President said to put it back in the box and nail it tight.'"

In trying to mute speculation about Soviet steps to make Cuba a nuclear launching pad, Kennedy had advice from Rusk and Bundy that Moscow had never risked deploying nuclear missiles outside of the Soviet Union. The intelligence bureaus of the State and Defense departments as well as their Army, Navy, Air Force, and National Security Agency counterparts agreed. The U.S. Intelligence Board, the coordinating agency or clearinghouse for all intelligence estimates, concurred. "There is no evidence that the Soviet government has ever provided nuclear warheads to any other state," Bundy advised.

When news of the Soviet buildup in Cuba, leaked by Senate Republicans, made headlines in the United States, Bundy urged Kennedy to hold a press conference drawing a "sharp distinction between what is now going on and what we would not tolerate." In short, the Soviet shipments did not add up to offensive weapons and posed no direct threat to the United States or any Latin American country, which is what Kennedy unequivocally told a press conference on September 13.

As with CIA and military miscalculations about the Bay of Pigs invasion, the various national security agencies were misinformed about past Soviet behavior and current Soviet intentions. Between January and May 1959, Moscow had set up nuclear missile launchers in East Germany and deployed warheads under Soviet control. In August, however, apparently eager to make sure that their capacity to strike all of Western Europe with the missiles in East Germany did not undermine prospects for an Eisenhower-Khrushchev Paris summit in the spring, Moscow dismantled the sites. Although the deployment had registered on Western intelligence services, none of them had a full report on the Soviet action until the beginning of 1961, and by then the missiles were long gone. Since none of the intelligence services could offer a satisfactory explanation for the missiles' initial deployment and subsequent removal, they concluded that the temporary deployment was an aberration not worth serious consideration.

Neither Kennedy nor anyone in his White House knew about the deployment, nor did anyone in the intelligence services come forward in September 1962 to report it or suggest that the Soviets might be replicating their action. U.S. intelligence officials knew that shipping missiles to Cuba would be much harder to disguise than the deployments to East Germany; it was another reason to discount the 1959 episode as of no significance in assessing current events. Since Moscow had no indication that the West knew about the 1959 deployment, the Soviets could hope that similar shipments to Cuba would also go undetected. The failure of U.S. intelligence agencies to tell the White House about the East German deployment was a blunder on par with earlier misjudgments on Cuba. The failure to report the seemingly inconsequential 1959 episode may have resulted from the belief that it was of no real importance, but it can also be assumed that it grew out of a desire to defend an agency for which the new administration already had questionable regard.

Yet whatever the sources of the omission, it certainly colored assumptions about current events in Cuba. Even when information came in on September 21 from spies that twenty medium-range ballistic missiles, which could hit targets nearly eleven hundred miles inside the United States, had arrived on the island, intelligence analysts marked the report as only "potentially significant," believing that Moscow was incapable or unwilling to take such action.

For Kennedy and his advisers in September and early October 1962, a Soviet buildup in Cuba, including offensive weapons posing a direct threat to U.S. territory, was an unwanted challenge. And not simply because it would compel consideration of an attack on Cuba that would undermine the Alliance for Progress; it would also threaten defeat in the November elections by Republicans blaming the president for having failed to topple Castro and head off grave dangers to the homeland.

The civil rights crisis in Mississippi in September also discouraged Kennedy from compounding his difficulties by describing events in Cuba as a national peril. Throughout 1961 and into 1962, civil rights had been a source of ongoing irritation to him rather than an opportunity to reform historic wrongs. He had refused to make good on his promise to integrate public housing with a stroke of the pen and he avoided tensions with segregationists by appointing five southern racists from Alabama, Georgia, Louisiana, and Mississippi to federal judgeships. In November 1962, Kennedy would finally issue an executive order integrating public housing, but his slowness to act supported Martin Luther King's observation that Kennedy lacked the "moral passion" to fight hard for racial equality.

At the beginning of October, one consequence of the Mississippi troubles was to keep public discussion of Cuba to a minimum. The mess at the state university in Oxford was bad enough without now conceding that the White House had failed to anticipate Soviet aggression in Cuba. During the night of the Mississippi crisis, Kennedy was angered by a tip that *New York Times* columnist James Reston was publishing an article saying that Kennedy was more eager to meet with the Soviets than they were to meet with him. Distressed at the thought of how weak it would make him look when tied to the Mississippi embarrassment and news of missiles in Cuba, Kennedy told aides, "We ought to knock it [Reston's claim] down tonight. That's just kicking Reston right in the balls, isn't it?" he asked, pleased at the thought of showing some toughness.

In refocusing his attention on Cuba, Kennedy hoped that he would not be dealing with a major crisis that could threaten a war with Russia. His eagerness not to face an emergency reflected itself in resistance to a McCone recommendation for additional U-2 flights. In an October 5 meeting of national security officials, Mc-

Cone described the likelihood of Soviet ground-to-ground missiles as more a "probability than a mere possibility." Bundy, who remembered Kennedy as "always edgy about McCone," disputed McCone's conclusion and saw little need for new U-2 flights over Cuba. Rusk, speaking for the president, opposed more flights as threatening to produce a crisis with Moscow. Moreover, Kennedy, who told Bobby that he thought McCone was "a real bastard," directly pressed McCone not to publicize his views on Cuba for fear it would inject the island into the campaign. As it was, press reports that Khrushchev had described the United States as "too liberal to fight" to defend Berlin or oust Castro had provoked congressional Republicans to demand that the president respond with tough talk.

Kennedy was convinced that Republican assertions in September and early October about Soviet missiles in Cuba were no more than campaign rhetoric. At the end of August, when Senator Kenneth Keating, a New York Republican, had complained that the administration had no plan for countering Soviet missile bases in Cuba, Kennedy had dismissed him as a "nut." Despite assurances from Khrushchev and Ambassador Anatoly Dobrynin in Washington that Soviet installations on the island were strictly for defense, the pressure on Kennedy to ensure that McCone and Republican critics were wrong convinced him to authorize a U-2 flight over Cuba for October 14.

The result shocked Kennedy and all his advisers—except for McCone, who had accurately assumed that the Soviets were building medium-range missile sites. On the evening of October 15, while Bundy was hosting a dinner party at his home, Ray Cline, the CIA's deputy director of intelligence, phoned to report that new U-2 photos conclusively revealed four medium- and two intermediate-range installations along with twenty-one crated bombers capable of carrying nuclear weapons. The intermediate-range weapons could travel 2,100 miles and reach America's most populous cities, as well as Washington, D.C.

Bundy's response was surprising. He decided to wait until the next morning to bring Kennedy the bad news. He believed it would be better for Kennedy, who had had a long day on the campaign trail, to get a good night's sleep before confronting the greatest crisis of his presidency. But Bundy knew that they were now facing a potential disaster and that the president should be the first to know. One can speculate that Bundy also wanted the time to reflect on his own failure to have measured accurately what they were dealing with and how he could provide wiser counsel. He concluded that no one, except for McCone, whose anti-Soviet ideology led him always to expect the worst from the Kremlin, could have assumed that Khrushchev would be so reckless. And so the new challenge was to assess how they could achieve any sort of reasonable agreement on the missiles with someone as seemingly erratic as the Soviet first secretary.

McNamara was unable to explain how the administration could have been so shortsighted about Moscow's decision to risk putting offensive weapons in Cuba. But as in the Bay of Pigs fiasco, the experts had again fallen short of providing wise counsel in the weeks and months preceding the discovery of the missiles. At the start of the crisis, McNamara was less concerned about the misreading of Soviet intentions by American intelligence than about the danger of a nuclear war. "The Pentagon is full of papers talking about the preservation of a 'viable society' after nuclear conflict," he told Schlesinger. "That 'viable society' phrase drives me mad." He added: "A credible deterrent cannot be based on an incredible act." Having shared in the failure to anticipate Soviet actions in Cuba, McNamara saw it as essential to find sensible answers in the coming confrontation.

When Bobby Kennedy heard the news, he rushed in to see Bundy. After being shown the photos, he was incensed at the lies Khrushchev and Dobrynin had told them. Pacing back and forth in Bundy's small office, he began pounding a fist in his hand and cursing, "Oh shit! Shit! Shit! Those sons a bitches Russians."

The president was equally outraged when Bundy gave him the

bad news at eight in the morning. Still in bed wearing pajamas and scanning the morning papers, Kennedy had just finished reading a front-page *New York Times* story headlined "Eisenhower Calls President Weak on Foreign Policy." Bundy's news seemed certain to encourage further attacks on Kennedy's leadership: "He [Khrushchev] can't do that to me," he exclaimed. His response spoke volumes about the preoccupation with domestic politics that had contributed to his reluctance to believe Khrushchev was putting offensive weapons in Cuba. True, it was easy to assume that Khrushchev would not be so reckless. After all, he knew that the United States had a huge military advantage over the Soviet Union and that a confrontation could be a prelude to a losing war. No one in the White House could quite imagine that he was ready to take such a gamble. So it was a sinking feeling for Kennedy to realize that he had misread events again on Cuba and had been so poorly served by almost all his advisers.

The presence of the missiles now opened up a two-front conflict for Kennedy. There was the domestic political fallout from having been slower than the Republicans to anticipate the Soviet aggression. McCone, who Bobby later complained made certain that congressmen and senators saw him as blameless for the administration's belated recognition of the missile threat, encouraged the view that the intelligence failure did not come from the CIA. Bobby countered with the argument that if McCone was so worried about the Soviet threat, why did he go off to Europe on a honeymoon? Bobby also asserted that McCone never put any warnings in writing, which is misleading. The contemporary record shows that McCone was assertive and emphatic about his view of the danger. But no one in the White House inner circle—the president, Bundy, McNamara, Rusk—had much liking for the self-righteous, doctrinaire McCone and that made it easier for them to set aside his predictable warnings about Cuban perils.

However, the political tension between the White House and the CIA was decidedly muted alongside much more compelling concerns about how to remove the missiles from Cuba. Despite the poor record of his advisers on Cuba, Kennedy now turned to them for help. While he was determined to respond critically to anything his advisers told him, Kennedy's strategy now was to broaden the group of consultants in order to ensure the widest possible judgments on how to end the Soviet threat peacefully, if possible. First, before he even addressed the external danger, he needed to guard against a domestic explosion of war fever, which meant hiding the crisis for as long as possible from the press and public. The *New York Times* story, however, had included a quote from Eisenhower, saying that on his watch "no threatening foreign bases were established." It immediately heightened fears that the Soviets were building a missile base in Cuba, and this seemed likely to agitate demands in Congress and the press for a military response. A public clamor for quick action would inject a sense of urgency into White House discussions that Kennedy considered unhelpful in finding a measured response to the Soviet challenge. Consequently, he tried to counter any suggestion that a crisis was brewing by following a normal schedule.

Shortly before noon, he gathered thirteen officials in the White House Cabinet Room to inform them of the missiles and begin a response. An expert from the National Photographic Interpretation Center and the CIA's assistant director of photographic interpretation set up easels and explained what the group was seeing on the enlarged photos, which looked like nothing more than a lot of indecipherable shapes and smudges. The president sat in the center on one side of an oblong table, flanked by Rusk and McNamara. Bobby Kennedy sat across at a discreet distance from his brother, sometimes leaving his seat to pace nervously about the room, giving a false impression of being a bit outside the inner circle.

Lyndon Johnson sat across from Kennedy uncharacteristically silent and stifling any hint of resentment at having been so largely ignored by the White House during his almost two years as vice president. Jackie Kennedy remembered that Johnson "never wanted to make any decision or do anything that would put him in any position. . . . Jack would say you can never get an opinion out of Lyndon at any cabinet or national security meeting. He'd just say, you know, that he agreed with them—with everyone—or just keep really quiet."

Although Kennedy had largely consigned Johnson to the outer fringes of the administration, he had been sensitive to his imperious nature and had dispatched Johnson on trips abroad to temporarily boost Johnson's standing and sense of importance. But Johnson's barnstorming on these visits, Jackie recalled, "embarrassed" the president. She thought that Kennedy had a "steadily diminishing opinion" of his vice president. She said that Kennedy "grew more and more concerned about what would happen if LBJ ever became president. He was truly frightened at the prospect." In the midst of the current Cuban crisis, however, Johnson was too experienced a politician to be excluded from the administration's inner circle.

Bundy, acting CIA director Lieutenant General Marshall Carter, Taylor, who had become chairman of the Joint Chiefs in October 1962, Secretary of the Treasury C. Douglas Dillon, Undersecretary of State George Ball, Deputy Secretary of Defense Roswell Gilpatric, and Deputy Undersecretary of State for Political Affairs U. Alexis Johnson were the other participants in the initial Executive Committee, or ExCOM, as it was called, discussions. All of them listened intently as Kennedy explained the terrible burden of responsibility facing them. No one in the room doubted that what they said and the president did could affect the world's well-being forever. Others would eventually be included in the deliberations, but for the moment, the group gave Kennedy a broader base of opinion than he had previously consulted.

While Kennedy was eager to hear any counsel the assembled experts had to offer, he initially focused on two questions: Why was

Khrushchev taking a gamble that could end in such a disaster? And second, how much time did they have before the problem became public knowledge? Kennedy made clear his eagerness to keep the crisis under wraps for the time being.

On his first question, Kennedy answered himself by saying, "It must be that they're not satisfied with their ICBMs," referring to intercontinental ballistic missiles. Taylor agreed, saying that missiles in Cuba would reduce the U.S. strategic advantage over the Soviet Union. Rusk chimed in: Khrushchev "knows that we have a substantial nuclear superiority, but he also knows that we don't really live under fear of his nuclear weapons to the extent that he has to live under fear of ours." By threatening the United States with medium-range missiles in Cuba, Rusk believed, Khrushchev was creating a power balance. As for any leeway they might have before public discussion and agitation of the Cuban threat erupted, they speculated that it could be anywhere from two days to a week.

Kennedy welcomed the observations on Khrushchev's motives and the timing of any release of information, but he emphasized that initial discussions should focus on what to do about the missiles. "We're certainly going . . . to take out these missiles," he said. But the hard question was not just how to do it, but how to do it without getting into a catastrophic nuclear war.

The meeting, which lasted for only an hour and ten minutes, touched off a divide between advocates of prompt action and those urging caution before resorting to armed attacks. As the principal military spokesman in the room, Taylor made the case for relying on the country's armed forces to combat the Soviet threat. McNamara, his ostensible boss, took up the argument for not rushing into a conflict. Given what we know now about Kennedy's ultimate response to the crisis, McNamara may have been speaking for the president. For while Kennedy taped all the ExCOM conversations, there are no records of what he may have said to his defense secretary before the meeting or whether they even spoke; it seems likely, however, that some kind of pre-meeting conversation revealed Kennedy's determination to avoid precipitous action at the same time he found a means to compel a Soviet retreat from building an offensive base in Cuba.

McNamara began his remarks by questioning whether the missiles were ready to be fired. He urged more U-2 flights to obtain a clearer picture of the weapons' state of readiness. Taylor doubted the wisdom of such caution: He believed the Soviets could fire the missiles "very quickly," and emphasized the great importance of a surprise attack. He recommended simultaneous air strikes against airfields and nuclear sites and a naval blockade to bar any more missile deliveries. A decision on whether to invade could wait until after they completed the initial attacks.

McNamara warned against hasty action. He thought that the nuclear warheads might not yet be in Cuba. And even if they were already there, he doubted the wisdom of hitting them from the air. Not only because they had no assurances that bombing attacks would take out all the missiles but also "because, I think the danger to the country in relation to the gain that would accrue would be excessive." In short, an air raid could lead to retaliatory strikes against the United States with nuclear weapons that would touch off a devastating all-out conflict. "It could be a very heavy price to pay in U.S. lives for the damage we did to Cuba."

Picking up on McNamara's warning, Kennedy asked Taylor, "How effective can the takeout be?" Taylor answered, "It'll never be a hundred percent," and predicted that they would need to continue air raids for as long as they believed necessary. McNamara then asked, "Should we precede the military action with political action? I would think the answer is almost certainly yes." And the approach should be directly to Khrushchev and as soon as possible.

Listening to the discussion, Rusk tried to find a middle ground between McNamara and Taylor. Acknowledging that the administration couldn't sit still, he cautioned that they had to take account of their many allies and the likelihood that any measures they took would affect them. He wanted Kennedy to bring the Organization of American States (OAS) into the discussion and possibly have them press Castro to understand that Khrushchev was using him and that he would be wise to break with Moscow. To meet Taylor's pressure for a military response, Rusk suggested calling up reserves, reinforcing Guantánamo, promoting guerrilla operations in Cuba, urging allies to suspend all trade with Cuba, and telling Khrushchev that he was risking a serious crisis. Rusk also speculated that Khrushchev might be trying to trade the missiles in Cuba for concessions on Berlin.

No one was persuaded by Rusk's concern about consulting allies. Bundy predicted that they would object to U.S. action to remove the missiles, "saying that if they can live with Soviet MRBMs [medium-range ballistic missiles], why can't we?" And the Germans, mindful of a possible Soviet interest in a trade, would complain "that we were jeopardizing Berlin because of our concern over Cuba." Kennedy didn't like the idea of warning allies because, as he said, it was tantamount to "warning everybody. And obviously you can't sort of announce that in four days . . . you're going to take them out." He asked that they figure out who to tell about their plans beyond de Gaulle, who they assumed would be entirely supportive. Rusk's comments particularly bothered Bobby Kennedy, who already considered him "rather a weak figure." In September, when Rusk had canceled U-2 flights over Cuba out of fear that they would be shot down and cause an international incident, Bobby savaged him in a meeting: "What's the matter, Dean?" he asked. "No guts?"

During the meeting, Bobby was uncharacteristically reticent, perhaps reflecting his brother's reluctance to commit himself to anything until he had a chance to fully reflect on his options. But when the president summed up the initial discussions, describing the options as bombing the missile sites and launching a blockade, Bobby raised the possibility of invading Cuba to prevent the Russians from resupplying the island and rebuilding the missile sites after air raids.

Like his brother, Bobby reflected the two poles of their thinking. They could not leave the missiles in place and they needed, both for the sake of America's hemisphere and national security and the president's domestic political standing, to get rid of Castro. At the same time, they believed that decisive action against Castro and the Soviets in Cuba could provoke a nuclear war that would produce global devastation and an end of the United States as they knew it.

Nonetheless, they felt compelled to give prime consideration to a military response. As the morning meeting ended and they made plans to reconvene that evening, Kennedy asked that the Joint Chiefs be there and that additional reconnaissance flights be scheduled promptly. It was evident to all that they had entered a grave crisis, which they remained eager to hide from the public by inconspicuously leaving the White House from the East Gate rather than the west entrance.

In the five hours before the group reconvened in the Cabinet Room, the participants worked to sort out the issues confronting them. Following lunch with Libya's crown prince, Kennedy asked Adlai Stevenson, who had come down from New York, to join him in the White House living quarters, where he briefed him on the missiles. The choice, Kennedy told him, was between a military strike and finding some other means to remove the threat. Predictably, Stevenson urged against a precipitous attack that could close off a peaceful solution. Meanwhile, Bobby Kennedy convened Mongoose planners at the Justice Department, chided them for having fallen so short in effective counters to Castro, and hinted that the president might be getting ready to unleash the U.S. military against Cuba. He emphasized the urgency of combating Castro's threat by declaring his intention to hold daily half-hour meetings until he considered it no longer necessary.

McNamara met with the Joint Chiefs at the Pentagon, where they dismissed the likelihood of a Soviet nuclear response to any U.S. action. They also pressed the case for comprehensive air raids that would not only eliminate the missile sites but also cripple Castro's air force and anything else that posed a threat to U.S. territory. They viewed a "surgical strike against the MRBM sites alone" as posing an "unacceptable risk"; it would leave Castro free to use his air force against Florida's coastal cities.

Kennedy was not ready to decide on any course of action yet, but the choices before him, as shown by what Stevenson and Bobby Kennedy were advising, seemed clear enough: precede military action with diplomatic and political initiatives to save the peace or strike quickly at the Soviet missile installations before they were fully operational and the Republicans could criticize the White House for having failed to preempt an increased communist threat to the country's security. Each choice carried substantial risks: Discussions with the Soviets and Cubans would give them the chance to strengthen Cuba's defenses against attacks and increase threats to Berlin or some other vulnerable Western target, while a surprise attack could lead to a wider war and all the catastrophic losses a nuclear conflict seemed certain to produce.

The same group of advisers, now including Ted Sorensen and Edward Martin, the State Department's assistant secretary for inter-American affairs, met for an hour and twenty-five minutes in the Cabinet Room beginning at 6:30. Sorensen was present not for advice but in order to know what he might need to include in a speech. Marshall Carter reported that they now had U-2 photos showing that the Soviets were building sites for between sixteen and twenty-four MRBMs that could be ready to launch in two weeks or even sooner. Remembering his reticence in challenging the CIA before the Bay of Pigs operation, Bundy pressed for assurances that the CIA was not mistaken about what they were dealing with. Carter reaffirmed his assessment, and McNamara, who said that he had tried unsuccessfully to prove that this was a misreading of the evidence, supported Carter's conclusion.

Rusk and Martin now weighed in with a proposal that Kennedy ask an intermediary to advise Castro that he was risking the destruction of his regime. Castro should also be told that the Soviets were playing him for a fool: They were ready to swap the missiles and his government for concessions on Berlin. Moreover, the emissary should hint that if Castro had trouble compelling the Soviets to dismantle the missile sites, the United States would be prepared to help him. Rusk acknowledged that any message to Castro might trigger strengthened defenses around the missiles rather than pressure on the Russians to retreat from installing offensive weapons. Rusk then warned that attacking the missiles might touch off upheavals in six Latin American countries with active communist parties. In addition, he predicted that Moscow would respond to any U.S. attack on Cuba with threats against NATO allies that would undermine the alliance unless the United States gave them advanced notice of military strikes against the missiles.

Unwilling to decide on a course of action yet, Kennedy offered no response to Rusk's and Martin's remarks. Instead, he asked what the Chiefs were proposing. McNamara and Taylor made clear that the Chiefs believed that a limited assault would provoke reprisal attacks on the United States. Only a full-scale air campaign, which would last five days, seemed the best option for ensuring the national security and would leave them time to decide whether an invasion should follow.

McNamara tried to blunt the Chiefs' recommendation by suggesting that they first consider political steps, as Rusk had proposed. Nor was he persuaded by the Chiefs' call for military action, which he warned would trigger a Soviet military response. It would require a partial mobilization in anticipation of what the Soviets might do. An invasion following air strikes would compel a large-scale mobilization and a declaration of national emergency. He saw an alternative between the Rusk and Chiefs proposals: a "declaration of open surveillance," which meant imposing "a blockade against offensive weapons entering Cuba," coupled with a warning that "we would be prepared to immediately attack the Soviet Union in the event that Cuba made any offensive move against this country."

McNamara shared the existing belief that the best way to prevent a nuclear war was through deterrence or the understanding by Moscow that any attack on the United States would bring the virtual annihilation of the Soviet Union. However, early in his tenure as defense secretary, he had concluded that a nuclear conflict was simply impermissible. He had read a study Eisenhower had commissioned on nuclear conflict, which assumed that a war would destroy both sides. He then advised Kennedy that "the President never initiate, under any circumstances, the use of nuclear weapons." His conversations with U.S. military chiefs about an all-out conflict had convinced him that their idea of unleashing America's missiles and bombers against Russia was "just absurd." It was clear to him from the start of the Cuban crisis that a military response was the prelude to a disaster.

Thinking out loud about the suggestions before him, Kennedy summarized the dilemma: He thought that publicly revealing the missile sites without attacking them would demonstrate restraint on the United States' part and put the burden on the Soviets not to deepen the crisis. On the other hand, going public would foreclose a surprise air assault and make it more difficult to destroy the sites. Moreover, he didn't think that Castro would "suddenly back down. I don't think he plays it that way." As for informing Khrushchev, Kennedy believed that he had already made clear to the Soviet leader his determination to prevent the installation of offensive weapons in Cuba. He saw justification in striking without a political overture: Khrushchev "initiated the danger, really hasn't he?" he rhetorically asked. "He's the one that's playing God, not us."

Kennedy's hard line triggered discussion of a possible nuclear attack. Rusk and Bundy doubted that Khrushchev would be so reckless. Yet Kennedy was not so sure: "We certainly have been wrong about what he's trying to do in Cuba," he said. "Not many of us thought that he was going to put MRBMs on Cuba." Because no one could say with any certainty what Khrushchev intended, Bundy thought the more important question was whether they actually needed to destroy the missiles: What is the real impact on the position of the United States of MRBMs in Cuba? he asked. How much does this change the strategic balance? McNamara replied that the Chiefs said, "Substantially." His view, however, was "Not at all." Most U.S. intelligence experts thought the ground-to-ground missiles in Cuba would give Moscow a military advantage, but had doubted that the Soviets would have been willing to risk the heightened tensions in relations with the United States. Taylor acknowledged that MRBMs in Cuba meant only "a few more missiles targeted on the United States" rather than some dramatic reduction in our security.

Everyone agreed, however, that it had political repercussions, although Kennedy thought it could lead to an even larger buildup in Cuba, which would create strategic concerns. Still, he didn't believe it was the greater military threat that mattered: "It doesn't make any difference if you get blown up by an ICBM flying from the Soviet Union or one from 90 miles away," he said. "Geography doesn't mean that much." Nonetheless, it would give Khrushchev the ability "to squeeze us in Berlin" and use the missiles as leverage in Latin America: If we faced trouble in Venezuela and made noises about sending in troops, Bobby Kennedy predicted, Castro could threaten us with the missiles. "It makes them look like they're co-equal with us," the president observed. "They've got enough to blow us up now anyway. . . . After all, this is a political struggle as much as military."

Yet he could not discount the need to get the missiles out of Cuba. The question was how to do it. Kennedy rejected the Chiefs' call for a large-scale air attack. It would create a "much more hazardous" crisis. Bundy agreed. He saw "political advantages . . . of the small strike. It corresponds to 'the punishment fits the crime' in political terms. We are doing only what we warned repeatedly and publicly we would have to do." As for timing of any air assault, they agreed that it could happen in four or five days, on Saturday or Sunday.

McNamara didn't think the timing and extent of an attack

was as important as a discussion of its consequences. Taylor advised that the Chiefs might dig in their heels against a limited air strike: "they would prefer taking no action. . . . It's opening up the United States to attacks which they can't prevent." Kennedy was not convinced: The larger attack they preferred would increase "the chances of it becoming a much broader struggle." He feared "the dangers of the worldwide effect."

Bobby Kennedy sided with the Chiefs. Unlike McNamara and Bundy, he focused not on the prospect of a nuclear war but the more narrow consequences for his brother. As was evident from his demands for more aggressive Mongoose action, he remained eager to bring down Castro and wash away the Bay of Pigs failure. Moreover, he was angry at being lied to by Georgi Bolshakov, a Soviet intelligence officer at the Washington embassy, who had become the Kennedys' back channel to Khrushchev. Bobby could imagine that Moscow had also deceived Bolshakov, an amiable, corpulent military bureaucrat who was little more than a go-between. But Khrushchev had directly told the president as well that there were no surface-to-surface missiles in Cuba. "It had all been lies, one gigantic fabric of lies," Bobby later said. And so he assumed that even if they persuaded the Soviets to remove the missiles, there would be nothing to prevent them from building new missile sites in six months. He spoke in favor of a full-scale attack followed by an invasion that could end Castro's regime and the prospect of a renewed threat. He said, "We should just get into it, and get it over with, and take our losses." He suggested creating a provocation at Guantánamo "or whether there's some ship that . . . you know sink the *Maine* again or something."

On the surface, it seems like a reckless disregard for the sort of consequences that McNamara feared. But Bobby was largely reflecting his brother's current outlook, though with the sort of vehemence that expressed the Kennedys' determination to best opponents. The president did not want to come down clearly on any side of the argument—at least not yet—and used Bobby as a sounding board. He understood that if he made his judgment evident in a group meeting, it could close off discussion by advisers reluctant to challenge the president's thinking. Yet when McCone talked to Kennedy privately on the morning of October 17, he had the impression that the president "leaned toward prompt military action" and instructed him to brief Eisenhower on developments. McCone reported that the ex-president was ready to "support any decisive military action."

On Tuesday evening of the sixteenth, Kennedy had heard enough and ended his part in the meeting by asking his advisers to keep discussing alternatives until Thursday morning, when he planned to confer with them again. In the meantime, he would follow his announced schedule of a meeting at the White House with the German chancellor on Wednesday morning followed by a political visit to Connecticut to support Democratic congressional candidates. The public was still to be kept in the dark about the emerging crisis.

Kennedy's determination to maintain a business-as-usual image carried over to a White House dinner in honor of Charles E. Bohlen before he went to Paris as ambassador. Isaiah Berlin, the British philosopher and historian and a friend of Schlesinger's, was one of the guests. Berlin recalled later that Kennedy "was very amiable, in a jolly mood, which was very extraordinary" on the day he had been alerted to the missiles in Cuba. "The sangfroid which he displayed, an extraordinary capacity for self-control on a day on which he must have been extraordinarily preoccupied, was one of the most astonishing exhibitions of self-restraint and strength of will which I think I've ever seen."

After Kennedy left on the evening of the sixteenth, the advisers continued discussing the options. McNamara was the most outspoken and unambiguous. He believed it essential to consider more fully the consequences of the major alternatives. He decisively favored what he called "the political approach" or "a non-military action." The missile sites presented Kennedy with "a domestic political problem," not "a military problem." The objective should be "to prevent their use." And this could be done by daily twenty-four-hour surveillance and a "blockade" to prevent the import of any additional offensive weapons. McNamara proposed an announcement of an "ultimatum" to Khrushchev warning him that "if there is ever any indication they're to be launched against this country, we will respond not *only* against Cuba, but we will respond directly against the Soviet Union with a full nuclear strike."

The recorded conversations give us a good idea of how the president and his advisers were responding to the crisis, but there are no records of what Kennedy's advisers thought and felt as they retired for the evening to the privacy of their homes. One can only imagine the tension all of them experienced as they considered the gravity of the problem facing them. If they agreed to steps that led to war with Moscow, it was almost too awful to contemplate. Their decision, or, more to the point, their advice and Kennedy's decision could decisively affect hundreds of millions of lives. None of them could escape the sense of responsibility for what they would urge Kennedy to do. And if Kennedy was not entirely mindful of what the crisis could bring, Stevenson hammered home the potential horror they faced with a letter to him saying, "The means adopted here have such incalculable consequences that I feel you should have made it clear that the existence of nuclear missile bases anywhere is negotiable before we start anything" (*negotiable* was double underlined). Stevenson added, "blackmail and intimidation *never*, negotiation and sanity always."

Only Kennedy has left a record of sorts on how the burden of governing was affecting him. We know that he shared McNamara's forebodings about the horrors of a nuclear war and that the responsibility of the decision weighed heavily on him. His medical records kept by Dr. Janet Travell, one of his principal physicians, give us some understanding of how the crisis took its toll on him. As a rule, he relied on antispasmodics to control a spastic colon; antibiotics to combat urinary tract problems and sinusitis; and hydrocortisone, testosterone, and salt tablets to manage his Addison's disease. During the crisis, Travell increased the amounts of the last three to ensure that his Addison's or adrenal problems did not get out of hand and sap his energy or reduce the flow of adrenaline and capacity to concentrate. When he had to speak publicly to the country and world about the crisis, for example, he relied on additional amounts of the hydrocortisone and salt tablets to prepare him for the challenge.

During this time, Jackie Kennedy asked the president's gastro-enterologist to stop the antihistamines he was taking for food allergies. She complained that they were having a "depressing action" on him and asked that the doctor prescribe a medication that would produce "mood elevation." The physician put him on a small dose of Stelazine, an antipsychotic drug that was also pre-scribed to control anxiety, which was what it was supposed to do for Kennedy. A decided improvement in Kennedy's emotional state allowed him to get off the drug after only two days.

The recorded conversations with his advisers give no indication of someone overwhelmed by current pressures and suggest that the medicines allowed him to function as effectively as any president hoped to in a grave crisis. The fact that Kennedy had hidden his health problems from the public may have been essential in helping him win a very close election for the White House. Happily, in 1962, as far as anyone can tell, Kennedy's health troubles did not reduce his capacity to muster the necessary energy and act sensibly in the Cuban Missile Crisis.

By Thursday morning, October 18, when Kennedy met with his advisers again, they had identified four possible actions to remove the missiles: an ultimatum to Khrushchev followed by an attack if he failed to take the missiles out of Cuba; an unannounced air raid against only the missile sites; a message telling Khrushchev that the United States was establishing a naval blockade around Cuba; or a large-scale air strike followed by an invasion.

Douglas Dillon and George Ball weighed in with more elaborate and passionate memoranda underscoring the divide among advisers and the momentous consequences of the president's decision. Dillon saw no room for negotiations and little alternative to military action. He opposed any request to Khrushchev for talks, arguing instead for a blockade and demands for the removal of the missiles. An immediate air strike should follow a Castro refusal. He warned that the nation's survival depended on the prompt elimination of the Soviet weapons in Cuba. Why the United States could not live with the missiles left in place was left unsaid.

Ball disagreed. He thought the missiles made little strategic difference. He compared a surprise air offensive to Japan's sneak attack on Pearl Harbor, which had justified war crime trials against Japanese leaders. Such an attack on Cuba would bring condemnation from world opinion as a violation of American traditions and professed moral standards. Ball urged a blockade that he believed would cripple and bring down Castro's government.

As the advisers convened in the Cabinet Room at 11:10 A.M., up-to-date reconnaissance photos showed intermediate-range ballistic missile (IRBM) sites, which had twice the range of MRBMs and carried warheads of roughly twice as much yield. Kennedy was told that these more powerful missiles brought the continental United States, except the Pacific Northwest, within range. Rusk and McNamara reacted to the news with anger and assertions that military strikes might need to replace the diplomatic track they had favored earlier. Remembering the appeasement of the thirties, Rusk worried about "the effect on the Soviets if we were to do nothing." But he feared that the Soviet response to air raids and a possible invasion of Cuba would lead to a dangerous escalation. So, everything considered, he favored discussions with Khrushchev, who "might realize that he's got to back down." It could "prevent a great conflict."

McNamara was more supportive of military action now than Rusk. He said that he had conferred with the Chiefs and shared their conviction that a full invasion might be in order. Kennedy asked McNamara why the new information changed the recommendation. McNamara still believed that the missiles in Cuba did not change the military equation between the United States and the Soviet Union. But he was willing to back the Chiefs' call for action out of political considerations. If they didn't act, how could any of America's allies continue to trust us; or expect Khrushchev to take U.S. deterrence seriously; or expect domestic public opinion to back the administration? While Taylor endorsed McNamara's support of strong measures, he disputed his assumption that it was strictly or largely a political matter. He shared the Chiefs' view that Khrushchev was turning Cuba into "a forward base, of major importance to the Soviets."

Kennedy was not convinced by the advice urging prompt air strikes and a possible invasion—either for military or political reasons. He discounted McNamara's assertion that the use of force was essential to hold the alliance together. Most allies do not see Cuba as a serious military threat, he said: "They think we are slightly demented on this subject." They would see an air attack on Cuba as "a mad act by the United States." Kennedy was also skeptical about the wisdom of landing U.S. troops in Cuba: "Nobody knows what kind of success we're going to have with this invasion," he said. "Invasions are tough, hazardous," as the Bay of Pigs had demonstrated. He leaned toward some kind of diplomatic initiative; he wanted to know what would be the best method of quick communication with Khrushchev. And the more important decision was, "What action we take which lessens the chances of a nuclear exchange, which obviously is the final failure."

If Kennedy needed support for a political initiative before they resorted to armed force, he found it in the advice of two American Soviet experts, Charles Bohlen and Llewellyn Thompson. The fifty-eight-year-old Bohlen was an American aristocrat. The offspring of a privileged family, he was schooled at St. Paul's and Harvard. After travels abroad, including a few months in China, he decided on a Foreign Service career with a focus on the Soviet Union. Years of service in the Moscow embassy had led Eisenhower, against Secretary of State John F. Dulles's wishes, to make him ambassador from 1953 to 1957. Tensions with Dulles over Bohlen's support for accommodations with the Soviet Union had forced him to leave Moscow, but it gave him standing with Democrats who admired his courage in standing up to right-wing Republicans, including Joe McCarthy, who had failed to block his appointment to Russia. Before Bohlen sailed in October 1962 for Paris, where Kennedy had made him ambassador to France, the president had consulted him about Khrushchev's motives and how he thought the crisis with Moscow could be resolved.

As he left Washington, Bohlen sent the president a letter saying that the missiles had to be forced out of Cuba by either diplomatic or military means. But diplomacy should be first; this advice reflected Bohlen's long-standing belief in a shared Soviet-American desire to avoid a war. A message to Khrushchev was an essential first step. Bohlen did not think it would impede later possible military steps. An attack without a prior diplomatic initiative would provoke a war with Cuba that would antagonize America's allies. "I feel very strongly that . . . a limited, quick action," he wrote, "is an illusion and would lead us into a total war with Cuba on a step-by-step basis which would greatly increase the probability of general war." The letter echoed Kennedy's fears and strengthened the president's resolve to find an alternative to military action.

Bohlen's departure for France had provoked some debate. He persuaded Kennedy that his staying in Washington would alert the press to the crisis and that Llewellyn Thompson, ambassador to Moscow, who had comparable expertise on the Soviet Union and held similar views to his, could speak for both of them. Bohlen's departure, however, infuriated Bobby Kennedy, who later complained that "Chip Bohlen ran out on us—which always shocked me. . . . That wasn't necessary; he could always have postponed it.

We said he could fly over, but he decided to leave this country in a crisis . . . when he had been working with all of us for such a long period of time." But the president disagreed with his brother's assessment and let Bohlen go to Paris.

Kennedy was also content to have Thompson as his principal adviser on likely Russian reactions to U.S. initiatives. Thompson had served as Bohlen's successor in Moscow and had been brought back from the embassy in June to become the State Department's principal Soviet expert. The fifty-eight-year-old Thompson was respected as a long-serving diplomat without a political agenda. As the U.S. expert who knew Khrushchev better than anyone else, he had a keen sense of his potential reactions to various policies as well as the likely response of others in the Soviet Union, including their interest in restraining Khrushchev. Thompson recalled for Kennedy an incident in 1960 during the U-2 crisis, when Soviet generals made clear to Thompson that Khrushchev was acting rashly. Thompson thought that Khrushchev might again be at odds with his military chiefs and that negotiating proposals might pressure him into conciliatory talks.

Thompson unequivocally supported a blockade. He believed that it would prevent the shipment of additional weapons to Cuba and would ultimately compel Khrushchev to dismantle the existing sites. Kennedy agreed that a blockade seemed unlikely to provoke a nuclear war, but he worried that Khrushchev would move against Berlin. Thompson was convinced that Khrushchev wanted to negotiate. If the United States bombed the missile sites, he said, Khrushchev would retaliate by taking "out one of our bases in Turkey . . . and then say: 'Now I want to talk.'" Khrushchev's "whole purpose of this exercise is to build up to talks with you, in which we try to negotiate out the bases."

Bobby saw problems with a blockade. It not only posed a threat to Berlin, but was also "a very slow death," with dangers in stopping and examining Russian ships and shooting down Russian planes that tried to land in Cuba. Whatever you do, Thompson advised, he urged Kennedy to make it as easy as possible for Khrushchev to back down. He thought Bobby's point was weakened by the likelihood that negotiations during the blockade would deter both sides from aggressive action.

McCone, who had sat silently, now reported that Eisenhower wouldn't support anything short of a military response. It was McCone's trump card for trying to force Kennedy into the sort of actions favored by the Chiefs. But it was clear to Thompson and Bobby that, while Kennedy had his doubts about the effectiveness of a blockade, he was not ready to risk a nuclear war with a full-scale assault. Thompson reinforced the president's reluctance by predicting that if the United States killed Russians in an attack, it would mean war. And Bobby, reflecting his brother's doubts, declared, "I think George Ball has a hell of a good point." "What?" Kennedy asked, eager to hear Ball's argument again. Bobby replied that the world would ask, "What kind of a country we are. . . . We did this against Cuba." We had consistently decried the threat of a Soviet first strike. "Now, . . . we do that to a small country. I think it's a hell of a burden to carry." Rusk agreed: It would be like carrying "the mark of Cain" on our brows. McNamara concurred, and Thompson said that it was essential that we not reject negotiations, which step would make a war inevitable.

After two days of discussion, they were still without a plan. Although McNamara acknowledged that no one had an ideal solution, he thought they needed to settle on a clear-cut diagram. McNamara tried to sum up their choices as the meeting came to an end: One was prompt military action, and the other was the slow move toward armed attacks, but only after setting up a blockade accompanied by an ultimatum to Khrushchev to remove the missiles. McNamara leaned toward the second option in the belief that it would not shatter any of the country's alliances and might facilitate an exchange in which the United States removed its missiles from Turkey and Italy at the same time Khrushchev took his out of Cuba.

A series of evening meetings made clear that Kennedy favored a blockade. Robert Lovett, whom Kennedy had brought into the discussions out of an eagerness to hear from the most experienced people he knew, suggested that they follow the State Department's legal adviser's suggestion that they call it a quarantine, which would define the action as more of a defensive measure than an act of war. As notes Kennedy made after these talks showed, Lovett especially influenced him: His long experience in government and reputation for moderate good sense helped sway Kennedy. By contrast with Acheson, who urged prompt military action, and Bundy, who opposed either an attack or a blockade as likely to cause the loss of Berlin and divide NATO, Lovett thought the blockade was the best way to resolve the crisis, with force as a last resort.

It is striking that Kennedy had not directly consulted the military chiefs before deciding to introduce a blockade. Should the blockade fail, he would resort to military steps, and so needed the Chiefs on board for that, but, assuming that they would be single-minded in their call for attacks, he held them at arm's length. Moreover, his memories of the naval officers he had seen in action during World War II and their advice before the Bay of Pigs had deepened his distrust. The Army's slow response to the Mississippi violence had added to his doubts about the military's competence. After the Army's failure to act quickly, Kennedy said, "They always give you their bullshit about their instant reaction and their split-second timing, but it never works out. No wonder it's so hard to win a war." It wasn't until the morning of October 19 that Kennedy finally brought the Chiefs into the discussion, but only for forty-five minutes.

The meeting confirmed his assumption about their views. At the start of the discussion, Taylor said that the Chiefs were agreed on a surprise air strike followed by surveillance to assure against further threats and a blockade to prevent shipments of additional weapons. Kennedy responded by telling the Chiefs that he saw no "satisfactory alternatives" but considered a blockade the least likely to lead to a disastrous nuclear war.

LeMay responded forcefully in opposition to anything but

direct military action. Moreover, he dismissed the president's observation that if the United States hit the Soviet missiles in Cuba, they would respond by taking Berlin. On the contrary, he said, hitting the missiles would deter the Soviets, and a failure to destroy the offensive weapons in Cuba would encourage Moscow to move against Berlin. "This blockade and political action, I see leading into war," he added. "It will lead right into war. This is almost as bad as the appeasement at Munich. . . . I just don't see any other solution except direct military intervention right now." Admiral George Anderson, the Navy chief of staff, General Earle Wheeler, and Marine Commandant David Shoup voiced the same conclusion: "The full gamut of military action," as Wheeler put it.

LeMay then commented on "the political factor," which, he said to Kennedy, "you invited us to comment on . . . at one time." Reminding Kennedy that he had "made some pretty strong statements . . . that we would take action against offensive weapons, I think that a blockade and political talk would be considered by a lot of our friends and neutrals as being a pretty weak response to this. And I'm sure a lot of our own citizens would feel that way, too. In other words, you are in a pretty bad fix at the present time." Offended by LeMay's bluntness and suggestion that he was acting like Britain's Chamberlain, Kennedy asked: "What did you say?" "You're in a pretty bad fix," LeMay replied, refusing to back down. Masking his anger with a contrived laugh, Kennedy said, "You're in there with me."

After Kennedy, McNamara, and Taylor left the meeting, the tape recorder caught the Chiefs attacking Kennedy. Shoup told LeMay: "You pulled the rug right out from under him." "What the hell do you mean?" LeMay asked. Shoup explained: "I agree with you a hundred percent," adding that escalation by small steps was a terrible idea. "If somebody could keep them from doing the goddamn thing piecemeal. That's our problem. You go in there and friggin' around with the missiles. You're screwed. . . . Either do this son of bitch and do it right, and quit friggin' around." Wheeler saw no chance of it: "It was very apparent to me," he said, "that the political action of a blockade is really what he" wants.

Kennedy was also angry. When deputy defense secretary Roswell Gilpatric saw him after he left the meeting, he thought the president "was just choleric. He was just beside himself, as close as he ever got." Kennedy then told Kenny O'Donnell, "These brass hats have one great advantage in their favor. If we . . . do what they want us to do, none of us will be alive later to tell them that they were wrong."

While Kennedy had concluded that a blockade was his best option for removing the missiles without a war, he wanted to ensure a consensus that precluded any public dissent by his advisers, especially the military chiefs, who could wound him politically if the blockade failed to remove the missiles and he had to resort to air attacks and possibly an invasion. The Chiefs could paint him as hesitant to use force and complain about losses resulting from the absence of surprise. Determined to keep the public in the dark until he rather than someone in Congress or the media revealed the crisis, Kennedy left on a campaign trip to the Midwest on Friday, October 19. He instructed Bobby to "pull the group together" to allow him to say later that all hands supported the blockade.

As Kennedy campaigned in Illinois and Ohio, his advisers met at the State Department, where they debated the choice between an air strike and a blockade. When a tentative commitment to a blockade was described as the current state of thinking, Taylor dissented, saying the Joints Chiefs shared his view. Bundy declared his shift from the previous day favoring non-action to air strikes, which he considered much more likely to remove the missiles than a blockade. Acheson predictably chimed in with a plea for a decisive air strike: They needed to understand that they were now dealing with an irresponsible "madman." "We had better act and act quickly," he warned. Dillon and McCone agreed, and Taylor predicted that imposing a blockade would mean abandoning an air assault or at least one that could be highly effective. McNamara said that he would order preparations for a prompt air attack but continued to prefer a blockade.

Bobby Kennedy now made clear what the president wanted. Grinning with perhaps the satisfaction of knowing that he was giving marching orders to a group of men unaccustomed to taking rather than giving direction, Bobby explained that he had spoken to his brother that morning and that the president saw no room for a surprise attack. It would evoke memories of Pearl Harbor. A blockade would make clear the administration's determination to get the missiles out of Cuba, but it would also "allow the Soviets some room for maneuver to pull back from their over-extended position." After some further discussion, Bobby agreed that a blockade could be a first step with an air strike in reserve if the Soviets did not take out the missiles.

Despite his show of confidence, Bobby called his brother and persuaded him to return to Washington on Saturday instead of Sunday to hammer home what he wanted. Pretending to have a cold, Kennedy returned to Washington to attend an afternoon National Security Council meeting at the White House. The session, the longest yet of the discussions, lasted two hours and forty minutes and included twenty-two officials, among them the president, Bobby, and all the principal advisers from the CIA, the Defense, State, and Treasury departments as well as Taylor, Bundy, and Sorensen. The discussion was essentially a rehash of now-familiar arguments, with Taylor pressing for full-scale air strikes and the president reiterating his preference for a blockade, with air attacks against only missiles and missile sites if the Soviets refused to remove the offensive weapons.

At a second NSC meeting lasting more than two hours the following day, the focus shifted to a presidential address in which Kennedy intended to demand "nothing less than the ending of the missile capability now in Cuba." He agreed, however, to use the word "quarantine" instead of "blockade" to avoid comparisons with the 1948 Soviet disruption of land traffic into Berlin. He also directed that a letter to Khrushchev be prepared saying how perilous the Soviet leader's actions were and how eager the United States was "to resume the path of peaceful negotiation."

On Monday, October 22, Kennedy implemented his decision to establish a blockade around Cuba: He instructed that "everyone should sing one song in order to make clear that there was now no difference among his advisers"; formally established an ExCOM of the NSC with him as chairman to meet every morning at ten in the Cabinet Room until the crisis ended; met with congressional leaders at the White House to explain his actions; and sent Khrushchev a letter with a copy of a speech he would make that evening. The letter explained why he was establishing a blockade: It was "the minimum necessary to remove the threat to the security of the nations of this hemisphere." In choosing a blockade, he "assumed that neither you nor any other sane man would, in this nuclear age, plunge the world into war which it is crystal clear no country could win and which could only result in catastrophic consequences to the whole world, including the aggressor."

The initial Soviet response was discouraging and even frightening. Khrushchev replied on October 23 that Kennedy's statement of the problem represented a "serious threat to peace and security of peoples." He described the blockade as "aggressive actions against Cuba and against the Soviet Union" and insisted that the weapons in Cuba were "exclusively for defensive purposes." Kennedy replied that evening describing "the current chain of events" as the result of Moscow's "offensive weapons" in Cuba, asked that Khrushchev issue "the necessary instructions to your ships to observe the terms of the quarantine," and expressed "concern that we both show prudence and do nothing to allow events to make the situation more difficult to control than it already is."

That evening, at the end of an NSC meeting, the president and

Bobby talked for ten minutes about the coming confrontation with Khrushchev. "How does it look?" Bobby asked. "Looks like hell— looks real mean, doesn't it?" Kennedy replied. "But . . . there is no other choice. If they get this mean on this one, it's just a question of where they go about it next." Bobby agreed. But it wasn't just the Soviet threat that needed answering as a way to avoid another Munich; the Congress also worried them: Without the quarantine, Bobby said, "You would have been impeached." Kennedy thought that was right and feared that after the elections the House would try to impeach him anyway for having been slow to respond to the Soviet aggression. But the larger concern was "the great danger and risk in all of this," which he saw as "a miscalculation—a mistake in judgment." Having recently read Barbara Tuchman's *The Guns of August*, a searing account of how "the Germans, the Austrians, the French, and the British . . . somehow seemed to tumble into war . . . through stupidity, individual idiosyncrasies, misunderstandings, and personal complexes of inferiority and grandeur," he feared that while "neither side wanted war over Cuba," they could find themselves in a conflict for "reasons of 'security,' 'pride' or 'face.'" Kennedy was determined not to repeat the German chancellor's response in 1914 to the question, "How did it all happen?" Which was: "Ah, if only we knew."

October 24 was a day of near despair followed by hope. Bobby recorded that at a meeting the previous night with Soviet ambassador Dobrynin, when he asked if Soviet ships heading for Cuba would try to run the blockade, Dobrynin assumed they would. U.S. readiness for a war had been increased from Defense Condition 3 to DEFCON 2, a prelude to a general war. The Strategic Air Command was put on a nuclear alert: Land- and submarine-based missiles were poised to attack, as were the country's 1,400-plus bombers loaded with nuclear weapons aimed at preselected Soviet targets. "In fifteen years of intercepting U.S. military messages," historians Aleksandr Fursenko and Timothy Naftali point out, "the Soviet military intelligence service may never have seen anything like this."

McCone reported at the morning's ExCOM meeting that the Russians were making rapid progress on the intermediate- and medium-range missile sites. Numerous Soviet ships were heading toward the island, including submarines and three possibly carrying missiles. The Soviets were also bringing their "military forces into a complete state of readiness."

Bobby recalled that the Wednesday ExCOM meeting "seemed the most trying, the most difficult, and the most filled with tension. . . . I sat across from the President. This was the moment we had prepared for, which we hoped would never come. The danger and concern that we all felt hung like a cloud over us all. . . . These few minutes were the time of greatest worry by the President. His hand went up to his face & covered his mouth and he closed his fist. His eyes were tense, almost gray, and we just stared at each other across the table. Was the world on the brink of a holocaust and had we done something wrong? . . . I felt we were on the edge of a precipice and it was as if there were no way off."

There were also hopeful signs of a Soviet retreat. As McNamara discussed plans for intercepting the Soviet vessels, McCone was handed a message saying that six Soviet ships in Cuban waters had either stopped or reversed course. The blockade seemed to be persuading the Soviets to back away from a confrontation. Rusk whispered to Bundy, who was sitting next to him, "We are eyeball to eyeball, and I think the other fellow just blinked." At an afternoon meeting with the president, Rusk said that the Kremlin's public silence about the missiles in Cuba meant they were trying to avoid a war scare. He also thought it significant that Khrushchev had sent a telegram to the British philosopher and pacifist Bertrand Russell, saying: "The Soviet Union will take no rash actions, will not let itself be provoked by the unjustified actions of the United States. We will do everything which depends on us to prevent the launching of a war."

Yet the crisis was far from over. Kennedy wanted to be sure that there were no plans to grab any of the Soviet ships. McNamara thought not, but Bobby and Rusk asked if the Navy was instructed not to pursue the retreating vessels. Mindful of how some unplanned event could trigger a conflict, Kennedy sent McNamara to the Navy's operations center in the Pentagon to make sure that ship commanders on quarantine duty strictly followed his orders to let the Soviet vessels retreat without incident. Navy chief of staff Admiral George Anderson was unhappy about the visit from McNamara and Gilpatric, which he saw as unneeded civilian interference. McNamara's questions about Navy's plans for stopping ships provoked Anderson to answer that the Navy had been doing this since John Paul Jones and he saw no reason to explain long-standing procedures. He waved a copy of the Navy regulations manual at McNamara and urged him to read it. "I don't give a damn what John Paul Jones would have done," McNamara exploded. "I want to know what you are going to do now." McNamara left in a huff, declaring, "That's the end of Anderson." (After the crisis, Kennedy forced his retirement and made him ambassador to Portugal.) It also deepened Kennedy's distrust of his military advisers. If he was going to avert a disaster, part of his challenge was to keep control of headstrong subordinates.

But even if Kennedy could rein in the men under his command, he could not control the Soviets. On the night of the twenty-fourth, he received Khrushchev's reply to his message of the day before counseling prudence and conformity to the quarantine. Khrushchev described Kennedy's actions as tantamount to an "ultimatum" that "flung a challenge at us," threatening Russia with "force." He denounced Kennedy's motives as "hatred for the Cuban people and its government" and "considerations of the election campaign in the United States." It was "the folly of degenerate imperialism." The United States was engaged in "an act of aggression which pushes mankind toward the abyss of a world nuclear-missile war." Soviet ships and forces would "protect our rights" on the high seas. It was a terrifying moment; Kennedy could only imagine that the two nations were on the brink of a disastrous war.

But Schlesinger passed along a message from Averell Harriman, whose familiarity with Soviet affairs convinced him that Khrushchev was desperate to find a way out of the crisis. "The instructions to Soviet ships to change course; the message to Bertrand Russell; and his obviously premeditated appearance last night at an American concert in Moscow" were signals that "the worst mistake we can possibly make is to get tougher and escalate." At 1:59 in the morning of the twenty-fifth, Kennedy sent Khrushchev a firm but conciliatory note: The crisis was the result not of anything the United States had done, but of the Soviet decision to place offensive weapons in Cuba, despite clear warnings against doing so, and then repeated lies about their actions. Kennedy expressed "regret that these events should cause a deterioration in our relations. I hope that your Government will take the necessary action to permit a restoration of the earlier situation." Kennedy did not budge on the quarantine.

Kennedy sent his reply without consulting all the members of the ExCOM or any of the Joint Chiefs, who were not to do anything without explicit orders from him. In fact, only Bundy, Rusk, and Sorensen seem to have been involved in drafting Kennedy's response, as was made evident when Bundy read both Khrushchev's letter of the twenty-fourth and the president's reply to the ExCOM meeting on the morning of the twenty-fifth. In moving forward without a prior full-scale discussion, Kennedy was signaling that the group had reached a consensus and emphasizing that he alone would make the final decisions on any actions that could trigger a war. The exchanges at the morning conference underscored these conditions. Everyone was in agreement that they should accept a proposal from U Thant, the U.N. secretary-general, to hold off on a confrontation at sea to allow discussions that might resolve the crisis. The fact that all Soviet ships with possible additional missiles aboard had reversed course and turned away from Cuba had made the decision easy.

The crisis, as Kennedy made clear in a conversation that evening with British prime minister Harold Macmillan, was anything but over. Kennedy reported that Khrushchev described U.S. behavior as "piratical," intended to resist it, and had "the means of action against us." While the Soviets were retreating from a collision by turning around their ships with "sensitive cargo," Kennedy saw it as only a first step in settling the larger issue, which was to persuade Moscow to stop building the missile sites and remove all offensive weapons from Cuba.

On Friday morning, October 26, the odds of achieving a Soviet stand-down seemed very long indeed. With U.S. newspapers featuring front-page stories about a buildup of American forces in the southern United States in preparation for an offensive against Cuba, and Kennedy at the morning ExCOM meeting saying that the missiles had to go, the likelihood of military action seemed very great. And time was running out: "We can't screw around for two weeks" and wait for them to finish building these sites, Kennedy said. The quarantine itself wasn't going to get the missiles out. It was only going to prevent additional missile shipments to Cuba. "We're either going to trade them out, or we're going to have to go in and get them out ourselves." After the morning meeting, Kennedy told the British ambassador that with the Soviets pushing to complete the construction of the missile sites, the United States could not wait much longer before taking action. During a brief meeting with intelligence officials at noon, Kennedy said that he saw only two ways to get rid of the missiles—through diplomatic discussions, which he didn't believe "will be successful," or air strikes followed by an invasion, which would likely trigger the firing of the Soviet missiles.

If he had to resort to military action, however, he was determined to make sure it would be the result of his conscious decision and not some misstep or miscalculation by a subordinate. During the afternoon, when Lincoln White, the State Department's press spokesman, told journalists at a daily press briefing that the

United States would be ready for "further action" if necessary, it added to the war scare. Kennedy was furious. He called White and told him, "That's the sort of stuff that's got to come from me and the White House. Christ, we're meeting every morning on this to control this, the escalation." The press was now saying that "further action is going to be taken," and Kennedy feared that this coverage was increasing the pressure for escalation. So, he told White, "You have to be goddamn careful." Otherwise, we will "find ourselves getting out of control."

Meanwhile, all this talk of action and escalation, including four cables from the Soviet Embassy in Washington warning that Kennedy was on the verge of going to war, frightened Khrushchev. On October 25, after receiving Kennedy's latest unyielding letter, he told Kremlin colleagues that he wished to end the crisis. He planned to tell Kennedy that if he promised not to invade Cuba, he would remove the missiles from the island.

At 4:30 in the afternoon, Rusk called the president to report that U Thant thought that Moscow was open to an exchange in which they would remove the missiles from Cuba if the United States pledged not to invade. The Canadians reported hearing the same thing. Kennedy was entirely receptive, telling Stevenson that he was eager for prompt agreement on Khrushchev's terms: a no-invasion pledge in return for removing the missiles. At the same time, Rusk reported that Moscow's KGB officer at the Washington embassy had contacted ABC reporter John Scali to say that Khrushchev would be interested in the exchange proposed to U Thant at the U.N.

And then at about nine in the evening, a letter from Khrushchev reached the White House with confirmation of Soviet eagerness for a deal. Khrushchev praised Kennedy's understanding of the situation. He declared his love of peace: "War is our enemy and a calamity for all of the peoples." He assured Kennedy that his concerns about "offensive weapons in Cuba were groundless." But "let us not quarrel now," he added. "It is apparent that I will not be able to convince you of this." And though he went on for several pages making the case against any Soviet aggressive intentions and declaring his sole interest in defending Cuba from an American assault, he ended by asking Kennedy to promise not to invade Cuba or support any sort of forces planning an invasion; then the missiles would disappear from Cuba.

Although an end to the crisis now seemed more likely, a settlement remained elusive. When Kennedy met with his advisers the next morning, they continued to see obstacles to an agreement. Six Soviet and three satellite ships were heading toward the quarantine line and work at the missile sites was proceeding night and day. Also, newspapers were reporting that Khrushchev had released his October 26 letter to Kennedy, which included an offer to swap the missiles in Cuba for U.S. Jupiter missiles in Turkey, a condition not mentioned in his previous night's private letter. Was Khrushchev changing his position? And because the Turks and NATO allies would see any such trade as abandoning them, while neutral observers would consider it a fair deal, how could Kennedy answer Khrushchev's public proposal? All the advisers, including Bobby Kennedy, argued against letting Khrushchev conflate the two issues. But Kennedy was more focused on ending the current crisis, saying that a lot of people would see the Turkish-Cuban swap as a rather reasonable position. Kennedy and Stevenson, who had been brought into the discussion, urged that they ignore the Turkish part of the bargain and focus on Khrushchev's narrower proposal for a U.S. pledge tied to elimination of the Cuban missiles.

Kennedy now suggested that the White House announce that it was dealing with several different, complicated proposals that required consideration over a period of time. Kennedy told the Ex-COM that if the Soviets insisted on the Turkish-Cuban exchange, it would remain difficult to resist. People all over the world would see it as crazy for the United States to fight a nuclear war over keeping its missiles in Turkey. Kennedy ended the morning meeting with a reminder to the group that the Jupiters were dispensable. Whatever his advisers might think, Kennedy was ready to trade the missiles in Turkey for those in Cuba and an end to a crisis that threatened a disaster.

The discussion continued in an afternoon session lasting almost four hours. Between meetings, the Joint Chiefs had urged the president to order a massive air strike against Cuba the following day, Sunday, or at the latest Monday, with an invasion to come shortly after. At the same time, reports arrived of a near clash between Soviet and U.S. planes off Alaska, where a U-2 had strayed into Soviet airspace. In addition, U.S. reconnaissance flights over Cuba had, for the first time, been fired on and a U-2 had been brought down with the death of the pilot. The Chiefs saw the crisis as an opportunity to hit back at the Soviets. What was the use of America's military advantage if they didn't exercise it when they had the chance?

The ExCOM focused on how to respond to Khrushchev's private and public pronouncements. Kennedy did not want simply to ignore or reject his public proposal about the Jupiters. Instead, Kennedy suggested that they urge Soviet suspension of work on the Cuban missile sites and assurances that the missiles already in place were being made inoperable. Then the United States would be prepared to discuss removing the missiles from Turkey. "We're not going to get these weapons out of Cuba unless . . . we're going . . . to take our weapons out of Turkey . . . now that he made that public," Kennedy told the ExCOM. Thompson and McCone disagreed. Thompson thought the public proposal on Turkey was a way of pressuring Kennedy to accept the private proposal on exchanging the dismantling of the missiles for a non-invasion pledge, and McCone believed that "the important thing for Khrushchev . . . to say" was: "I saved Cuba."

At eight in the evening, after additional discussions at the Ex-COM meeting, Kennedy replied to Khrushchev in a letter that put aside the Turkish question. He welcomed Khrushchev's desire for a prompt solution and stated his willingness not to invade Cuba and to end the quarantine if Khrushchev dismantled the missiles and returned them to Russia. Kennedy hoped this could be done in a couple of days, when they could begin having broader arms control discussions.

The letter was to be hand-delivered to Ambassador Dobrynin by Bobby Kennedy at the Justice Department. Although Kennedy asked Bundy, McNamara, Rusk, Ball, Gilpatric, and Thompson to join him in instructing Bobby on what to tell Dobrynin, Kennedy believed that among all his advisers only Bobby could be entirely trusted to act on his instructions. Bobby was to leave no doubt in Dobrynin's mind that a settlement was essential on the terms described in the president's letter and that any further delay would trigger U.S. military action. At the same time, Bobby was to make clear that the United States was ready to remove the missiles from Turkey, but that this could only follow a settlement over Cuba, and the commitment to do this had to be kept secret. Bobby closely followed the scripted directions in his meeting with Dobrynin, saying that Khrushchev had at most twenty-four hours to end the crisis by dismantling the missiles. In return, the United States would not permit an invasion of Cuba from American soil. When Dobrynin asked about the Jupiters in Turkey, Bobby explained that "there could be no quid pro quo," but that in four or five months "these matters could be resolved satisfactorily."

The Turkish part of the arrangement, which had been so much the focus of the ExCOM's discussions after Khrushchev had included it in his public proposal for a settlement, was the product of Rusk's advice. He had urged Bobby to say that "the president was determined to get them out and would do so once the Cuban crisis was over." Kennedy quickly signed on to this and the agreement that knowledge of this commitment would be limited to those in the room.

Kennedy had ample reason to complain about Rusk's caution and passivity in managing the State Department and his failure to propose fresh initiatives in foreign policy. But Kennedy gratefully embraced his judgment on how to finesse the Turkish issue.

He then asked Rusk to join him in a secret plan to eliminate the Turkish problem if Khrushchev insisted on including the Jupiters in any Cuban deal. Rusk was to ask Andrew Cordier, former U.N. undersecretary and dean of Columbia's School of International Affairs, to request that U Thant publicly propose the simultaneous elimination of Turkish and Cuban missiles. Kennedy believed that it would be much easier for him to accept this arrangement if the initiative came from the U.N.

Kennedy was spared a renewed discussion of a Turkish-Cuban bargain when Khrushchev agreed to the conditions Bobby had put before Dobrynin. Increased tensions over the destruction of the U-2 and intelligence describing growing pressure on Kennedy to launch an attack on Cuba persuaded Khrushchev that he needed to reach a settlement before he was driven into a war. To guarantee a quick response to Kennedy's latest proposal, Khrushchev instructed that a letter to Kennedy be read on Moscow radio. It declared: "In order to eliminate as rapidly as possible the conflict which endangers the cause of peace . . . the Soviet Government, in addition to earlier instructions on the discontinuation of further work on weapon construction sites, has given a new order to dismantle the arms which you described as 'offensive,' and to crate and return them to the Soviet Union." He also took note of the president's promise not to invade Cuba or to permit such an attack from any other country in the Western Hemisphere. In a follow-up secret message, Khrushchev noted the president's commitment to remove the Jupiter missiles from Turkey in four or five months.

Kennedy and his civilian advisers breathed a huge sigh of relief. They had faced the possibility of a nuclear exchange with the Soviets as the ultimate failure. Jackie Kennedy reflected the depths of their fears when she told her husband that she and the children wanted to die with him, if it came to that. Despite her reluctance to leave, he sent her and the children away to the safety of a bomb shelter. He then invited Mimi Beardsley to spend the night of October 27 with him at the White House. She witnessed Kennedy's "grave" expression and "funereal tone" that evening, when he told her that "I'd rather my children be red than dead." He never could have said that publicly; it would have been seen as defeatist, a readiness to surrender to Moscow rather than fight them. But it revealed his conviction that almost anything was better than a nuclear war.

With the immediate crisis at an end, Kennedy cautioned everyone not to gloat, and Jackie Kennedy remembered at the end of the crisis "thinking of the Inaugural Address—'Let's never negotiate out of fear'—because I thought how humiliating really for Khrushchev to have to back down. And yet, somehow Jack let him do it with grace and didn't rub his nose in it." On the contrary, Kennedy, as he said to some dinner guests after the crisis, "wondered how he would now get on with Khrushchev; he wondered if this humiliation cost Khrushchev too much; he wondered if something ought to be done to save his face and what, if so, he could do."

Kennedy also counseled against thinking that everything was settled. The United States still had to verify that the missiles were being dismantled and shipped back to Russia. Moreover, the Joint Chiefs remained convinced that the Soviet threat would not disappear without an attack on Cuba that could neutralize the island. They sent the president a memo describing Khrushchev's response as a delaying tactic "while preparing the ground for diplomatic blackmail." Unless there was "irrefutable evidence" that the Soviets were removing the missiles, the Chiefs recommended the full-scale air strike and invasion that had been planned for Sunday or Monday. Taylor, who had faithfully reflected the Chiefs' views throughout the crisis, now separated himself from their advice.

Kennedy ignored the Chiefs' recommendation, but a few days after the crisis ended, he met with them as a gesture of regard for their help. They were openly contemptuous: A talking paper they had prepared for the meeting, according to Taylor, was "condescending and full of platitudes: 'we were saying, Now see here, young man, here is what we think you ought to do.'" Although they put the paper aside, Anderson and LeMay made their contempt for Kennedy's leadership clear: "We have been had!" Anderson said. LeMay called the settlement "the greatest defeat in our history," one that could only be remedied by a prompt invasion. Kennedy was "absolutely shocked" by their remarks and "stuttering in reply." Soon after, *Washington Post* editor Benjamin Bradlee heard from him "an explosion . . . about his forceful, positive lack of admiration for the Joint Chiefs of Staff." Kennedy would later tell John Kenneth Galbraith, "Ken, you have no idea how much bad advice I received in those days."

But he could not simply disregard their advice: "We must operate on the presumption that the Russians may try again," he told McNamara. And when Castro refused to allow a U.N. inspection of Cuba and posed a continuing threat of subversion in other hemisphere countries, Kennedy maintained plans to bring him down. An invasion, however, was off the table. As Kennedy told McNamara on November 5 about an invasion, "Consider the size of the problem, the equipment that is involved on the other side, the Nationalists' fervor which may be engendered, it seems to me we could end up bogged down. I think we should keep in mind the British in the Boer War, the Russians in the last war with the Finnish and our own experience with the North Koreans." Not to mention the broken pledge about an invasion, which would have brought condemnation not just from Moscow but also around the world.

What lessons did Kennedy take away from the crisis about the Soviets, his military, and his closest advisers? Khrushchev had made a mistake in putting the missiles in Cuba, and he knew it and had backed down. More to the point, he understood that Soviet national security interests were not clearly at stake and that world opinion would not have been on his side in a war. And so he relented and pulled the missiles out of Cuba. But, as Kennedy told Schlesinger, if Soviet interests were directly at stake, the outcome would have been different.

As for the U.S. military, Kennedy justifiably considered them irresponsible: They wanted an invasion and that "would have been a mistake—a wrong use of our power. But the military are mad," he told Schlesinger. It is impossible to say whether an invasion would have provoked a nuclear exchange with the Soviets. But it is clear that they had tactical nuclear weapons ready to fire if U.S. forces had invaded the island. Whether they would have fired them is unknowable, but the risk was there and certainly great enough for firings to occur in response to an invasion. As Bundy said later, "recognition that the level of nuclear danger reached in October was unacceptably high for all mankind may be the most important single legacy of the Cuban missile crisis."

Kennedy also told Schlesinger that the military wanted to invade. "It's lucky for us that we have McNamara over there," who had initiated the idea of a blockade or quarantine as an alternative to quick military strikes. Kennedy could have described McNamara as a counterbalance to Acheson and Taylor, who had consistently backed the Chiefs' pressure for action, until their startling call for implementing plans for the air assault and invasion after Khrushchev had agreed to remove the missiles. Their insistence on action before diplomacy was given a chance was distressing enough. But after Khrushchev had agreed to remove the missiles? It is no wonder that Kennedy kept the Chiefs at arm's length during the crisis, never giving them a regular presence in the ExCOM discussions, which Kennedy could justify with Acheson briefly and Taylor consistently at the table.

Kennedy said nothing about Bundy, who had run an erratic course during the crisis, careening from suggestions for passivity to calls for military action. He "did some strange flip flops," Bobby Kennedy recorded. "First he was for a strike, then a blockade, then for doing nothing because it would upset the situation in Berlin, and then, finally, he led the group which was in favor of a strike." Schlesinger said that at "one time he was a hawk and another time he was a dove." Jackie Kennedy recalled that "Bundy in the missile crisis, when you think of that great mind, in the beginning he wanted to go in and bomb Cuba. And at the end, he wanted to do nothing. So, if you'd been relying on that great intelligence, look where we'd be?" As the missile crisis ended, Bundy himself, speaking of the advice offered at the ExCOM meetings, said "some had been hawks and some had been doves, but today was the day of the doves."

Rusk had been a cautious but steady presence throughout the discussion. He described his function as trying "to keep the group from moving too far or too fast." Bobby Kennedy privately described him as "playing the role of the 'dumb dodo' for this reason." From the perspective of 1965, when Bobby was already critical of Rusk for his ties to Lyndon Johnson and escalation of the Vietnam War, Bobby ungenerously described Rusk during the missile crisis as having "a virtually complete breakdown mentally and physically." Although Kennedy never held Rusk in the same high regard as McNamara, Rusk was, in fact, a voice of reason in the crisis that helped Kennedy resist the rash urgings of the military Chiefs. Llewellyn "Tommy" Thompson was especially valuable in helping Kennedy take Khrushchev's measure. Bobby Kennedy remembered him as "tremendously helpful. . . . He made a major difference. The most valuable people during the Cuban crisis were Bob McNamara and Tommy Thompson," Bobby said.

Bobby should have included himself. He was the president's closest confidant during the crisis. Dobrynin recalls that he and Bobby had almost daily conversations, usually lasting about two hours, between one and three in the morning, with no one else present. Although he thought that Bobby sometimes overdramatized the tensions between the president and his military chiefs, Dobrynin believed that "in general he rather correctly reflected the tense mood inside the White House"; it gave Khrushchev a realistic sense of the crisis they were facing. At their decisive meeting on the night of October 27, Bobby made clear that "a lot of unreasonable people among American generals—and not only generals—were 'spoiling for a fight.'" Dobrynin had "no doubt that my report of this conversation turned the tide in Moscow."

Most important, Bobby was less a thoughtful commentator during the ExCOM's deliberations than an instrument of his brother's ideas and intentions. Where Kennedy needed to stay somewhat above the debate over finding a way through the crisis, Bobby could freely state his brother's views and at times openly announce that he was declaring what the president wanted done. It was an essential role that allowed Kennedy to provide the sort of effective leadership that carried the country and the world to a peaceful resolution of the most dangerous Cold War conflict between the United States and the Soviet Union.

o o o o

"Mankind Must Put an End to War"

In December 1962, Kennedy was much happier about his presidential performance than he had been at the end of 1961. A national approval rating of 74 percent and a Gallup poll describing him as the most admired man in the world were causes for considerable satisfaction. The second-best showing in the November midterm elections by a party in a hundred years, and the conviction of three-quarters of U.S. voters that Kennedy would win reelection, were additional reasons for an upbeat mood at the White House.

Kennedy relished the prospect of running in 1964 against Arizona senator Barry Goldwater, the darling of ultraconservatives. This group had won 109 of the 176 Republican House seats in the recent elections and had become a dominant force in the Republican Party. They were enthusiastically promoting Goldwater's presidential nomination, but Kennedy saw him as a relatively easy mark. Goldwater's antagonism to Social Security and seeming readiness to consider fighting the Soviet Union—"We should think about lobbing one into the men's room of the Kremlin," he jested—frightened people. Why, even Dave Powers could beat him, Kennedy joked. They would all get to bed much earlier on election night than they had in 1960 if Goldwater were his opponent, Kennedy quipped.

A big reelection victory would allow Kennedy to overcome his disappointing legislative record. All of his major initiatives—a tax cut, federal aid to education, and Medicare—were stalled in Congress, while southern resistance ruled out even asking for a civil rights law. But a second term, supported by a large national majority for him and congressional Democrats, seemed likely to produce a domestic legacy rivaling Franklin Roosevelt's New Deal. Moreover, with almost two-thirds of the country optimistic about the economy and a like number hopeful after the resolution of the missile crisis that the United States could achieve a peaceful settlement of differences with Russia in the future, Kennedy could imagine a truly distinguished presidency.

And after that, he could picture Bobby succeeding him—the first two-brother presidential dynasty in American history. On January 4, when Bobby gave a compelling speech at the National Archives commemorating the centennial of the Emancipation Proclamation, liberal activist Joe Rauh passed a note to Schlesinger saying, "Poor Lyndon." What do you mean? Schlesinger asked him. "Lyndon must know he is through," Rauh replied. "Bobby is going to be the next President."

Only Vietnam cast a shadow over the administration's prospects. True, a test ban treaty and the 1964 elections remained unfinished projects, but favorable outcomes seemed within reach. By contrast, Kennedy received several reports about Vietnam in December 1962 that troubled him. Yes, Diem and Nhu were optimistic about the eventual success of Strategic Hamlets in defeating the Viet Cong, but they could not imagine a successful end to the fighting for at least three years, if even then. Roger Hilsman warned that although Diem and Nhu saw "the tide turning against insurgency and subversion . . . this optimism was premature." A State Department working group on Vietnam thought that bringing U.S. journalists in Saigon on board was the biggest remaining problem in ensuring sustained U.S. support for a successful war effort. Even then, South Vietnam's survival was less than certain: Averell Harriman, Deputy Undersecretary U. Alexis Johnson, Hilsman, Deputy Secretary of Defense Roswell Gilpatric, and Michael Forrestal, the National Security Council's point man on Vietnam, thought that the war could be won, but only through increased use of U.S. air support and a sustained effort for an uncertain period of time. Whether the Congress and the public would remain steadfast if the required sacrifices became too great or lasted too long was open to question.

After visiting Vietnam with other senators in November, where he had been seven years before, Mike Mansfield gave the president a more discouraging report. Nolting and Harkins acknowledged the challenges in battling the communists, but they told Mansfield that they had every hope of winning. Diem, with whom Mansfield had a cordial relationship dating from the 1950s, and the Nhus, whose influence and control over Diem did not escape Mansfield, also saw better days ahead.

But Mansfield did not trust either the embassy's or the palace's optimism. Eager to escape the Potemkin village atmosphere of officials in Saigon, Mansfield arranged to have a four-hour lunch with more critical American correspondents. The move greatly irritated American officials, who saw the journalists as a major part of their problem. Their reporting angered Diem and the Nhus and made them harder to deal with. Diem and the Nhus also thought that the journalists were undermining morale among the diplomats and troops serving in Saigon as well as threatening to turn key Washington policymakers against substantial long-term support for the war. Some in the U.S. military and embassy tried to strong-arm the reporters into getting "on the team." Nolting told them, "Stop looking for the hole in the doughnut." He told Washington that his assignment was being "badly hampered by irresponsible, astigmatic and sensationalized reporting." The Marine general in Saigon belittled the journalists as weak-kneed liberals who cried at the sight of dead bodies. Although the four reporters—Halberstam of the *New York Times*, Neil Sheehan of

UPI, and Peter Arnett and Malcolm Browne of the Associated Press—did not describe a hopeless conflict, they astutely warned of disaster from the Diem government, which they believed incapable of beating the Viet Cong.

Mansfield saw little that had changed since his last visit, despite billions of American dollars and the increased presence of U.S. military advisers. He heard lots of hopeful talk about Strategic Hamlets, but the French had also been full of unwarranted optimism about new concepts. As far as he could tell, the Viet Cong still controlled the countryside. He thought that defeat of the communists would take "an immense job of social engineering, dependent on great outlays of aid on our part for many years" and a much more effective government in Saigon.

He believed it was worth giving the Hamlets program a chance to work, but warned that if they failed, the United States would face the possibility of "a truly massive commitment of American military personnel and other resources—in short going to war fully ourselves against the guerrillas—and the establishment of some form of neocolonial rule in South Vietnam. That is an alternative I most emphatically do not recommend. . . . We are not, of course, at that point at this time. But the great increase in American military commitment this year has tended to point us in that direction." Mansfield warned against a rapid slide into a quagmire. He urged a reassessment of U.S. interests in Southeast Asia, which might discourage any expansion of U.S. commitments to Vietnam's civil war. He shared the English view that "[e]very people have a right to their own Wars of the Roses." The failure of the present effort should persuade the administration to seek neutralization of the region.

A front-page *New York Times* story by Halberstam on December 3 describing Mansfield as "Cool on Vietnam War" angered Kennedy. In a discussion with Mansfield at his Palm Beach retreat in December, he told the senator that his assessment did not tally with what he was hearing from administration subordinates. Afterward, Kennedy told Kenny O'Donnell that "I got angry with Mike for disagreeing with our policy so completely, and I got angry with myself because I found myself agreeing with him." He was more irritated with his military and national security advisers for once again urging him to take on what might prove to be an impossible challenge. This could be worse than the Bay of Pigs. The failure there was a quick disaster, but this could be a slow, drawn-out fiasco that could cost American lives and significant amounts of money and undermine him politically at home and abroad.

In public, he tried to maintain a measured posture on America's part in the conflict. At a December 12 press conference, a journalist described "a good deal of discouragement about the progress" in Vietnam despite a year of stepped-up aid, and asked Kennedy's assessment. The president acknowledged that "we are putting in a major effort" with about ten or eleven times the number of men there than a year ago and a lot of equipment. He also described "a number of casualties and great difficulty . . . in fighting a guerrilla war. . . . We don't see the end of the tunnel," he said, but he didn't think conditions were worse than a year before.

Events in January of the new year deepened Kennedy's private skepticism about rescuing South Vietnam from a communist takeover. On January 3, 1963, in a pitched battle at Ap Bac, a village thirty-five miles from Saigon in the rich Mekong Delta, which the government saw as vital to its survival, a Viet Cong battalion—a force of some two hundred men—inflicted a stunning defeat on Diem's forces. Despite a four-to-one advantage in troops backed by artillery, armor, and helicopters, the South Vietnamese performed miserably. According to a U.S. adviser's after-action report, they showed "a reluctance to incur casualties, an inability to take advantage of air superiority, and a lack of discipline." In short, the government's men could not match the enemies' willingness to take losses and fight for something they believed in. The battle also cost the lives of three American advisers, including an Army captain "out front pleading with them [the South Vietnamese] to attack." Five U.S. helicopters ferrying troops into battle were shot down. When the defeat became headline news in the United States, the Joint Chiefs, instead of acknowledging that military advisers had had little success in turning South Vietnamese troops and their commanders into an effective army, criticized the reports as distorting "both the importance of the action and the damage suffered by the US/GVN forces."

To refute the press accounts, which stirred complaints that the Kennedy administration had been hiding the truth about Vietnam from the public, General Harkins and Admiral Felt attacked the newsmen as irresponsible for saying that the South Vietnamese forces didn't and won't fight and for ignoring victories that they asserted were occurring more frequently. To bolster the view that the war was going much better than journalists said, the Chiefs sent Army Chief of Staff General Earle G. Wheeler to lead a team of Pentagon officers on an inspection trip to Saigon to report on future prospects for the war. A career Army officer, the fifty-four-year-old Wheeler had served in a variety of senior staff positions, but had limited experience as a field commander and deferred to those who did. He seemed unlikely to take issue with what Harkins thought, which is what the Chiefs wanted. If the United States wasn't winning on the battlefield, at least it could encourage perceptions that it was until the tide turned against the Viet Cong.

Because Mansfield's report had so troubled him, Kennedy earlier had directed Hilsman and Forrestal to make a fact-finding trip to Vietnam to see if anything could be done to improve chances of a successful outcome. On the ground for ten days beginning December 31, the two kept extensive notes on what they found, which was the basis for a lengthy report they sent Kennedy on January 25. They saw clear evidence of improvement in the war from the previous year and substantial optimism at the U.S. Embassy and among the military about long-term results in the fighting. Harkins and Nolting were decidedly optimistic.

But the "negative side of the ledger," Hilsman and Forrestal reported, "is still awesome." They were hard-pressed to understand the excessively positive official outlook: "Things are not going nearly so well as the people here in Saigon, both military and civilian, think they are." They agreed "that we are probably winning, but certainly more slowly than we had hoped. At the rate it is now going the war will last longer than we would like, cost more in terms of both lives and money than we anticipated, and prolong the period in which a sudden and dramatic event would upset the gains already made." They suggested the need for a plan better coordinating civilian and military actions; providing some long-range thinking about future conditions in South Vietnam; better understanding of how to conduct a counter-guerrilla war; and means of persuading Diem to win greater popularity among his countrymen. It was hardly a prescription for how to win the war—just another call to find the means to do so.

Kennedy at once read the battle at Ap Bac and the Hilsman-Forrestal report as evidence that Galbraith, Ball, and Mansfield were right about the nearly impossible task of defending an unpopular regime in Saigon against a highly motivated insurgency. When he read press reports of visits to Vietnam by senior U.S. military and civilian officials that increased impressions of American involvement in the war, he complained to Forrestal on January 25: "that is exactly what I don't want to do."

On February 1, Forrestal reminded Kennedy that he would meet that afternoon with General Wheeler as well as Rusk, McNamara, Taylor, Harriman, and McCone to hear Wheeler's report on Vietnam. Forrestal listed seven questions Kennedy might want to ask the group, including whether a new ambassador should replace Nolting, whether the administration was being firm enough with Diem, and whether the United States should reconsider the way it was conducting military and political operations.

After the meeting, Forrestal apologized to Kennedy for the "complete waste of your time." The meeting was supposed "to provide you an opportunity to initiate action on some of the problems in South Vietnam" listed in his earlier memo. "The rosy euphoria generated by General Wheeler's report made this device unworkable." Speaking for the Joint Chiefs, but especially Taylor, who even more than McNamara saw the conflict as his war, Wheeler tried to counter the negative impressions of Hilsman and Forrestal and the journalists. He described a situation in South Vietnam that had changed from near desperation to likely victory. The United States was winning and only needed to stay the course. Because he and the Chiefs did not want to attack Kennedy's civilian advisers directly, Wheeler said the greatest problem was the reporting of U.S. journalists in Saigon, who had done "great harm" by encouraging public and congressional concern the United States was locked into a losing effort. It was transparent to Forrestal and Kennedy that the military was spinning its wheels. No doubt they believed their own rhetoric, but Wheeler's performance only deepened Kennedy's conviction that they weren't to be trusted.

For the next three months, from the beginning of February to the end of April, while he focused on nuclear test ban negotiations, finding some fresh approach to Cuban problems, and made plans for a European trip, Kennedy gave limited attention to Southeast Asia. Four passing mentions in his January State of the Union address, led by an optimistic assessment that the "spear point of aggression had been blunted in Vietnam," underscored his wish to put the conflict at the bottom of his priorities. He left it to his military and civilian officials and the journalists in Saigon to argue about how the United States could defeat the communists and withdraw from Vietnam. Kennedy allowed the problem to fester rather than confront a hard decision to expand U.S. involvement or shut it down. His hope was eventually to withdraw from Vietnam with at least the appearance, if not the actuality, of victory. It was something of a pipe dream, but simply walking away from Vietnam did not strike him as a viable option—for both domestic political and national security reasons.

On March 6, when a reporter at a press conference asked his reaction to a Mansfield recommendation for a reassessment of U.S. Asian policy and a possible reduction of aid, he said, "I don't see how we are going to be able, unless we are going to pull out of Southeast Asia and turn it over to the Communists . . . to reduce very much our economic programs and military programs in South Viet-Nam, in Cambodia, in Thailand. I think that unless you want to withdraw from the field and decide that it is in the national interest to permit the area to collapse, . . . it would be impossible to substantially change it particularly as we are in a very intensive struggle in those areas." He asserted that if the communists controlled all of Southeast Asia, it would jeopardize India and all of the Middle East. "I don't see any prospect of the burden being lightened for the United States in Southeast Asia in the next year if we are going to do the job and meet what I think are very clear national needs." He said nothing, however, about what might come after 1964, signaling by his silence the hope that by then they might be able to ensure Vietnam's autonomy or at least the appearance of it and bring home American advisers.

While Kennedy stood aside, the debate over Vietnam continued. On one side were the determined optimists: Nolting and Harkins in Saigon and the Chiefs led by Taylor in Washington, with unflinching support from Wheeler and Marine General Victor H. Krulak, special assistant to the Chiefs for counterinsurgency. They believed winning in Vietnam was an essential predicate to beating back communist insurgencies in developing countries. As important, they thought that the expanded U.S. involvement of the past year had established the conditions for victory. And the path to success was not through political reforms imposed on Diem, but through military action. As Wheeler had said in a speech at Fordham University in November 1962, "it is fashionable in some quarters to say that the problems in Southeast Asia are primarily political and economic rather than military. I do not agree. The essence of the problem in Vietnam is military."

Civilian advisers, led by Harriman, Hilsman, and Forrestal, were no less eager to defeat the Viet Cong insurgency. But they were less certain about the outcome and were convinced that political rather than military initiatives were the key to success: Pressuring Diem and the Nhus to introduce political reforms was essential to Saigon's long-term stability. They also believed that South Vietnam's armed forces needed to take greater direction from U.S. military advisers in order to ensure more aggressive action.

American journalists in Saigon represented a third side in the debate. They did not think that a Diem government could outlast the communists. It was a corrupt regime sponsoring an Army led by handpicked mandarins more committed to preserving their privileges and insulating their troops from battlefield losses than to sacrifices in the service of victory. These correspondents made every effort to expose the weaknesses in South Vietnam's government and army. They saw their job as telling the truth about an unpopular autocratic regime and an army that wouldn't fight. They were not, however, urging a withdrawal from Vietnam, but a change in government and military actions that could bring victory.

The argument in Washington and Saigon over how to win the war and escape from Vietnam intensified in early February when Forrestal suggested to Harriman that the United States broaden its contacts with noncommunists. Forrestal saw it giving the United States a more independent position in Vietnam and increasing our alternatives if a change of government became desirable. He thought that the embassy should make clear to Diem that U.S. interests meant "a friendly attitude towards all his people" and a full airing of differences with him.

Harriman passed Forrestal's letter along to Nolting, who bristled at the suggestion that "we are living in cocoons here, dealing only with GVN officials and deliberately cutting ourselves off from other Vietnamese elements." Nolting warned that encouraging oppositionists could "stimulate revolution. . . . If the idea is to try to build up an alternative to the present government . . . I am opposed." He saw no alternative to Diem. Forrestal was scathing about Nolting's resistance. "It's about what I expected," he told Harriman, "since this is more a question of attitude than of making a case one way or the other. . . . Fritz tends to be more concerned about preserving the legitimate government than keeping in touch with the opposition." Schlesinger weighed in on Forrestal's side, telling him that Nolting's letter to Harriman "is one of the most dismal documents I have ever encountered." He suggested replacing Nolting in Saigon. What no one wanted to confront was that the bureaucratic infighting signaled America's involvement in a failing policy.

The one thing policymakers in Saigon and Washington could agree on was the destructive influence of the journalists on the war effort. When Hilsman held an informal meeting with American reporters during his January inspection trip, it turned into a shouting match. The journalists described Diem as hostile to his U.S. advisers, whom he was ignoring, and the South Vietnamese government as on the road to defeat. Hilsman, convinced that his firsthand experience with guerrilla operations in Burma during World War II gave him superior understanding, snidely dismissed the newsmen as "naïve." The important thing, he told them, was not for Americans "to be liked, but to be tough and get things done." The reporters had no quarrel with that prescription, but they were less accepting of Hilsman's conviction, or at least hope, that Diem could be pressured into reforming his government and convincing his army to fight more aggressively, as the Americans were advising.

The U.S. military was angrier than Hilsman was with the journalists. When Admiral Harry Felt arrived in Vietnam after the Ap Bac debacle, Sheehan greeted him at the airport with a request for a comment. "I don't believe what I have been reading in the papers," he said, before hearing from subordinates in Saigon. "As I understand it, it was a Vietnamese victory—not a defeat, as the papers say." Harkins, who was standing next to him, parroted his optimism. "Yes, that's right. It was a Vietnamese victory. It certainly was." Informed by an aide that his questioner was Sheehan, Felt said, "So, you're Sheehan. . . . You ought to talk to some of the people who've got the facts." Without missing a beat, Sheehan fired back, "You're right, Admiral, and that's why I went down there every day."

There were also dissenting voices within the ranks of the American military and civilian managers trying to chart a winning strategy in Vietnam. Among the most informed and outspoken critics was John Paul Vann. He was thirty-seven years old, an up-through-the-ranks lieutenant colonel who had seen more combat operations in Vietnam—participating in more than two hundred helicopter assault landings—than any other American. He was a fearless warrior, frequently exposing himself to dangers in attempts to motivate aggressive action by the Vietnamese troops he accompanied on missions. His frustration at the refusal of the ARVN to take advantage of greater numbers and superior arms to attack the Viet Cong boiled over into outspoken complaints. A highly critical report he wrote on Ap Bac angered his superiors, who wanted the whole command in Saigon to speak with one voice. Completing his tour of duty in Vietnam in April 1963 and assigned to the Pentagon, Vann tried to brief his superiors on Diem's and ARVN's failings, but Taylor, Wheeler, and Krulak, who shared the Saigon command's aversion to conceding Vietnamese imperviousness to U.S. advice, refused to give Vann a hearing. It was groupthink unworthy of such intelligent and competent leaders and a formula for defeat in the war.

The State Department also downplayed tensions with the Vietnamese that undermined prospects of a more stable and secure nation; it was too frustrating to concede that nation-building in Vietnam was beyond Washington's reach. As Hilsman told Rusk at the beginning of April, "the strategic concept . . . for South Viet-Nam remains basically sound. If we can ever manage to have it implemented fully and with vigor, the result will be victory." But no one in the department seemed to have reliable suggestions for implementing it or could even demonstrate that the overall plan would work.

In March, the General Accounting Office had circulated a draft report on aid programs for Vietnam that was severely critical of the Vietnamese government's failure to mobilize its resources. Nolting pressed the department to bury this "public chastisement of the GVN" as likely to encourage coup plotting, raise enemy morale, and reverse recent gains in the fighting. The department arranged to have the report classified and hidden from the press and public. Similarly, when a public affairs officer in the department took issue with the conviction that the correspondents in Saigon were a menace to the U.S. mission in the conflict, arguing that the reporters were better informed and had a clearer understanding of conditions in Vietnam than most U.S. officials in the country, the department shelved his report.

Robert G. K. Thompson, a highly decorated British officer who had been a central figure in defeating guerrilla insurgents in Malaysia and had become the head of a British Advisory Group on Vietnam, encouraged American illusions about progress defeating the Viet Cong. At an April I meeting with Harriman, Thompson reported that when McNamara had asked about reducing U.S. forces in Vietnam, Thompson had said that continuing gains in the fighting during 1963 would allow the withdrawal of some one thousand men.

Thompson's credentials in defeating communist insurgents in Malaysia convinced his American audience that he was an authority worth hearing. His conviction that Strategic Hamlets would make a difference in Vietnam won him an interview with Kennedy on April 4. Eager for good news that would chart out a relatively short timeline for U.S. victory, Kennedy embraced Thompson's vision of success. The French had lost in Vietnam, Thompson replied to a Kennedy question about their failure, because they did not have the

Strategic Hamlets, which were bringing the peasants to the government's side. Morale was on the upswing, Thompson reported, and continued progress by the summer would give Kennedy the option of withdrawing a thousand advisers by the end of the year. He also told Kennedy, who had asked about the quality of the political opposition, that it was "very poor," and that Diem was the only one who could win the war. By the close of the meeting, Kennedy was delighted to have such an upbeat assessment.

A three-and-a-half-hour meeting that night at the palace in Saigon between Nolting, who had agreed to resign at the close of his two-year tour in Saigon, and Diem gave the lie to all the optimism. Diem adamantly refused to accept American advice on funding for counterinsurgency programs that the embassy and U.S. military advisers believed were essential to winning the war. Diem's resistance perplexed Nolting. But Diem insisted that following the U.S. lead would undermine his authority by giving the appearance of Vietnam as a U.S. protectorate. Diem was indifferent to Nolting's warning that his resistance to U.S. advice would curtail American aid and reverse the gains made over the last eighteen months.

Still, Diem and the Nhus did not seem to care. On April 12, when Ngo Dinh Nhu spoke with an embassy official, he said that "it would be useful to reduce the numbers of Americans by anywhere from 500 to 3,000 or 4,000." He did not think that U.S. advisers in the provinces understood the difficulties of the local officials who were complaining about unwanted pressure from the Americans. They were creating a "sense of inadequacy and inferiority" among the Vietnamese. Mrs. Nhu publicly declared that the Americans were trying to make " 'lackeys of the Vietnamese.' "

An April 17 U.S. National Intelligence Estimate concluded that it was impossible "to project the future course of the war with any confidence. . . . Despite South Vietnamese progress, the situation remains fragile." The analysts doubted that Diem had the will or capacity to lead his country to victory. Nonetheless, the State Department's working group on Vietnam was willing to bet on the likelihood of gains in the fighting and proposed a substantial reduction of the more than sixteen thousand U.S. advisers in Vietnam by the end of 1963. The CIA expected the Saigon government to endorse the idea of a significant reduction of U.S. personnel; the agency said "the force is too large and unmanageable."

By the spring of 1963, Vietnam seemed to be primarily a burden with no clear solution. An evaluation of the Strategic Hamlets program described it as an "excellent" concept that was "seriously handicapped by a lack of understanding" and insufficient "will to put it into effect." To make it work would require "a psychological revolution in the way the Vietnamese Government and its officials operate." As for Diem, he remained resistant to reforms the embassy had been pressing on him since 1960. The State Department's Vietnam working group proposed a fresh discussion of what economic and fiscal pressures might force him to respond. They also wanted renewed consideration of contingency plans should Diem be ousted. Despite, or perhaps because of, these uncertainties, McNamara directed Pentagon planners to discuss a more rapid withdrawal of U.S. personnel from Vietnam than the numbers projected over the next three years. In a May 7 meeting, he told the president that he hoped they could pull out one thousand troops by the end of the current year and avoid the sort of commitment the United States had faced in Korea, which he saw as excessive and feared would occur from an unchecked growth of involvement in South Vietnam.

In May, if Kennedy needed any additional incentives to withdraw from Vietnam, he found them in the disturbing behavior and comments of Diem and Nhu. In Hue, in the central highlands, government repression of Buddhists, who were peacefully protesting a ban on displaying their flags on Buddha's birthday, embarrassed Washington as a demonstration of Diem's religious intolerance: a Catholic president repressing his country's Buddhist majority. A *Washington Post* story on May 12 reporting an interview with Nhu outraged congressmen and senators and raised new questions about why the United States was expending lives and money in Vietnam. Nhu said that he would like to see half of the U.S. troops in his country withdrawn; they were unnecessary and gave the communists propaganda. He said that many of the Americans who died were "cases of soldiers who exposed themselves too readily," suggesting that they were reckless. The State Department complained that Nhu's comments would "likely generate new and reinforce already existing US domestic pressures for complete withdrawal from SVN." In addition, "statement that some American casualties incurred because our advisors are daredevils and expose themselves needlessly likely to have very bad effect on morale US forces."

Managing Diem and Nhu had become a war within the war. Harkins was convinced that if Diem agreed to an all-out offensive in 1963, it would end the war. "The equipment is on hand, the units are trained, morale is high, and from all I can ascertain, the determination and will are present," Harkins told Diem. But doubtful that his government could survive a campaign with significant losses, which he feared would occur, Diem wouldn't budge. The State Department's working group discussed punitive cuts in aid to force Diem's hand on reforms and military cooperation. But they thought the result would be more private and public conflict with him. The outcome might then be a push for "a change in government. We don't want to blunder into that," they concluded.

Kennedy was caught between an eagerness to end U.S. involvement in what could be a long, draining war and the political liability from accusations that he had cut and run. On May 22, when a reporter asked his response to Nhu's call for a drawdown of U.S. troops, he replied: "We would withdraw troops, any number of troops, any time the Government of South Vietnam would suggest it. The day after it was suggested, we would have some troops on their way home. . . . We are hopeful that the situation in South Vietnam would permit some withdrawal in any case by the end of the year, but we can't possibly make that judgment at the present time." That decision would depend on "the course of the struggle the next few months."

With a *New York Times* headline announcing "Failure to Gain Clear-Cut Victories and Political Frustrations Hamper U.S. Involvement," Kennedy invited Mike Mansfield in for another conversation. According to Kenny O'Donnell, who sat in on part of the meeting, Kennedy told the senator that he agreed with his "thinking on the need for a complete military withdrawal from Vietnam." But he couldn't do it until after the 1964 election— otherwise "there would be a wild conservative outcry" against returning him to the presidency for a second term. After Mansfield left, Kennedy told O'Donnell, "In 1965, I'll become one of the most unpopular Presidents in history. I'll be damned everywhere as a Communist appeaser. . . . If I tried to pull out completely now from Vietnam, we would have another Joe McCarthy red scare . . . but I can do it after I'm reelected. So we had better be damned sure that I *am* reelected."

Because Kennedy had lost whatever confidence he had that Diem could be pressured into following America's lead in winning the war, he wanted contingency plans for a change of government. The plan, which won White House approval on June 6, was to be held in the greatest confidence lest it stir a firestorm of controversy from both Saigon and the communists accusing the United States of neocolonialism.

Pressure to dump Diem mounted with a failure to end Buddhist protests. On June 8, when Madame Nhu denounced the Buddhists as "reds' dupes," the tensions between the government and Buddhist leaders provoked a crisis in U.S.-Vietnamese relations. Rusk cabled the embassy: "Madame Nhu's intolerant statement has seriously weakened GVN's position as defender of freedom against Communist tyranny and has greatly increased difficulty of U.S. role as supporter of GVN." An embassy protest to Diem reflecting Rusk's complaint brought no retreat. And on June 11, when a photograph of a Buddhist bonze (priest) burning himself to death reached around the world, the *New York Times* reported that Diem's government and the war against the Viet Cong were in serious jeopardy. The embassy received word that Vietnam's air force chief of staff, a Buddhist, wondered why the United States did not seize this opportunity to overthrow Diem.

Rusk instructed the embassy to inform Diem that he needed to do whatever it takes to meet Buddhist demands. Relations between Washington and Saigon were nearing a breaking point; the United States was considering reexamining "our entire relationship with his regime."

Rusk's warning to Diem went forward without Kennedy's knowledge or approval. It was unusual for Rusk to act so independently. But he believed that he was expressing the president's wish to keep his distance from the whole Vietnam mess. Besides, he saw Kennedy as preoccupied with a crisis in Alabama, where Governor George Wallace was resisting the integration of the state's university, and racial tensions across the South seemed poised to erupt in widespread violence. After great reluctance to face down the segregationists, Kennedy had decided to put a comprehensive rights bill before Congress that would end segregation in all places of public accommodation.

Kennedy had tried to avoid a showdown over civil rights, viewing it as likely to refocus too much national attention on a fiercely divisive domestic issue that could jeopardize his reelection. But the repression of peaceful marchers in Birmingham by club-wielding police prodding snarling dogs to bite demonstrators, including children, joined by firemen assaulting marchers with high-pressure hoses that tore off their clothes, provoked outrage across the nation. A front-page *New York Times* image of a dog lunging to bite a teenager sickened Kennedy and convinced him to act. He now saw the segregationists as "hopeless, they'll never reform." They "haven't done anything about integration for a hundred years," he added, "and when an outsider interferes, they tell him to get out; they'll take care of it themselves, which they won't."

But Kennedy still held back on making civil rights reform a top priority. His resistance to investing too much of his energy and political capital in this domestic battle reflected itself in the slap-dash manner in which he asked Congress to act on civil rights and prepared on June 10 for an Oval Office speech to the nation. There was little discussion about the bill with anyone except Bobby Kennedy, who, as in the Mississippi crisis, managed the administration's response. Lyndon Johnson, in fact, was the member of the administration best able to help design and pass a landmark civil rights law. In 1957, as Senate majority leader, he had engineered the passage of the first major civil rights law since 1875, during Reconstruction. But neither Kennedy consulted him about the current crisis. LBJ learned about the president's plans from a *New York Times* article. He complained to Ted Sorensen that he never saw the bill, didn't know what was in it, nor had discussions with anyone on how to pass it. Johnson's concerns persuaded Kennedy to hold off sending the bill to Capitol Hill for a week. But he still failed to confer with Johnson about a concerted strategy for getting the bill approved.

The national address was also something of an afterthought. All Kennedy's speeches on big foreign policy matters had been the product of careful preparation. But his civil rights address was not ready until five minutes before he went before the cameras, and even then part of it had to be delivered extemporaneously. To his credit, Kennedy made a heartfelt appeal for what he described as a great cause: "We are confronted primarily with a moral issue," he declared. "It is as old as the scriptures and is as clear as the American Constitution. The heart of the question is whether all Americans are to be afforded equal rights and equal opportunities. . . . One hundred years of delay have passed since President Lincoln freed the slaves, yet their heirs, their grandsons, are not fully free. . . . And this nation, for all its hopes and all its boasts, will not be fully free until all its citizens are free. . . . I shall ask the

Congress of the United States to act, to make a commitment it has not fully made in this century to the proposition that race has no place in American life or law." Although given with clear emotion and sincerity, the speech was a last-minute pronouncement on a subject at the fringe of Kennedy's priorities, or more to the point, given with the understanding that it would alienate southern voters and jeopardize his reelection.

Kennedy was in fact frustrated at having to divert attention to civil rights, which led to his irritation at Rusk for implying that the United States might back a coup in Saigon. When he learned about Rusk's threat from a CIA Intelligence Checklist on June 14, he instructed his military aide to tell the State Department to stop making such pronouncements. "The President noticed that Diem has been threatened with a formal statement of disassociation," a White House memo recorded. "He wants to be absolutely sure that no further threats are made and no formal statement is made without his personal approval."

Kennedy had no special regard for Diem and his government. He saw the South Vietnamese leader's clash with the Buddhists as unacceptable and destructive to his regime. But he was reluctant secretly or overtly to back military opponents ready to overthrow him. As with the Bay of Pigs invasion, any coup would be laid at America's doorstep. A front-page *New York Times* story on June 14 saying, "U.S. Warns South Vietnam on Demands of Buddhists," had raised speculation that Washington was laying plans to get rid of Diem. In fact, Rusk sent word to South Vietnamese vice president Nguyen Ngoc Tho that the United States would be prepared to back him as Diem's replacement should it become necessary, but only if it were clear that the war could not be won with Diem in power. Kennedy believed that dumping Diem would mean greatly increased U.S. responsibility for Vietnam and diminished likelihood of a U.S. withdrawal in a timely fashion. He was not cate-

gorically opposed to a U.S.-sponsored coup, but he wanted to be
sure that Diem's replacement would be a major improvement and
would facilitate rather than reduce America's chances of escaping
from an unwanted war.

Despite Kennedy's injunction against further threats, Rusk,
Hilsman, and the State Department's working group continued
warning Diem that loss of U.S. support could follow unless he
resolved his Buddhist problem. They saw no choice if the United
States were to find a timely exit from Vietnam, as Kennedy wanted.
They cited growing congressional resistance to aid, declaring that
"a profound sense of irritation" was damaging prospects for long-
term cooperation.

The department's Intelligence Bureau warned that a coup
would force Kennedy and the United States to confront a grave
dilemma. An uprising would create difficult choices: If Washington responded with silence or a refusal to take sides, Diem would
conclude that it was a U.S.-sponsored coup and become entirely
unmanageable if he held on to power. If the coup succeeded without American help, the United States would face considerable
hostility from the new administration. In brief, regardless of who
prevailed, a coup in which Washington held its hand could be a
disaster for American policy.

American demands on Diem to settle the Buddhist crisis enraged him. He was angry at being told how to run his affairs and
was "suspicious that we were trying to undermine him." Diem
saw a request for approval of former Massachusetts senator Henry
Cabot Lodge, Jr., as Nolting's replacement as signaling a new
American policy. He saw Nolting as someone he could bend to his
purposes, but Lodge? "They can send ten Lodges," he said, "but I
will not permit myself or my country to be humiliated, not if they
train their artillery on this Palace." The State Department tried to
assure Diem that there was no change in policy: The embassy was
instructed to tell him that the United States was trying to save him
from a disaster and to advise him that Lodge was a conservative

Republican eager to defeat the communists and intent on expanding U.S. cooperation with Saigon.

Kennedy's eagerness to find some way out of Vietnam partly rested on his wish to keep the situation from adding to his difficulties with the Soviets. Above all, after the missile crisis, he thought that Khrushchev might be more receptive to a test ban. He was right. Having gone to the brink of a nuclear conflict, Khrushchev had a heightened sense of urgency about reining in the dangers of an apocalyptic war. On November 12, Khrushchev sent the president a letter saying that if they were to "draw the necessary conclusions from what has happened up till now," they could agree "that conditions are emerging . . . for reaching an agreement on the . . . cessation of all types of nuclear weapons tests." As a prelude to further negotiations, Kennedy asked that a committee of national security experts, including White House science adviser Jerry Wiesner, evaluate the results of the recent U.S. and Soviet atmospheric tests. In December, when the committee told Kennedy that little of importance had been learned from the recent tests and that they wouldn't affect the strategic balance, Kennedy was eager to make a new push for a ban.

On December 19, Khrushchev gave substance to his suggested cessation with a letter saying that the time had come to end all nuclear tests. Because they had not found an acceptable way to monitor underground tests, Khrushchev proposed that they focus on controlling explosions in outer space, the atmosphere, and underwater. He was willing, however, to consider a few on-site inspection stations with two to three inspections a year. Kennedy promptly replied that he hoped they were now starting down the road to an agreement.

Khrushchev's response: not so fast. Where he was willing to consider between two and four on-site inspections, Kennedy insisted that it had to be between eight and ten, a substantial decrease from earlier demands for between twelve and twenty. Nor could they agree on where detection stations should be located. The acrimony over these differences and other issues between them became so sharp that Bobby Kennedy refused to carry a message from Khrushchev to the president that Ambassador Dobrynin handed to him on April 3, 1963. But he couldn't resist giving Kennedy the gist of Khrushchev's twenty-five-page letter: Khrushchev described himself as tired of hearing about Kennedy's problems with the Senate. "The United States is run by capitalists who are interested only in war profits. They were the ones that were dictating policy. If President Kennedy was not as concerned about the Rockefellers and these capitalists then he would take this step for world peace. Further, who did we think we were in the United States trying to dictate to the Soviet Union?"

Although Khrushchev clearly was not yet ready to compromise, Kennedy, bolstered by the conclusion of his science advisers that little would be gained by additional atmospheric tests, was more determined than ever to find a way to overcome the barriers to a ban. With advisers predicting that continual testing would make it cheaper to build nuclear weapons, increase the likelihood of proliferation, and heighten prospects of a nuclear war, Kennedy saw the need for an agreement as irresistible. At news conferences in February and March 1963, Kennedy voiced fears that the absence of a ban would allow irresponsible governments to acquire nuclear weapons and increase chances of a general nuclear conflagration. He told reporters that he was "haunted by the feeling that by 1970, unless we are successful [in banning tests] there may be 10 nuclear powers instead of 4, and by 1975, 15 or 20 . . . I regard that as the greatest possible danger and hazard."

Prime Minister Macmillan, who was determined to achieve a test ban, urged Kennedy to work with him toward some formula that could break the stalemate with Russia. He thought "a personal message to Khrushchev . . . or perhaps some emissary such as Averell or your brother Bobby" might make a difference. Writing from Moscow, Llewellyn Thompson advised that Khrushchev's principal interest in a ban rested on inhibiting Chinese and West German acquisition of nuclear weapons. On April 15, seizing on Thompson's insight, Kennedy and Macmillan sent Khrushchev a joint letter asserting that agreement on nuclear testing might allow them to move rapidly to prevent proliferation of nuclear power.

But Khrushchev was still unprepared to compromise, and the Joint Chiefs also resisted the JFK-Macmillan initiative. Before Kennedy heard back from Khrushchev, Taylor, speaking for the Chiefs, expressed renewed doubts about the likelihood of effective verification arrangements. Ignoring the view of Kennedy's science advisers, he warned that the United States could maintain its nuclear weapons superiority only through continued testing in the atmosphere and underground. A report from the U.S. Embassy in Moscow was equally discouraging: Khrushchev's reaction to the Kennedy-Macmillan letter was "entirely negative." Thompson explained that Khrushchev was committed to additional tests and currently saw discussions as interfering with these plans. In addition, intensifying tensions with China's communists deterred Khrushchev from talks that could become a propaganda barrage against him. In a lengthy letter to Kennedy in May, Khrushchev repeated and underscored familiar arguments against inspections and for more comprehensive arms control arrangements. Khrushchev, however, left the door open to further test discussions, saying he would be happy to receive high-level representatives of the United States and Great Britain in Moscow for talks.

On May 30, Kennedy responded that he was eager to send a notable representative to Moscow during the summer to bridge the gap in their views. But he felt that something more dramatic needed to precede the arrival of any envoy if the stalemate were to be broken. His impetus partly came from a conversation with Norman Cousins, the editor of the *Saturday Review of Literature* and an outspoken peace activist, who had met with Khrushchev in Russia in April and then with the president to report on their conversation. Cousins described Khrushchev as eager for peace but under pressure from militants in the Kremlin. Kennedy sympathized with Khrushchev's predicament. He said that he had "similar problems . . . the hard-liners in the Soviet Union and the United States feed on one another, each using the actions of the other to justify his own position." Cousins urged Kennedy to handle the problem by offering a "breathtaking new approach toward the Russian people, calling for an end to the cold war and a fresh start in Soviet-American relations."

Cousins's urgings meshed with Kennedy's own determination to move beyond stale Cold War tensions, which kept the world on edge and threatened to trigger another crisis, like the one over Cuba, that could end in a world cataclysm. Slated to speak at the American University commencement on June 10, Kennedy decided to deliver a "peace speech" that would announce a unilateral suspension of atmospheric testing, a new push for a test ban, and, more broadly, a plea for "a genuine, lasting peace: . . . Not merely peace for Americans, but peace for all men; not merely peace in our time but peace for all time." It was not to be the sort of elusive Wilsonian dream of "a sudden revolution in human nature," but a peace resting on "a gradual evolution in human institutions." Because "our problems are man made," Kennedy said, "therefore they can be solved by man." But this would mean a dramatic shift in outlook: "enmities between nations . . . do not last forever." The aim must be not so much to eliminate differences but to "make the world safe for diversity." As fellow inhabitants of the same planet, the United States and Russia needed to guard the world against war for the sake of all "our children's future. . . . We are not helpless before that task" of world peace, he declared. "Confident and unafraid, we labor on—not toward a strategy of annihilation but toward a strategy of peace." The speech was one of the great state papers of American history.

In preparing the address, Kennedy had limited the discussion to a handful of White House advisers, including Sorensen, who did the principal drafting, as well as Bundy, Schlesinger, and Tom Sorensen, Ted's brother, an official at the United States Information Agency. McNamara, Rusk, and Taylor were only told about the speech two days before Kennedy delivered it. He did not want predictable quibbling from the principal national security officials, who would surely object to so unconventional and idealized a call for changes in the country's perspective on foreign affairs. His caution was borne out by instant objections from Taylor that the Joint Chiefs could not endorse a unilateral suspension of atmospheric tests. Congressional Republicans, the mainstay of conventional thinking, dismissed the speech as "a soft line that can accomplish nothing . . . and a dreadful mistake." In England, however, it was hailed as an extraordinary landmark pronouncement, and Khrushchev called it "the best speech by any President since Roosevelt."

The missile crisis had rekindled Kennedy's interest not only in avoiding an arms race and a nuclear war but also in muting difficulties over Cuba. The end of the crisis made clear to Kennedy and some of his advisers that Cuba had become a major impediment to more productive foreign policy initiatives.

But Castro, some of Kennedy's most militant counselors, and U.S. domestic politics made Cuban relations a source of continuing contention. Castro was furious at Khrushchev for giving in to Kennedy and promising to remove the missiles in exchange for a verbal non-invasion pledge. As a consequence, he refused to allow U.N. inspectors to enter the island. Castro's obstructionism made Kennedy's advisers distrustful of Khrushchev's commitments. After all, he had lied about having offensive weapons on the island, so why shouldn't they suspect that he was using Castro as a stalking horse for keeping the missiles in Cuba? Rusk, McNamara, Taylor, McCone, and Dillon pressed Kennedy to resume reconnaissance flights. They also pressured him into insisting that Soviet IL-28 bombers, which could carry nuclear bombs, be added to the list of offensive weapons removed from the island.

Kennedy had originally opposed listing the aircraft as among the offensive weapons the Soviets were to eliminate. Compared to the missiles, they seemed like an inconsequential threat. Kennedy's advisers, however, insisted that they now be included—not only because they could carry nuclear bombs from Cuba to targets in the United States, but also because political opponents could use them to attack the administration for being careless about national security. But adding the IL-28s to the missiles threatened to unhinge the settlement with Khrushchev and rekindle the crisis. Kennedy continued to see the bombers as an insignificant danger, but once he appreciated how they could be converted into a political liability, he insisted on their removal. It took numerous exchanges between Moscow and Washington, including several interventions by Bobby Kennedy with Dobrynin at the Soviet Embassy, before a final agreement was reached on November 20: Kennedy would announce an end to the quarantine, and the Soviets would remove the bombers along with the missiles. Kennedy also agreed not to press for on-site inspections, convinced that U.S. surveillance of Soviet ships carrying the missiles and planes from Cuba would suffice to demonstrate whether Khrushchev was keeping his word. His objective, he told Khrushchev on November 21, was to push Cuba to the side, so that they might move on to other issues.

Yet Kennedy could not ignore the possibility, as his military Chiefs were warning, that the Soviets might try again to sneak offensive weapons onto the island. Nor could he overlook newspaper reports that the Soviets might be hiding missiles in caves or that they might try to build a naval base, which could give them, in Kennedy's words, "a near parity with us if we should once again blockade." Khrushchev dismissed the stories about hidden missiles in Cuba by saying, "We do not live in the caveman age to attach great significance to the rumors of this sort." But given how badly Kennedy had been burned by Khrushchev's earlier deceptions, he felt compelled to see the island as an ongoing threat to U.S. security. Specifically, having ignored warnings from McCone before

October that the Soviets were turning Cuba into an offensive base, he decided to take renewed warnings from him more seriously. At the beginning of December, McCone pointed to evidence of increased air defenses, which suggested that the Soviets might be planning to put more missiles in Cuba. McCone also urged attention to comments by Che Guevara, Castro's spokesman, that Cuba intended to continue communist subversion in Latin America and that the Soviets saw the island as a communist sanctuary in the Western Hemisphere.

For many of Kennedy's advisers, overturning Castro's regime remained an appealing prospect. An administration memo of December 3 declared, "Our ultimate objective with respect to Cuba remains the overthrow of the Castro regime and its replacement by one sharing the aims of the Free World. . . . All feasible diplomatic, economic, psychological and other pressures" were to be brought to bear on behalf of this goal. Amazingly, the Joint Chiefs described themselves as ready to use "nuclear weapons for limited war operations in the Cuban area." They rationalized the reliance on such overkill by directing that "collateral damage to nonmilitary facilities and population casualties will be held to a minimum consistent with military necessity."

Surely the Joint Chiefs knew that their assumption about limiting damage was nonsense. A 1962 Air Force pamphlet on "The Effects of Nuclear Weapons" tried to whitewash the consequences of using just a one-megaton nuclear bomb, never mind a sixteen-megaton weapon. The pamphlet acknowledged that radiation exposure was likely to cause hemorrhaging that would produce "anemia and death." The authors also explained that "if death does not take place in the first few days after a large dose of radiation, bacterial invasion of the blood stream usually occurs and the patient dies of infection."

What the pamphlet neglected to say was that a one-megaton blast would kill—"vaporize"—everyone within a six-square-mile area. "Outside this circle," a 1999 report explained, "the light from the explosion will blind people ten miles in every direction." People seeing the explosion fifty miles away "will have a large spot permanently burned into their retinas." And most people hundreds of miles from the blast will die later, suffering an agonizing death. A one-megaton nuclear explosion would "create a firestorm that can cover 100 square miles. It will melt everything in its radius." Dropping a one-megaton nuclear bomb on Cuba would have made the island a living hell.

While Kennedy did not veto the Chiefs' radical plan for a nuclear attack, he had no intention of ever acting on it, especially since he knew that curbing "collateral damage" seemed like a pipe dream—more a way to justify using these ultimate weapons than a realistic assumption. In April 1961, he had already dismissed talk by the Joint Chiefs about using nuclear weapons against communist forces in Southeast Asia as ridiculous. In a meeting with the Joint Chiefs and McNamara on December 5, 1962, Kennedy questioned the value of building so many nuclear bombs: "What good are they?" he asked. "You can't use them as a first weapon yourself. They are only good for deterring . . . I don't see quite why we're building as many as we're building."

But if he saw no place for nuclear bombs in Laos or for toppling Castro, he did not rule out an invasion of Cuba or at least planning for one. As he told the National Security Council on December 10, "domestic public opinion, including congressional opinion," would exert such "great pressure on the Government in the next few months" that we "must not go too far down the line on no-invasion assurances."

On December 27, in a meeting with the Joint Chiefs at his Florida retreat in Palm Beach, Kennedy told them, "We must assume that someday we may have to go into Cuba, and when it happens, we must be prepared to do it as quickly as possible." He wanted them to think about plans for an invasion that might come "one, two, three, or four years ahead." Whatever the reality of his intentions, such a directive would discourage leaks from the Chiefs about his willingness to live with Castro, which could then become a political liability.

Two days later, Kennedy spoke to forty thousand Cuban exiles in Miami's Orange Bowl. He had been urging that Cuban émigrés settle in different parts of the United States. He hoped to limit the concentration of exiles in Florida, where they could exert pressure on a Washington administration by bloc voting in a presidential election; it could give the state to any candidate supporting their demands for an aggressive policy to oust Castro. But with the Cubans congregated in Miami and a presidential election less than two years away, Kennedy felt compelled to appease them with an appearance. Rusk, Bundy, and Kenny O'Donnell had all urged him not to speak, arguing that it would raise doubts in Moscow about the sincerity of his non-invasion pledge. Domestic politics, however, which included keeping his Chiefs convinced that he was serious about getting rid of Castro, took priority.

But more was at work here than domestic politics. Kennedy had a profound sense of guilt toward the Bay of Pigs captives in Castro's prisons. Cardinal Cushing of Boston remembered that when he and Kennedy talked about the prisoners, "It was the first time I ever saw tears in his eyes." He was determined to free the 1,113 Cubans captured at the Bay of Pigs, agreeing, despite the opposition of some in the United States, to swap millions of dollars in medicines for them. Bobby Kennedy was especially committed to winning their freedom: "We put them there and we're going to get them out," he said to an aide. He told his brother that he could not turn his back on the brave men who had done his bidding in the invasion.

Jackie Kennedy, who shared the stage with the president and spoke in Spanish to the crowd, lauded the brigade members for their bravery. She remembered Kennedy as sincerely moved by the plight of the prisoners and their families and felt obliged to free the captives. Jackie recalled the occasion as "one of the most moving things I've ever seen. All those people there, you know, crying and waving, and all the poor brigade sitting around with their bandages and everything." Some of them were obviously malnourished and sickly. Kennedy was carried away by the emotions of the moment, and when some of the brigade returnees presented him with a flag they had carried with them to the beaches at the Bay of Pigs, Kennedy declared in unscripted remarks, "This flag will be returned to this brigade in a free Havana." The crowd responded with shouts of "Guerra, Guerra, Guerra!" It aroused suspicions in Havana and Moscow that Kennedy would not honor his non-invasion pledge.

There were good reasons for suspicions. By January 1963, Kennedy and his advisers saw themselves starting over in their dealings with Cuba. The island had absorbed far more of their attention than they had ever thought likely. But they still felt compelled to devise a plan that could oust Castro and neutralize the island as a source of tension with Moscow. Bundy advised Kennedy that he was setting up a new interdepartmental group with members from State, Defense, the CIA, and Nick Katzenbach from the Justice Department representing Bobby. "The time is ripe for such a re-organization," Bundy told the president, "because we seem to be winding up the negotiations in New York [on the missiles], the prisoners are out, and there is well nigh universal agreement that Mongoose is at a dead end."

But what was to replace Mongoose? Sterling Cottrell, a deputy assistant secretary of state in the Bureau of Inter-American Affairs, was to become a new coordinator with responsibility for covert operations. But the problem of organization, Bundy emphasized, was less important than how to design an effective plan. Bundy favored reducing the CIA's role and developing ways to communicate with moderates in the Cuban government, "including even Fidel himself," who they thought might be at odds with hard-line communists.

Kennedy liked Bundy's plan, and when he discussed Soviet-American relations with Vasily Kuznetsov, Moscow's deputy foreign minister, at the White House on January 9, he urged him to remove Soviet troops and armaments from Cuba; it would further relax tensions over the island and lessen areas of disagreement with the Soviet Union. Kennedy emphasized that the United States had no intention of invading Cuba and explained that his speech to the exiles in Miami signaled nothing more than a hope for a change in Cuban conditions. But he also described speeches by Castro and Che Guevara urging "armed struggle in Latin America" that would take "power from the hands of the Yankee imperialists" as provocative and a source of ongoing friction.

Kennedy wanted to reduce chances for an incident with Castro's government by replacing U-2 surveillance flights over the island with reports from the representatives of friendly countries stationed in Havana and from visitors to Cuba with access to members of the Cuban government. CIA and Pentagon officials were unhappy with the possibility that Kennedy was ready to live with or reach some accommodation with Castro. They favored no letup in the battle to bring down Castro. The Pentagon wished to plan for undisguised, full-blown military support of any anti-Castro uprising. The Chiefs continued to believed that "we had missed the big bus" by not destroying Castro's government during the missile crisis.

Kennedy was torn between proposals for passive acceptance of Castro and plans for renewed efforts to eliminate his regime. On one hand, he was mindful of how close they had come to a nuclear war over Cuba and was determined to avoid another such crisis. Moreover, in January, when one of Castro's trusted advisers told James Donovan, a New York lawyer who was in Havana negotiating the release of Americans in Cuban prisons, that they should discuss reestablishing diplomatic relations and Castro invited Donovan to return in March "to talk at length . . . about the future of Cuba and international relations in general," Kennedy was interested in the possibility of some sort of rapprochement.

On the other hand, Kennedy was wary of the political pressure that Republicans, led by Senator Kenneth Keating of New York, were generating over the failure to bring down Castro's communist government. At an NSC meeting on January 25, Kennedy complained that "Keating was alleging that there is now in Cuba ten times as much [Soviet] equipment as there was." But when McCone publicly acknowledged that Moscow was not "withdrawing their sophisticated equipment from Cuba" and told a Senate committee that "there is about twice as much Soviet equipment in Cuba as there had been prior to the Russian buildup," Kennedy could not ignore demands for renewed efforts to oust the communists from the island. The anti-Castro hawks constantly reminded the press and public that the communists were now "only ninety miles from our shores." Kennedy also worried that Castro continued to promote subversion in the hemisphere and might succeed in creating other communist regimes in the region, which would threaten U.S. interests and become a new point of Republican attack on his administration.

Kennedy told the NSC that "the time will probably come when we will have to act again on Cuba." He was interested in the possibility of using the island as a counter to threatened Soviet control of Berlin. "We should be prepared to move on Cuba if it should be in our national interest," he said. "The planning by the US, by the Military in the direction of our effort should be advanced always keeping Cuba in mind in the coming months and to be ready to move with all possible speed. We can use Cuba to limit Soviet actions just as they have had Berlin to limit our actions."

At an NSC meeting on January 25, the issue of Cuba provoked heated discussion between Llewellyn Thompson and John McCone. Thompson complained that Cuba was eclipsing larger foreign policy goals. An American obsession with Cuba was taking precedence over more important relations with the Soviet Union: specifically, efforts to negotiate a nuclear test ban treaty and a possible chance to take advantage of a developing Sino-Soviet split. Thompson acknowledged that domestic politics, particularly congressional agitation about Cuba, stood in the way of more rational calculation, but he urged the need to educate congressmen about the country's greater interests. Taylor doubted the likelihood of altering Congress's focus. He complained at the next week's NSC meeting that congressional hearings on the defense budget had turned into an investigation on Cuba. "Most of the time Secretary McNamara has spent on the Hill was taken by Cuban questions rather than military budget problems," he told Kennedy.

McCone was pleased to see Congress maintain a steady drumbeat about Castro's dangerous ties to Moscow. During congressional testimony, he foresaw the continued presence of Soviet troops in Cuba as well as sophisticated military equipment. He speculated that Castro wanted Russian troops as an insurance policy against an internal revolt encouraged by U.S. subversion. In White House meetings, Secretary Dillon weighed in with objections to unrealistic hopes of wooing Castro away from the communists, urging instead a renewed commitment to overthrow his government. Bobby Kennedy sided with the hawks pressing the struggle against Castro. He acknowledged that another invasion of Cuba was currently out of reach, but thought that the United States needed to work more effectively with Cuban brigade members by encouraging them to use sabotage to weaken the communists.

Kennedy, trying to square national security concerns with domestic political pressures, sought a middle ground between his advisers' conflicting policy suggestions. Eager to keep Cuba from poisoning Soviet-American relations, he postponed orders that barred all U.S. flagships from carrying goods to Cuba and that closed U.S. ports to ships involved in trade with the island. He also rejected using balloons to drop propaganda leaflets over Cuba and refused to authorize landing groups in Cuba for intelligence gathering unless it was essential. Kennedy also wanted military and civilian officials to mute their complaints about the continuing presence of 4,500 Soviet troops in Cuba. He thought it useful to suggest that the Soviets might be acting as a check on possible reckless actions by Castro. At the same time, he was responsive to political pressure to remove Soviet forces from the island. On February 5, he gave in to congressional pressure to encourage U.S. allies to reduce trade with Havana and agreed to have reconnaissance overflights of Soviet ships leaving Cuba to track their departing troops.

It was clear to Kennedy that he could not ignore a domestic political problem over Cuba, which was partly the result of McCone's press leaks that the situation in Cuba was more ominous than the White House believed. McCone mainly directed his fire at McNamara, who closely reflected the president's wishes. Bundy noted that McCone "was something between concerned and angry" and feared that the conflict could become "the first big, internal, high-level personality clash of this administration. . . . McCone is afraid of the military situation in Cuba while McNamara is not." Bundy was not unmindful of other internal conflicts between Sorensen and O'Donnell, Bobby Kennedy and Chet Bowles, Bobby and Johnson, but Cuba was an explosive issue that would command widespread public attention should the extent of their internal arguments about Cuba become fully known.

Because Rusk was much more cautious about coming down on either side of the argument, McCone felt that he could enlist him as an ally in a campaign to stiffen Kennedy's resolve. In mid-February, he told Rusk, "I am growing increasingly concerned over Soviet intentions in Cuba." He saw fresh signs of Soviet plans to reintroduce "an offensive capability." He believed it "highly dangerous" for the intelligence community to make "categoric" judgments about what they were dealing with unless there were "penetrating and continuing on-site inspection."

By March, Kennedy was finding it impossible to keep the lid on administration infighting over Cuba. He told McCone that "an attempt . . . to drive a division within the Administration, most particularly between CIA on the one hand and State and Defense on the other . . . worried him and he hoped we could avoid any statements on the Hill, publicly or to the Press, which could exacerbate the situation." McCone saw "no reason for all the furor" but described the problem in the CIA as resulting from a concern with "an inhibiting policy" limiting overflights of Cuba. Kennedy acknowledged that he had been "one of those who did not think the Soviets would put missiles in Cuba," but, without conceding any compelling need for a more aggressive program of U-2 missions, he urged McCone to "minimize" their "internal problem" and "not permit it to get into an interdepartmental row."

At an NSC meeting on March 13, Kennedy was principally concerned with congressional pressure to drive out Castro. McCone reported that he had fended off such demands for anti-Castro measures by explaining that once Soviet troops had left the island, they could look forward to a military coup serving U.S. aims. Kennedy wanted to defend the White House from congressional demands for action if all the Soviet troops didn't leave Cuba. He said that "we should protect ourselves as best we can" by emphasizing efforts to isolate Cuba, mainly through trade and shipping policies and pressure on Latin American governments to prevent students and subversives from going to Cuba.

Bobby Kennedy urged his brother to go beyond these actions. He wanted him to think of ways that the United States could facilitate the military uprising McCone had predicted. If there was "evidence of any break amongst the top Cuban leaders and if so, is the CIA or USIA attempting to cultivate that feeling? I would not like it said a year from now that we could have had this internal breakup in Cuba but we just did not set the stage for it." When Kennedy failed to offer any reply, Bobby asked: "Did you feel there was any merit to my last memo?" He added, "In any case, is there anything further on this matter?" Again, Kennedy didn't answer: He wanted no additional aggressive action that could trigger a new crisis with the Soviets.

But he had limited control over Cuban problems. The crosscurrents were a constant source of irritation that exasperated him. He found himself caught between the Cuban exiles supported by

McCone and the CIA, and White House advisers sympathetic to Soviet complaints that Cuba kept getting in the way of reducing tensions with the United States. On March 18, Cuban exiles describing themselves as Alpha 66 attacked Soviet ships and installations in Cuba. Although the State Department, with Kennedy's approval, condemned this hit-and-run raid as doing more to strengthen than weaken Castro's government, Castro and the Soviets blamed Washington for facilitating it.

The raids opened a new round of arguments about Cuba. Kennedy complained to the NSC that despite State Department condemnations, Havana and Moscow refused to believe that the exiles weren't supplied by Washington, even if they weren't launching the raids from U.S. bases. Kennedy wanted to cut off their supplies or at least advise them not to attack any of the Russians on the island. He thought the raids would then "draw less press attention and arouse less acrimony in Moscow."

McCone defended the exiles. It "would be extremely difficult to control them because they are brave men fighting for the freedom of their country," he told Kennedy at an NSC meeting. Besides, "the raids would cause trouble inside Cuba and would discredit Castro in Latin America if he were unable to prevent them." The Soviets then might have second thoughts about backing Castro. Should the United States try to stop them, it would provoke widespread press criticism as well as congressional objections. He suggested officially condemning the raids while not barring the raiders from using the United States as a base. Rusk saw nothing but trouble in McCone's recommendations. But characteristically walking a middle ground, he suggested that the administration plan the raids if they were worthwhile. Dillon, who had consistently sided with advocates of militant action toward Cuba, wanted simply to accept the raids as impossible to control.

Kennedy was unconvinced by McCone and Dillon. He said, "These in-and-out raids were probably exciting and rather pleasant for those who engage in them. They were in danger for less than an hour. This exciting activity was more fun than living in the hills of Escambray, pursued by Castro's military forces." He then read a dispatch describing the Soviet protest, underscoring the damage the raids were doing to relations with Moscow. At the same time, he didn't want the public to think that the United States was "prosecuting Cuban patriots." And he hoped to blunt domestic criticism by letting the exiles strike at Cuban but not Soviet targets. In short, he was eager to make this issue disappear from Soviet-American relations as fast and fully as possible, but without provoking a public outcry against appeasing communists.

Under pressure from the CIA and the Florida exile community, the administration continued covert operations against Castro and did little to halt raids striking Soviet as well as Cuban targets. Kennedy acknowledged that sabotage aimed at disrupting the Cuban economy and promoting Soviet-Cuban tensions and exile attacks would not bring down Castro. Nevertheless, he felt compelled, if simply to answer domestic pressures and boost the morale of the exiles, to maintain some kind of "noise level" in anti-Castro operations.

As a consequence, Cuba remained a serious irritant in relations with Moscow. On April 3, Soviet ambassador Dobrynin visited Bobby Kennedy at the Justice Department, where he presented a twenty-five-page paper describing Khrushchev's anger toward the president for allowing Cuba to remain an obstacle to improved Soviet-American relations. Khrushchev complained that the United States was dismissive of Soviet concerns, that McNamara was making "warlike pronouncements," and that "the United States was only interested in . . . building up their efforts to dominate the world through counter-revolutionary activity. . . . We should understand that we could not push the Soviet Union around." For Khrushchev, the raids and sabotage were like rubbing salt in open wounds. The retreat in Cuba had humiliated him: The Chinese snidely described his response to the quarantine as "capitulationism" and reported Kennedy as saying, "I cut his balls off." Press stories saying that Castro described Khrushchev as a man with "*No cojones* [balls]" and a "*Maricón* [homosexual]" enraged the Soviet leader.

Khrushchev wasn't the only one who was angry. The message and Dobrynin's verbal presentation incensed Bobby Kennedy. He considered them "so insulting and so rude to the President of the United States" that he refused to accept the message or transmit it. He said that he had never "insulted or offended" Dobrynin or "his country or Mr. Khrushchev" and considered such a message distinctly unhelpful in advancing their relations. As a principal architect of the anti-Castro campaign, Bobby took the Soviet rebuke personally and refused to give any ground, despite his brother's desire to keep Cuba from impeding an improvement in Soviet-American relations. It seems likely, however, that Kennedy endorsed Bobby's hard line, believing that he could not allow impressions either in Moscow or the United States that Khrushchev could address him in abusive language.

Had Kennedy's advisers, including his brother, been of one mind about forcing Cuba off the front pages by ending the raids and subversion, Kennedy might have dropped the anti-Castro campaign. True, the exiles would have howled at his betrayal of their fight against communism, and congressional Republicans would have echoed their complaints. With approval ratings of between 66 and 76 percent, however, he was in a strong position to resist their pressure. But with a divided administration, including a CIA and Joint Chiefs of Staff warning not to be fooled again by Khrushchev, and a need for a cooperative Senate should he win agreement to a nuclear test ban treaty under negotiation, he found it impossible to shelve plans for a change of regimes in Cuba.

In response to all the cross pressures, Kennedy moved in two directions at the same time. On April 9, he rejected CIA and U.S. Army proposals for radio broadcasts spreading anti-Castro propaganda in Cuba, except for some inciting Cubans to strike out against Soviet personnel in Cuba, and those messages only if they could not be traced back to the United States. He also approved the training of exiles for sabotage in Cuba and for advance attacks within the island should they ever decide to invade. Simultaneously, Kennedy expressed great interest in Castro's statement to James Donovan, who was discussing the release of U.S. citizens imprisoned in Cuba, "that relations with the United States are necessary" and he "wanted these developed." There is little doubt that Kennedy wanted a resolution of the distracting troubles with Cuba. As he wrote Khrushchev on April 11, "I have neither the intention nor the desire to invade Cuba. . . . The pressures from those who have a less patient and peaceful outlook are very great—but I assure you of my own determination to work at all times to strengthen world peace." As he had told Bundy after the missile crisis, indulging "his taste for understatement, 'nobody wants to go through what we went through in Cuba very often.'" Bundy, who was reinforcing the president's wish to find some way out of the Cuban dangers, added, "We must make it our business not to pass this way again."

On the same day Kennedy wrote Khrushchev, the committee charged with planning Castro's fall reviewed "Black Operations" and "Sabotage Targets" that would meet the president's demand for "some action" and "a program which will show continuous motion," but without clear evidence of direct U.S. involvement. On April 15, when McCone met with Kennedy in Palm Beach, he described two possible ways to resolve the Cuban problem: Woo Castro away from the Russians and get him to follow a more benign path in Cuba, or use every possible pressure to get the Soviets out of the island, followed by means to topple Castro. Kennedy "suggested the possibility of pursuing both courses at the same time."

In the coming months, Kennedy's advisers were distinctly unhelpful in suggesting ways to make Cuba a less compelling concern in the reach for some sort of accommodation with Moscow. The U.S. National Archives are awash in boxes of files between 1919 and 1941 about war debts. An argument over how to collect the money owed to the United States and whether to tie it to the reparations payments dictated by the Versailles Treaty proved to be a debate devoid of a solution. Except for Finland, none of the debtor countries ever repaid the loans and the issue disappeared with the onset of World War II.

The discussion of how to deal with Castro proved to be every bit as sterile. "The elimination of Castro was a requirement," McNamara told the NSC on April 23, as it considered "A Sketch of the Cuba Alternatives." But he did not see how "our present policy" would bring the desired result. We needed to "create such a situation of dissidence within Cuba as to allow the U.S. to use force in support of anti-Castro forces without leading to retaliation by the USSR." McNamara's convoluted language reflected the emptiness of the plan, if one can call it that. "Are we keeping our contingency plans up to date?" Kennedy asked McNamara at the end of the month. McNamara assured him that they were. But the "small-scale sabotage" they were doing was producing "no real change in the situation." In fact, Castro was gaining ground in Latin America. Since the communists could not stand being made fun of, a State Department adviser said, maybe they could "destroy Castro's halo" with ridicule. Desperate for a solution to the Castro problem, the planners reverted to wondering whether "an attack on a United States reconnaissance aircraft could be exploited" to end Castro's regime. It was a variation on the stale get-them-to-attack-Guantánamo idea, which McNamara raised anew with the standing committee on Cuba at the end of May.

The advisers were as much at sea as ever. "All of the courses of action discussed were singularly unpromising," the NSC agreed at a May 28 meeting. Bundy saw no date certain when "we could overthrow Castro" and acknowledged that none of their planning seemed very promising. But McCone refused to confess failure. It suggested that his CIA was incapable of meeting the challenge to national security supposedly posed by Castro. His prescription: Increase economic hardship in Cuba, continue sabotage, and maybe the Cuban military would overthrow the communists and restore relations with the United States. And Bobby Kennedy, as out of ideas as the rest of the advisers, said that "the U.S. must do something against Castro, even though we do not believe our actions would bring him down."

And so the CIA just continued to hammer away at the same proposals—sabotage and propaganda—that had yielded no constructive results before. At a June 19 White House meeting about covert policy toward Cuba, McCone told Kennedy of "the importance and necessity for continuous operations." Although they "would create quite a high noise level" and although "no single event would be conclusive," they should not be abandoned.

No one at the CIA, above all McCone, thought to ask: Couldn't we live with Castro? Was his regime really more than a "thorn in the flesh," as Fulbright had described it? When Castro spent a month in Moscow in May, McCone believed it "was inspired by a Russian desire to forestall any effort by the U.S. to negotiate with Castro." After Castro returned from Russia, McCone dismissed "conciliatory" statements by him toward the United States as meaning that the Cuban had tied himself more closely to Russia than ever, and that "any reconciliation with the U.S. would have to be on Castro's and the Soviet Union's terms," which would be entirely unacceptable. Any such agreement would enhance Castro's standing in Latin America and fuel his ambitions to spread communism in the hemisphere. Moreover, McCone warned that domestic politics made this "very dangerous" for the administration. The American people would oppose a rapprochement unless Castro disavowed subversion, broke with Moscow, and opened his country to U.S. inspection. In making his views clear to the White House, McCone was putting Kennedy on notice that the president could face a political firestorm from McCone if he gave up trying to end Castro's rule. However much Kennedy disliked McCone's message and his boorish insistence on reminding him of the dangers to the White House from not following his advice, Kennedy believed that he could not risk a dispute with his CIA chief over communism in the hemisphere.

McCone especially worried that ABC reporter Lisa Howard, who had recently had several interviews with Castro, might be facilitating discussions aimed at an accommodation with Havana. He told Bobby Kennedy that he could not "overemphasize the importance of secrecy in this matter," predicting that "gossip and inevitable leaks . . . would be damaging." He warned against "active steps . . . on the rapprochement matter," urging that if the issue surfaced, it should be described as "a remote possibility" alongside "various levels of dynamic and positive action."

Others in the administration were more sanguine about the prospects for some sort of reconciliation that could tone down the agitation over Cuba and remove it from the list of problems dogging Soviet-American relations. Averell Harriman "flatly disagreed" with McCone's reading of Castro's visit to the Soviet Union. Khrushchev had used it "to prove the success of Russian policy toward Cuba and to refute Chinese accusations that Khrushchev 'softness' toward the U.S. had produced no returns." Harriman also doubted that the Soviets would try to make Cuba a showcase for communism in the Western Hemisphere and that even if they wanted to, it would not work. Rusk also doubted the CIA's hardline approach to Cuba and saw "some opportunity of a rapprochement." Bundy was skeptical but did not think "we should say now that we would never talk to Castro."

Kennedy continued to find himself betwixt and between on how to deal with Castro and Cuba. Less than eighteen months away from his reelection, he wasn't going to take the political risks involved in an open effort at reconciling with Castro. Nor would he risk overt attacks on his regime that rekindled a crisis with the Soviet Union.

o o o o

"The Two of You Did Visit the Same Country, Didn't You?"

In the second half of 1963, Kennedy remained convinced that his paths to reelection and global peace passed through a quiet time on civil rights, an arms control agreement with Khrushchev, a resolution of difficulties with Castro, and some kind of settlement in Vietnam. It was an intimidating agenda. But remembering what he had told Bobby about countering the burdens of political life with humor, he enjoyed poking fun at critics accusing him of gambling with U.S. security. In a phone conversation on June 4, 1963, with his old Senate colleague and friend George Smathers, Kennedy told him that he was writing to tell a foreign policy hawk that his worries about an assault on the United States by the United Nations were overdrawn. While he did not think that the United States "would be attacked by the United Nations, Iceland, Chad or the Samoan Islands," he was ready to do his duty as commander in chief. The following month, when the seventy-nine-year-old Harry Truman called to offer support for a test ban treaty, Kennedy told him, "You sound in good shape." To Kennedy's great amusement, Truman replied: "The only trouble with me . . . is keeping the wife satisfied." "Well, that's all right," Kennedy assured him through his laughter.

Kennedy's first order of business as the summer began was building on his peace speech and Khrushchev's apparent receptivity to test ban talks. But could it be converted into a treaty fulfilling some of Kennedy's goals and then pass a cautious Senate?

Kennedy sent a stellar delegation to Moscow headed by Harriman. As Schlesinger said, "from the viewpoint not only of ability and qualification but of persuading the Russians we meant business, he was the ideal choice." Moreover, his work on Vietnam as an assistant secretary, offering more realistic assessments than the Joint Chiefs or embassy officials in Saigon, had put him in Kennedy's good graces. Because both sides were prepared for a quick resolution, the negotiations lasted only an amazingly brief ten days in July. After all the years of discussions and acrimony about reining in armaments, it was an astonishingly quick agreement. Even before Harriman had arrived in Moscow, Khrushchev had declared his interest in a limited test ban treaty that would require no on-site inspections and would forbid explosions in the atmosphere, outer space, and underwater. Although Harriman would have to maneuver his way around Soviet pressure for a nonaggression pact between NATO and the Warsaw Pact countries, he was able to convince Khrushchev and Gromyko that a pact could be the subject of separate future discussions.

The more difficult battle was to assure Senate approval of the treaty. Remembering Woodrow Wilson's blunder in excluding leading senators from the Versailles negotiations, Kennedy sent Rusk with a bipartisan group of senators to Moscow to participate in a signing ceremony. Khrushchev, mindful of the Senate's importance in completing the agreement, fêted the delegation at a signing ceremony in the ornate Catherine Hall of the Great Kremlin Palace, followed by "a gala luncheon featuring brandy, speeches, and a Soviet orchestra playing Gershwin's 'Love Walked In.'" The delegation then joined Khrushchev at his summer retreat for games and swimming, where Khrushchev drubbed Rusk at badminton.

Kennedy's principal problem was to mute objections from the Joint Chiefs. In June 1963, the Chiefs had advised the White House that "every limited test ban proposal they had reviewed . . . contained shortcomings 'of major military significance.'" They believed that a test ban would erode America's strategic superiority. In testimony before the Senate Armed Services Committee, Admiral Anderson stated that the Joint Chiefs opposed a comprehensive test ban agreement. LeMay followed up with the warning that without testing the United States could lose its military superiority. To achieve some kind of broad consensus, McNamara organized a series of interagency consultations, but the result was not harmony but a divide in which White House science advisers and the CIA disputed the Chiefs' view that a test ban could change the basic U.S.-Soviet balance.

In July, as Harriman prepared to leave for Moscow, the Chiefs declared a limited test ban "at odds with the national interest." Kennedy saw them as his largest domestic impediment to an agreement: "I regard the Chiefs as key to this thing," he told Senate Majority Leader Mike Mansfield. "If we don't get the Chiefs just right, we can . . . get blown." They have "always been our problem." To quiet their objections to Harriman's mission, Kennedy promised that they would have the chance to speak their minds in Senate hearings if there were a treaty. The restive Chiefs then drafted a statement, which they did not send to the White House, saying a limited test ban was militarily disadvantageous. To further blunt their opposition, Kennedy directly told them that they "should base their position on the broadest considerations, not just military factors." At the same time, he refused to include any military officers in the Moscow delegation and instructed that none of the cables reporting developments at the conference go to the Defense Department. "The first thing I'm going to tell my successor," Kennedy said in private, "is to watch the generals and avoid feeling that just because they are military men their opinion on military matters is worth a damn."

Bringing the Chiefs to support ratification or at least refrain from registering strong objections to the treaty required the inclusion of clauses that permitted the United States to resume testing if essential to the national safety, and intensive lobbying by the White House. One can only imagine the pressure Kennedy brought to bear on the unhappy Chiefs to reverse themselves. Republicans asked if the Chiefs had been brought under "rack and screw" to follow Kennedy's lead. They diplomatically denied suggestions that they were cajoled into changing their minds. (An eighteen-minute tape from July 19, 1963, of JFK and LeMay talking that might show otherwise remains closed.) In testimony before the Senate Foreign Relations Committee, however, LeMay could not resist planting doubts: He explained that Kennedy and McNamara had promised to maintain a test program in case changed circumstances required testing. "We have not, however, discussed with them what they mean by that—whether what we consider an adequate safeguard program coincides with their idea on the subject," LeMay said.

Getting the treaty approved by the Senate proved to be less problematic than the White House had feared. President Eisenhower's support of a test ban agreement, Kennedy's success in the Cuban Missile Crisis, which had greatly boosted public confidence in his handling of foreign affairs, public statements by him and President Truman describing the treaty as reducing the likelihood of a nuclear war, and favorable media coverage made clear to Senate Republicans and some conservative Democrats that opposition to treaty ratification would be politically unwise. In September, as the Senate deliberated ratification, 81 percent of the public approved of the treaty. "I can't believe this fellow would be so stupid as to vote against this treaty," Fulbright told Kennedy about one opponent. On September 24, the Senate ratified it by a vote of 80 to 19.

Although the agreement would not mark the end of the Cold War arms race or fulfill Kennedy's hopes of nonproliferation, it was an initial step in reducing East-West tensions and encouraging hopes that the world might avoid the catastrophe of a nuclear war.

As he was winning his fight for a test ban, Kennedy continued to struggle with questions about Castro and Cuba. He remained leery of the domestic political risks in reconciling with Castro. He also saw overt attacks on his regime that rekindled a crisis with Moscow as unacceptable. Secret raids, however, were another matter. In June and July, when the NSC proposed renewed sabotage, Kennedy agreed, but only if the administration could "flatly deny" any involvement. He also ordered a declaration notifying the Russians and Latin Americans that the United States would not allow another Castro in the hemisphere. It was excellent politics: The Russians could not object to an admonition against renewed tensions with Washington over communism in America's backyard, and it gave Kennedy cover against allegations of letting the communists expand their foothold in the region.

Still, no matter how much raids and sabotage might appear to be the work of exiles operating without U.S. support, few would believe that Washington was not involved. Exile incursions aroused suspicions, particularly in Havana and Moscow, that Kennedy remained intent on bringing down Castro, which, of course, was true. On September 10, Dobrynin told Kennedy that U.S. denials of complicity in saboteur landings in Cuba and attacks on the island's industrial facilities were simply not credible. "If such attacks continued," Dobrynin said, "this could only lead to a new crisis," which Khrushchev and the president clearly did not want. Cuba should not jeopardize the interests of both countries and world peace. Without acknowledging a U.S. part in the assaults, Kennedy reminded Dobrynin of "how deeply the Cuban problem was felt in the United States," indicating that the island remained a major domestic political threat to his administration. He was asking Khrushchev to understand that he could not see a good way to quiet the anti-Castro Cubans in Florida without agreeing to some of their operations and neutralizing their ability to give political opponents an issue that appealed to considerable numbers of Americans.

Despite Kennedy's plea to Dobrynin for greater understanding of the domestic pressures compelling his aggressiveness toward Cuba, Dobrynin's warnings about the dangers to U.S.-Soviet relations were not lost on him. He sent word through Bundy to the NSC of his interest in ending all exile raids on Cuba. It brought immediate pushback from Gordon Chase, the NSC's Cuba expert. Chase did not dispute the danger to Soviet-American relations from U.S.-based raids. But attacks from other points in the Americas might be hurting Castro. In addition, blocking external raids would agitate exile protests, provoke loud objections from anti-Castro activists everywhere, and give Castro something to crow about: "The U.S. has capitulated."

Kennedy, supported by Bundy and McNamara, was not convinced that the raids were worthwhile—except as a shield against a political uproar in the United States. He was ready to hear something new on how to mute his Castro problem. And so when William Attwood, the ambassador to the West African nation of Guinea since 1961 and current assistant to Stevenson at the U.N., suggested a different approach to Cuba, Kennedy was ready to listen. The forty-three-year-old Attwood, who had been editor of *Look* magazine and a speechwriter for Stevenson in both his presidential campaigns, had interviewed Castro in 1959 as he took power in Cuba. Convalescing in New York in 1963 from polio, which he had contracted while in Guinea, Attwood had persuaded the State Department to let him temporarily work at the U.N. In a report to the department, Attwood related that the Guinean ambassador to Cuba, who was in New York for the opening of the U.N.'s 1963–64 session, told him that Castro was "unhappy about his present dependence on the Soviet bloc; that he does not enjoy being a satellite; that the trade embargo is hurting him—

though not enough to endanger his position; and that he would like to establish some official contact with the U.S. and go to some length to obtain normalization of relations with us—even though this would not be welcomed by most of his hard-core Communist entourage, such as Che Guevara."

Attwood could not vouch for the accuracy of the ambassador's assessment of Castro's interest in improving relations, though he had similar reports from "neutral diplomats and others I have talked to at the U.N. and in Guinea." Consequently, he thought a rapprochement with Castro was worth looking into as a "course of action, which, if successful, could remove the Cuban issue from the 1964 campaign." He did not suggest "offering Castro a 'deal'— which could be more dangerous politically than doing nothing." But it might provide a means "of neutralizing Cuba on our terms," which he saw as getting Soviet troops out of Cuba, an end to Cuban subversion in other Western Hemisphere countries, and a Cuban pledge of neutrality in the Cold War. Because existing policy had almost no chance of toppling Castro, Attwood thought that exploring the possibility of a rapprochement made good sense. He asked permission to speak to Carlos Lechuga, the Cuban ambassador to the U.N., and to arrange a meeting with Castro, which would be hidden from the press.

The CIA immediately countered Attwood's suggestion with renewed recommendations for undermining Castro's government. In fact, they saw evidence that Castro was losing his hold on the Cuban people. It was time not for conciliatory actions, but for stepped-up pressure on Havana through a wider trade embargo and more sabotage. To appease administration hawks, Kennedy pressed the British to restrict shipping to the island. He told the British ambassador that their ships led free-world trade with Cuba and it was embarrassing his administration.

At the same time, Kennedy saw the wisdom of resolving the Cuban problem by diplomacy. He was more than skeptical of CIA recommendations on Cuba. During a meeting with Kennedy at the

White House on October 10, Foreign Minister Gromyko complained that the United States was subjecting Cuba to constant pressure and provocation. Kennedy conceded that there was no "benefit to the US from harassment. This would not unseat Castro and serve no useful purpose." To mute worries about Cuba in the United States and pressure on him to topple Castro, Kennedy asked that Khrushchev publicly announce the departure of all Soviet military personnel from the island. Remembering that the Joint Chiefs had emphasized how important it was to get Soviet troops out of Cuba—they described the Western Hemisphere as having been invaded, as long as Russian forces were in Cuba—Kennedy saw their departure as certain to ease the pressure on him to help exiles drive the communists from the island.

Ending agitation about Cuba by reaching a modus vivendi with Castro also appealed to Kennedy and several of his top advisers—Bobby Kennedy, a convert to his brother's softer line, Bundy, McNamara, and Stevenson. After Kennedy had given Stevenson approval to pursue Attwood's initiative, Bundy instructed Chase, his deputy at the NSC, to speak to Attwood about the results of a conversation with Lechuga. Attwood reported that Lechuga suggested an envoy travel to Cuba for a conversation with Castro, but Attwood, mindful of the political embarrassment to the White House from any leak of an administration approach to Havana, urged Lechuga to hold secret conversations in New York at the U.N. headquarters. Lechuga was not hopeful that Castro would let anyone speak for him outside of Cuba or that anything would come of conversations: He saw Castro as "too well boxed in by such hardliners as Guevara to be able to maneuver much."

Still, the president, Bobby Kennedy, Bundy, and Stevenson thought the reach for an accommodation worth pursuing. Bundy told Attwood that the president favored "pushing towards an opening toward Cuba" that could remove Castro from "the Soviet fold and perhaps wiping out the Bay of Pigs and maybe getting back to normal." When Castro agreed to see Attwood in Havana, Kennedy, according to Bobby, "gave the go ahead." Attwood planned to go in December or January. He was to insist on the exit of the Russian military in Cuba, a "cutoff of ties with the Communists by Cuba, and the end of the exportation of revolution." It was conceivable that Castro could expel the Soviet military and even call a halt to subversion in the hemisphere. But ending his ties with the communists seemed highly unlikely, even if it was conceivable that he could mute his connections to the Soviet bloc.

At the same time that Attwood prepared to go to Cuba, Kennedy agreed to see Jean Daniel, a French journalist at *L'Express*, who was traveling to Havana to interview Castro. Kennedy waved aside a discussion of Indochina in order to focus on Cuba. He began by acknowledging that U.S. policy had contributed to Cuba's "humiliation and exploitation" under Castro's predecessor, Fulgencio Batista. "I believe that we created, built and manufactured the Castro movement out of whole cloth and without realizing it," Kennedy said. "Batista was the incarnation of a number of sins on the part of the United States. Now we shall have to pay for those sins." He described himself as sympathetic to "the first Cuban revolutionaries." But Castro had betrayed the revolution by becoming "a Soviet agent in Latin America." Worse, he had almost caused a nuclear war between the United States and the Soviet Union. Kennedy ended the interview by telling Daniel: "Come and see me on your return from Cuba. Castro's reactions interest me."

On October 31, Attwood received word that Castro "would very much like to talk to the U.S. official anytime and appreciated the importance of discretion to all concerned." He was "willing to send a plane to Mexico to pick up the official and fly him to a private airport . . . where Castro would talk to him alone. . . . In this way there would be no risk of identification at Havana airport." It would also allow him to hide the conversation from anyone in his government opposed to a rapprochement with the United States. Concerned that an intermediary might misrepresent his views or

that he might leak the conversation to opponents of even prelim-
inary talks with a Washington representative, Castro "wanted to
do the talking himself." However, he would not rule out delegating
the responsibility to someone else "if there was no other way of
engaging a dialogue."

Like Castro, Kennedy found that some in his inner circle also
opposed any kind of reconciliation. The State Department's Office of American Republic Affairs insisted that renewed relations
with Cuba depended first on the island's separation from the Sino-
Soviet bloc, a repudiation of any affiliation with communism, and
the restoration of expropriated properties and free enterprise.
Predictably, McCone joined the State Department's hard line by
insisting at a November 12 meeting that the sabotage programs
and economic sanctions were weakening Castro. Tired of CIA
pressure about Cuba that led to no change, Kennedy pointedly
asked whether the sabotage program "was worthwhile and whether
it would accomplish our purpose." Rusk chimed in by discounting
the value of "hit-and-run sabotage tactics," though he "said we
must replace Castro" and favored the "infiltration of black teams"
and "internal sabotage" as well as economic sanctions. Rusk's on-
the-one-hand, on-the-other-hand waffling here underscored Kennedy's irritation with him. By the end of the meeting, the group,
which included CIA, national security, and State Department officials, agreed to continue "CIA sabotage operations." Fear of leaks
that could make him look weak on Cuba dictated that Kennedy
not squelch CIA ops.

It did not, however, deter Kennedy from encouraging additional
discussions with the Cubans about better relations. Kennedy sent
word to Attwood through Bundy that he did not see it as "prac-
ticable . . . at this stage to send an American official to Cuba." He
preferred that a Cuban official meet Attwood at the U.N. to learn
if "there was any prospect of important modification in those parts
of Castro's policy which are flatly unacceptable to us: namely, . . .
submission to external Communist influence, and a determined
campaign of subversion directed at the rest of the Hemisphere."

Bundy emphasized that a reversal of these policies might not be enough to alter the current estrangement, but without a dramatic shift in policy, a visit to Cuba by a U.S. representative would be pointless. Ever mindful of how a proposal for altering relations could be a political time bomb in the United States, Kennedy instructed Bundy to make clear "that we were not supplicants in this matter and that the initiative for exploratory conversations was coming from the Cubans."

Castro was not prepared to concede anything. Agreeing to Washington's demands would have been a repudiation of his own administration. Nonetheless, he remained eager to explore ways to end hostile relations, which were causing economic hardship on the island and threatening to destroy his government, or at least his hold on power. On November 18, Castro advised Attwood through an intermediary that the invitation to visit Cuba remained open and reiterated that it could be secretly arranged. Attwood diplomatically declined, explaining that an initial discussion in New York was essential to make clear that productive conversations would follow. The Cubans agreed to set an agenda for a meeting between Attwood and Lechuga in preparation for a later meeting with Castro.

The same day Castro responded to Attwood, Kennedy spoke in Miami about Cuba before the Inter-American Press Association. Kennedy's speech, which was written by Richard Goodwin in consultations with Bundy and Schlesinger, aimed to advance talks with Castro. Kennedy said that Latin America's problems "would not be solved simply by complaining about Castro, by blaming all problems on communism." But Castro and communism had alienated Cuba from the United States and other hemisphere countries. "It is the fact that a small band of conspirators has stripped the Cuban people of their freedom and handed over the independence and sovereignty of the Cuban nation to forces beyond the hemisphere. They have made Cuba . . . a weapon in an effort dictated by external powers to subvert the other American Republics. This, and this alone, divide us. As long as this is true, nothing is possible. Without it, everything is possible. Once this barrier is removed, we will be ready and anxious to work with the Cuban people." The message was clear enough: If the Cuban government distanced itself from Moscow and abandoned subversive efforts to promote communism in Latin America, a new day could rise in relations with the United States.

During three weeks in Havana, Jean Daniel had tried unsuccessfully to meet with Castro. On the night of November 19, the day after Kennedy's speech, Castro showed up unannounced at Daniel's hotel, where they spoke until four in the morning. Daniel reported his conversation with Kennedy. Castro asked Daniel to repeat three times what Kennedy had said about Batista and how Castro had almost caused a nuclear war. He acknowledged that Kennedy was someone with whom he could talk, calling him an "intimate enemy." At the same time, he denied any concern about the United States and wondered why Kennedy couldn't accept him as the United States had accepted Tito, Yugoslavia's communist leader, who had separated himself from Russia. Signaling his interest in a possible rapprochement with Washington, Castro said he had some hope that Kennedy might become "the greatest President of the United States" by being "the leader who may at last understand that there can be coexistence between capitalists and socialists." He ended the conversation by telling Daniel, "You are going to see Kennedy again, be an emissary of peace. . . . There are positive elements in what you report." Neither Kennedy nor Castro knew if they could find enough common ground to reduce tensions to the point where they could reestablish formal Cuban-American relations. But the recent exchanges suggested that things were certainly moving in a new direction.

During the summer and fall, as the test ban treaty came to fruition and exchanges with Castro raised hopes of better relations, Vietnam had become a greater administration problem.

Pressure on Diem to settle the Buddhist crisis and reassurances that the United States had only the best of intentions toward him were ineffective. On July 1, at a State Department meeting, George Ball, Harriman, Hilsman, and Forrestal concluded that a further outbreak of government tensions with the Buddhists, including another immolation of a bonze, was imminent. Nolting, who was in Washington, needed to return to Saigon at once and Lodge's tenure needed to be moved up from September to early August. They discussed urging Diem to separate himself from the Nhus, who were identified as the leaders in the campaign against the Buddhists. Should new rioting occur, the administration would need to make a strong public protest despite danger that it might trigger a coup.

Kennedy was in Rome on a triumphal four-nation trip. Having put his civil rights bill before Congress on June 10 and won an outpouring of approval from European audiences, especially in Berlin, where a crowd of perhaps one hundred thousand had roared their approval of his sympathetic pronouncement, "Ich bin ein Berliner," Kennedy was at the height of his presidency. In his absence, the State Department, convinced it was serving Kennedy's goal of a timely exit from Vietnam, advised Kennedy of its warning to Diem that further incidents involving the Buddhists would compel the United States to take its distance from him.

On July 4, after his return from Europe, Kennedy met with Ball, Harriman, Bundy, Hilsman, and Forrestal to discuss the ongoing crisis in Vietnam. They advised him that the Nhus were continuing to provoke tensions with the Buddhists and that fresh demonstrations aimed at overturning Diem were likely. The group discussed getting rid of the Nhus but didn't think it would be possible. They agreed that coup attempts were likely over the next four months, but couldn't say if any of them would succeed. If one did, they hoped it wouldn't produce chaos. Forrestal reported Marine General Victor Krulak as saying that in any case, the South Vietnamese would continue to fight the communists. Kennedy focused on when Lodge would be able to replace Nolting in Saigon. He was eager to get him there as soon as possible. His unspoken concern was that Lodge should be in his post if and when a coup occurred. Lodge could give Kennedy political cover should congressional Republicans complain that the White House had facilitated, if not orchestrated, Diem's demise and a possible collapse of the war effort.

What was Kennedy supposed to believe? The disarray in Vietnam was increasing and destabilizing the government. The journalists in Saigon were describing an untidy war and, according to the embassy, were "saying quite openly to anyone who will listen that they would like to see regime overthrown." They manifested "intense hatred of all things GVN." The government also saw them as "actively encouraging the Buddhists." A "swift unprovoked and violent attack by government plain clothes police" on journalists covering an otherwise peaceful Buddhist religious ceremony did not surprise the embassy, though it provoked a telegram of complaint from the correspondents to President Kennedy.

At the same time, Krulak, who had spent a week traveling in Vietnam during the last week of June to update the Chiefs, remained optimistic about Diem and chances of winning the war. He had mainly visited the provinces, where he took Vietnamese military officers at their word and reported that "the counterinsurgency campaign is moving forward on the military and economic fronts. There is reason for optimism in both of these areas." The Strategic Hamlets program was working well. Offensive operations had thrown the Viet Cong on the defensive. U.S. advisers were working effectively with their Vietnamese counterparts. While the Buddhist difficulties were serious, they had not adversely affected military operations. Nolting was equally hopeful that Diem would resolve problems with the Buddhists. A coup would be a disaster: It would split the country into feuding factions, and the communists would win control of all Vietnam. By contrast with Krulak and Nolting, the CIA field chief in Saigon saw a volatile situation that put Diem's regime in considerable peril. On July 9, when Forrestal sized up the situation for Bundy, he described a dilemma: No one was able to offer confident estimates of whether Diem would survive. People in Saigon were more hopeful than analysts in Washington. Forrestal suggested a period of fence sitting while they waited on developments.

At a July 17 press conference, where Kennedy focused on nuclear test ban negotiations in Moscow and the domestic economy, he steered clear of Vietnam until halfway through the session, when a reporter asked if the turmoil was impeding the war effort. "Yes, I think it has," Kennedy said. He regretted that it had arisen just as the military situation was improving. But he declared: "We are not going to withdraw. . . . For us to withdraw . . . would mean a collapse not only of South Vietnam, but Southeast Asia." The dominoes would fall and U.S. national security would diminish. In brief, he more than implied that U.S. withdrawal would depend on victory in the conflict or at least the appearance of victory. In the meantime, his policy toward the Buddhist crisis in Vietnam was one of watchful waiting.

At the end of July, a report from Robert Manning, the assistant secretary of state for public affairs, whom Kennedy had asked to visit Vietnam to assess the tensions between the government and the press corps, concluded that there was "an unbridgeable gap between the official and the correspondents' assessment of the Vietnamese situation." The embassy saw great progress in the war, while the journalists thought that the war was being lost as long as the Diem government remained in power. The two sides in the debate had nothing but contempt for each other. For Kennedy, it was essential that the press begin to put a better face on the war. Since he had committed himself publicly to saving South Vietnam, he could not end or sharply limit U.S. involvement in the conflict if the newsmen, who simply refused to conform to embassy wishes, kept writing about a Vietnam in disarray and a faltering war effort.

On August 5, when the press reported a renewed government crackdown on the Buddhists and another self-immolation, the embassy was instructed to warn Diem and the Nhus that they were courting disaster, with Washington poised to denounce their actions. The next day, Hilsman calculated the chances of a coup as fifty-fifty and the likelihood of success as also even. He told George Ball that the department was continuously reviewing its contingency plans and maintaining contacts with oppositionist elements in hopes of shaping the outcome of any uprising. Because Diem's continued control seemed to be reducing hopes of defeating the Viet Cong, the department thought better of a coup. The press corps was all over these latest developments, reporting the Nhus as driving the government into greater conflict with the Buddhists and pressing Diem to take decisive action against them. Madame Nhu's comment that "all the Buddhists had done was to 'barbecue a bonze' with 'imported gasoline'" had moved the department to ask Diem to send her out of the country. "No decisions are required from you at this time," Forrestal told Kennedy, "but you may wish to give some guidance during the next week when it becomes more clear what the real intentions of the GVN or the Buddhists are."

On August 15, Kennedy met with Lodge before he headed to Saigon. They discussed the possibility of Diem's ouster. Kennedy thought that he "was entering a terminal phase." While Kennedy was not averse to seeing Diem ousted, he wished to be sure that his successor would be a more cooperative and effective ally. The press, they agreed, was a serious problem. Kennedy saw relations between the embassy and journalists as the worst to be found in any foreign capital. He wanted Lodge to take charge of press relations, which was code for telling him to bring the newsmen under greater control. Lodge anticipated difficulties handling them. He would not lie to them, but he had no plans to cooperate with them. Kennedy saw the journalists as instinctively liberal and anti-Diem. Two years before, they had been dead wrong in predicting Diem's fall in six months, he said. The Associated Press photo of the monk burning himself alive had generated more emotion around the world than any other photo he could think of.

Kennedy, of course, could not control the press. And so on August 15, when Halberstam reported that the war was going badly, despite increased U.S. involvement during the last twenty months, and that Diem's forces were losing control of the Mekong Delta, the key area of the country, Kennedy's antagonism to the journalists increased. Krulak gave him some comfort by explaining that Halberstam did not understand that American strategy was working and that the greater concentration of communist troops in the delta was evidence that they were being compressed into the southernmost part of the country, where they would "rot." "If Halberstam understood clearly this strategy, he might not have undertaken to write his disingenuous article," Krulak told McNamara.

For all his antagonism to the press, Kennedy could not dismiss their stories as biased reporting. He did not trust what the military was telling him. Their advice on the Bay of Pigs, Laos, the missile crisis, and their response to the Mississippi civil rights crisis the previous year had soured him on their judgment and accuracy. At the same time, Mike Mansfield told him that "we are in for a very long haul to develop even a modicum of stability in Viet Nam." He reminded Kennedy that he needed to ask whether Vietnam had taken on "a highly inflated importance and, hence, talked ourselves into the present bind." Had the administration "moved what may be essentially a peripheral situation to the core of our policy considerations?" By contrast with Mansfield's advice to reduce our commitment, the Joint Chiefs urged a delay in a decision McNamara had made in May to withdraw one thousand troops by the end of the year and keep them there until conditions had stabilized in Vietnam.

On August 21, when reports came in of a government crackdown on Buddhist pagodas all over the country and a declaration of martial law by the military, the State Department and White House felt more engulfed than ever in the fog of war. The initial impulse was to denounce the government's actions, but with Lodge on his way to Saigon and questions in the air as to who had authorized the raids, Kennedy chose to hold back on any statement or action. During a meeting at the White House, Ball, Harriman, and Krulak agreed that caution and clarification were the best approach for the moment. Responding to the upheaval described in press reports instead of embassy accounts especially bothered Kennedy; he wanted to know what American officials in Saigon thought had occurred, not what journalists were telling their editors and readers. The first cable to Lodge asked him to find out what had happened and to report who was controlling events—Diem, the Nhus, or the South Vietnamese army.

Embassy dispatches over the next two days made clear that there had been no coup and that the Nhus were behind the anti-Buddhist campaign. Diem's national security adviser now told the U.S. Embassy to do everything possible to get rid of the Nhus. The Vietnamese army would turn against them if the United States made its opposition to them clear. Vietnamese generals gave the embassy the same message. Before he had acquainted himself with his staff and office or learned his way around Saigon, Lodge found himself in the middle of a palace intrigue. It was the price of U.S. involvement in Vietnam's civil war and its byzantine politics.

Lodge was uncertain about whom to trust and urged the department to hold off on making any major decisions. More disturbing was a memo from the public affairs counselor at the embassy to Lodge describing the split between the press and the American mission in Saigon as well as among members of the U.S. military command. The journalists and the mission were at odds over whether the war was going well or badly, but so were members of the military advisory group. The 18,000 or 19,000 Americans in the country "are torn by doubt," Lodge's adviser told him. No one knew whom to believe, including Kennedy, who had recently asked for a factual assessment of the fighting and prospects for a successful outcome.

On August 24, Forrestal advised Kennedy that the embassy had sorted out recent events in Saigon: "It is now quite certain that Brother Nhu is the mastermind behind the whole operation against the Buddhists and is calling the shots." The remedy was to remove Nhu from power—with or without Diem's approval. Harriman and Hilsman agreed, and they were consulting John McCone at the CIA on how to get this done. A cable to Lodge that night stated that the U.S. government could no longer live with the Nhus. Diem had to remove them or face a break with Washington. The United States would support new leadership. Lodge was to find the means to achieve the stated ends and the White House would back him "to the hilt."

When shown the cable and told that Kennedy had approved its dispatch, Taylor complained to Krulak that the action represented an end run around accepted procedure. It did "not give Diem adequate chance to do what we want." Taylor, who opposed dumping Diem, said that it reflected "the well-known compulsion of Hilsman and Forrestal to depose Diem and, had McGeorge Bundy been present [in Washington], he would not have approved the message."

On August 26, Kennedy met with his top eleven State, Defense, NSC, and CIA officials. Hilsman reported on the embassy's approach to principal Vietnamese generals to enlist them in the ouster of Nhu in collaboration with Diem or in Diem's removal as well. Kennedy now voiced doubts about encouraging or facilitating a coup. He wanted to know more about the generals who might replace Diem and said: "Diem and his brother, however repugnant in some respects, have done a great deal along the lines that we desire." His greatest concern was that they were responding to pressure from the *New York Times* and Halberstam in particular. "He's a 28-year-old kid," Kennedy said, and complained that he had been wrong in the past. Kennedy "wanted assurances we were not giving him serious consideration. . . . It was essential that we not permit Halberstam unduly to influence our actions." Kennedy was less concerned about Halberstam than about rushing into a coup that would deepen American involvement in Vietnam without any guarantees that a new government would fare any better in the civil war than the previous one.

Kennedy made his reservations more evident by asking Taylor, who he knew opposed displacing Diem, whether he thought coup plans could succeed. Taylor caustically answered "that in Washington we would not turn over the problem of choosing a head of state to the military." McNamara joined Taylor in raising questions about a coup, asking, "What exactly do we mean . . . by the term 'direct support'?" which Lodge was instructed to tell the generals the United States was prepared to provide. Kennedy then questioned the likelihood that Diem would allow his brother Nhu to "be ejected from the scene." McNamara worried that the United States would "ultimately suffer" if "a weak man got in the presidency," replacing Diem. Kennedy wanted to know what would happen if the United States had to continue living with Diem and Nhu. Hilsman thought it would be "horrible." Rusk declared, "Unless a major change in GVN policy can be engineered, we must actually decide whether to move our resources out or to move our troops in." Kennedy did not want to choose either step. He asked that Nolting, who was certain to defend Diem, be brought into the discussion. When Hilsman objected that "Nolting's views are colored, in that he is emotionally involved in the situation," Kennedy responded, "Maybe properly."

While the White House debated policy in Vietnam, Nguyen Khanh, the lead general in the coup discussions, saved Kennedy from a decision by telling the CIA station chief in Saigon that he wasn't ready to move. He intended to wait until Nhu gave him a clearer motive for an uprising: If Nhu tried for a rapprochement with Hanoi, it would give the generals legal grounds for action. Forrestal reported Khanh's inaction to Kennedy. He wanted a pretext for a coup, which they didn't have. The generals simply weren't ready to risk their lives and so it relieved Kennedy of a decision he preferred not to make.

But the problem of what to do about Vietnam wasn't going away. After the meeting on the twenty-sixth, Forrestal urged Kennedy to understand that a majority of Vietnamese believed that the repression of the Buddhists could not have occurred without American equipment, wanted Diem out, and looked to the United States for change. Moreover, the next day, a CIA report that the Vietnamese generals were now promising a coup in a week forced Kennedy to give further consideration to U.S. support. Forrestal told him that the generals wanted some clearer expression of U.S. backing. With another meeting scheduled for the afternoon of the twenty-seventh, Forrestal suggested that Kennedy close the meeting by saying that the United States could not support a government dominated by Nhu, preferred to keep Diem in power but would leave it to the generals to decide, and would wholeheartedly back whatever new regime emerged as long as it prosecuted the war against the Viet Cong.

The 4 P.M. meeting now expanded to sixteen advisers, again including all the leading State, Defense, and CIA officials as well as Robert Kennedy and Nolting. McNamara began the discussion by insisting that the president didn't need to decide anything that day. Kennedy at once indicated his inclination to avoid endorsing a coup, by asking Nolting whether the generals could carry off a successful rebellion. Nolting was scathing: The generals, he said, "haven't the guts of Diem or Nhu. They will be badly split. They do not have real leadership, and they do not control the predominant military force in the country." He had "grave doubts" about any effective action. McNamara echoed Nolting's reservations, saying the coup generals lacked sufficient forces. When Kennedy asked Nolting about Diem's reliability, he emotionally replied that "Diem had kept his promises. . . . Diem is not a liar and is a man of integrity." Hilsman, with equal fervor, disputed Nolting's characterization of Diem as honest. It angered Nolting, who had abandoned any pretense of objective reporting; he saw his reputation as a successful envoy tied to Diem's survival and success in the civil war.

But Kennedy was less interested in Diem's character than whether a coup might succeed, saying he saw no point in backing an abortive strike. Nolting assured Kennedy that the military could not carry off an effective rebellion. Kennedy was more concerned to know if Diem was capable of winning the war. Nolting didn't know—so he counseled patience, saying a coup could always come later. In the meantime, the United States had created a problem by telling some of the generals to remove Diem. Kennedy thought that the embassy could certainly delay action. Hilsman warned, "The longer we wait the harder it would be to get Diem out." Kennedy wanted to hear from Lodge and Harkins on the chances for a successful coup, or so he said. But he had already decided against immediately trying to change the government, saying "the generals interested in the coup were not good enough to bring it about."

The struggle over how to proceed continued for another forty-eight hours. On the twenty-eighth, cables arrived from Lodge and the CIA chief in Saigon asserting that delaying a coup would diminish chances of success, with the danger that Diem might arrest the generals plotting against him. Vietnamese vice president Nguyen Ngoc Tho told Lodge that a coup was essential. A continuation of current conditions would endanger U.S.-Vietnamese relations. The CIA station chief warned that Vietnam would be lost to the communists if the Diem government remained in place.

Meanwhile, back in Washington, a White House meeting at noon on August 28, attended by twenty officials, including Kennedy, erupted in a fierce argument. Taylor launched the debate by explaining that even though anti-Diem troops would outnumber loyal forces, a small number of tough Diem loyalists could beat them back. George Ball believed that a coup was necessary to rid the country of the Nhus to ensure victory in the war. The only question was how to arrange a successful coup. McNamara doubted that it could be done. Kennedy said, well then, it shouldn't go forward. Nolting restated his opposition to a coup, predicting that a new government would not do any better in the fighting. Ball, Harriman, and Hilsman countered that Diem had shown that he couldn't win the war. They saw no choice but to oust him. Kennedy wasn't ready to agree. He restated doubts that rebel generals could defeat Diem and asked for suggestions on how to build up the anti-Diem forces.

Hilsman and Harriman saw no choice but to go ahead. Without a coup, the United States would lose in Vietnam and would have to withdraw. Besides, they said, "We can't stop the generals now and they must go forward or die." Kennedy again demurred: He suggested that they go back to Lodge and Harkins and explain that Diem seemed to hold the balance of power and ask their advice on what to do. Nolting said that the president was right: "only Diem can hold this fragmented country together." Harriman now exploded in anger at Nolting, saying he had disagreed with Nolting from the beginning and that he was "profoundly wrong," adding that the stakes in this debate compelled him to be so blunt. Gilpatric later recalled that this was the worst "tongue lashing" he had seen in Kennedy's presence and doubted that anyone other than Harriman, with his seniority, could have gotten away with it. Bobby Kennedy saw the division in the room as a very disturbing fundamental break in his brother's government. Kennedy now adjourned the meeting to let tempers cool and provide time for some reflection before they reconvened at six.

At the evening meeting, Kennedy put any decision on hold by directing that cables be sent to Lodge and Harkins in Saigon saying that nothing had been decided in Washington. Instead, he wanted their judgment on whether the generals were ready to act, and if not, he recommended a temporary stand-down from a coup, or no coup at all. As Taylor now emphasized to Harkins, "We do not want to become involved in any coup which will not succeed."

Harriman saw the cables for what they were—an expression of Kennedy's doubts about promoting a coup that would inevitably draw the United States more deeply into Vietnam.

As the meeting ended, Harriman said to Kennedy, "I hope we are not giving any idea of wobbling on our course." Kennedy replied: "We have to make sense; we must not let the field feel that we are in any way heavy-handed, or obliging them to take actions which are not, in their good judgment, sound." A memo describing the exchange noted: "The President had some difficulty containing himself until everyone had left the room, whereupon he burst into laughter and said, 'Averell Harriman is one sharp cookie.'" Harriman, who was the strongest advocate of promoting and ensuring the success of a coup, understood that Kennedy wasn't eager or even willing to do it.

During all this debate over Vietnam, Kennedy struggled to keep black demands for equal rights under control. On June 22, as he was about to go to Europe, he had met with civil rights leaders at the White House. His agenda was to encourage them to contain demonstrations that might jeopardize the civil rights bill before Congress. He warned against a planned march on Washington that could turn some members of Congress against the bill, saying, "I'm damned if I will vote for it at the point of a gun." But the civil rights leaders believed that a peaceful demonstration would do more to energize and promote a law than undermine it. As the meeting ended, Roy Wilkins conveyed to Schlesinger "his sympathy for the President in view of the pressures playing on him, the choices he had to make, the demands on his time and energy." On August 28, the march of some 250,000 people was an affirmation of peaceful democratic expression of which, Kennedy told the march's leaders at a White House meeting that evening, "[t]his nation can be properly proud." While the march worked no miracles on the Hill, where Kennedy's legislative initiative stalled in the Senate, it momentarily quieted this most volatile domestic issue and allowed Kennedy to return to the crisis in Vietnam.

The pressure on Kennedy to give the go-ahead for a coup was unrelenting. On the twenty-ninth, in response to his request for an independent judgment, Lodge declared, "Any course is risky, and no action at all is perhaps the riskiest. . . . We are launched on a course from which there is no respectable turning back." It was already an open secret that the United States favored a coup. More to the point, Lodge warned that Diem could not win the war. Because Harkins believed that a coup might be unnecessary if Diem ousted Nhu and because McNamara saw no alternative to Diem, Kennedy instructed the embassy in Saigon to make a final effort to pressure Diem into dismissing Nhu. But doubtful that Diem would accept the advice, Kennedy confirmed the earlier decision to inform the Vietnamese generals of U.S. backing for a change of government. Nonetheless, if he had last-minute doubts that an uprising would be successful, he insisted on the freedom to change course. "I know from experience," he told Lodge, "that failure is more destructive than an appearance of indecision." Lodge, however, cautioned him that a coup could take on a momentum of its own, and "you may not be able to control it."

The problem, however, was not a runaway operation, but, as Lodge reported by cable on August 30, "inertia. The days come and go and nothing happens." At a State Department meeting that afternoon, Rusk and McNamara said that "the Generals were either backing off or wallowing." McNamara thought that they had never even had a plan. That night, Rusk cabled Lodge that prospects for a coup now seemed "very thin" but assured him "that highest levels in Washington are giving this problem almost full-time attention." At 2:39 A.M. on August 31, the CIA station chief reported that "this particular coup is finished. . . . Generals did not feel ready and did not have sufficient balance of forces." Lodge followed up with the contemptuous conclusion "that there is neither the will nor the organization among the Generals to accomplish anything." Kennedy's advisers saw no choice but to reopen discussions with Diem about how to win the war.

Later that morning, at another State Department meeting, which Lyndon Johnson attended with all the administration's top national security officials, Paul Kattenburg, an expert on Southeast Asia who had just returned from Vietnam, urged the group to understand that Diem could not win the war, and that if the United States continued on the same track it would be forced to leave the country in six months to a year. "He had known Diem for ten years," he said, "and did not think that Diem would ever take the steps necessary to correct the situation. . . . He suggested that it would be better for the U.S. to withdraw honorably." Secretary of State Rusk dismissed Kattenburg's remarks as "speculative": A pullout made no sense. Johnson agreed, and in the sort of colorful language that all who knew him well found familiar, declared, "We must . . . stop playing cops and robbers" and talking about a coup. "There were bad situations in South Vietnam. However, there were bad situations in the U.S. It was difficult to live with [Louisiana congressman] Otto Passman, but we couldn't pull a coup on him."

Recalling the meeting and discussion sixteen years later, Kattenburg thought the "whole group of them . . . absolutely hopeless. . . . There was not a single person there that knew what he was talking about. . . . They didn't know Vietnam. They didn't know the past. They had forgotten the history. . . . The more this meeting went on, the more I sat there and I thought, 'God, we are walking into a major disaster.'"

After more than fifty-eight thousand American troops had died in Vietnam and the North had seized the South despite the sacrifices in American blood and treasure, Kattenburg was proven right. And even then only McNamara and Bundy publicly acknowledged how wrong they had been. Walt Rostow, a principal proponent of the war as Johnson's national security adviser, never conceded the war was a failure, arguing that aside from Vietnam, the United States had saved the rest of Southeast Asia from communism. McNamara and Bundy tried to understand their misjudgments in the belief that it might head off similar future disasters. Bundy not only was self-critical, but he also passed judgment on other advocates of America's increased involvement in Vietnam. In an interview, Bundy said later that Lodge was the stupidest man he had ever dealt with in public life.

Lodge was now instructed to renew pressure on Diem to push the Nhus aside and reform his government in hopes of increasing his popular support. At the same time, Kennedy used TV interviews with CBS and NBC to put Diem on notice that he stood squarely behind the demand for changes in Saigon. Could Diem's government regain the support of his people? CBS's Walter Cronkite asked. Kennedy replied: "With changes in policy and perhaps with personnel I think it can." Kennedy left no doubt about his determination to win in Vietnam. He called the suggestions of withdrawal a "great mistake"; the people who advocated it were "wholly wrong." The United States had no choice but to defend Asia and understand that we were locked in a "desperate struggle against Communism." Leaving Vietnam would open the way to Chinese expansion in Southeast Asia and trigger greater threats to other Asian nations.

Nonetheless, Kennedy emphasized that it was up to the Vietnamese to do the fighting, and unless the Diem government generated popular support for itself, it was likely to lose the war. "In the final analysis," he said, "it is their war. They are the ones who have to win it or lose it." Anyone who thought about his comments had to be puzzled. If success in Vietnam was so crucial to U.S. national security and Diem was in jeopardy of losing the war, could the United States ultimately avoid full-scale involvement? Or was Kennedy signaling that if Diem didn't reform he would back a coup that brought to power a government that would fight more effectively? It was an unacknowledged contradiction that Americans and Vietnamese were left to consider. Whether Kennedy purposely created this uncertainty or simply was expressing his own inner struggle about what he might do was unclear.

In suggesting that it was Vietnam's responsibility to fight the war, Kennedy had considerable hope that it might yet rise to the challenge with limited American help. Taylor told him that military operations in Vietnam for August were encouraging, despite Saigon's political disputes. Progress was also continuing with the Strategic Hamlets program, with 76 percent of the rural population under its protection, which said nothing about whether it was effective in defeating the Viet Cong.

As the summer ended, Kennedy's strategy was to keep up the pressure on Diem to end or at least greatly reduce the Nhus' power and limit the press stories about tensions between Saigon and Washington and pessimistic reports about the outcome of the war. Kennedy saw negative news accounts forcing him toward a choice between using U.S. forces and abandoning Vietnam, or encouraging a coup that might lead to victory or who knows what. He instructed government press officers to stay off TV and turn down calls from journalists requesting interviews. At the same time, he directed the embassy in Saigon not to initiate further contacts with the Vietnamese generals, but to be responsive to any initiative from them. He did not want the generals to think "that the U.S. had backed off" or excluded a coup from its plans for defeating the communists.

Nonetheless, Vietnam remained a muddle without a solution. On the morning of September 10, General Krulak and Joseph Mendenhall, a State Department Asian expert, reported to Kennedy on a four-day visit they had just made to Vietnam. Krulak described a war that was moving in the absolutely right direction and was going to be won. The impact on the war effort from the current tensions between the government and the Buddhists were at most "small": The ARVN units under American direction were "worrying about the Viet Cong and not about politics or religion," he said. Mendenhall saw a different universe: "a virtual breakdown of the civil government in Saigon as well as a pervasive atmosphere of fear and hate arising from the police reign of terror and the arrests of students. The war against the Viet Cong has become secondary to the 'war' against the regime." He concluded "that the war against the Viet Cong could not be won if Nhu remains in Vietnam." Krulak countered, "The battle was not being lost in a purely military sense."

An astonished and frustrated Kennedy asked: "The two of you did visit the same country, didn't you? . . . How is it that you get such different—this is not a new thing, this is what we have been dealing with for three weeks. . . . I'd like to have an explanation what the reason is for the difference." Kennedy didn't know what to believe or, more important, what to do. He had pressed the case in public for Diem to introduce political reforms and to convince U.S. congressional and public opinion that this was a conflict we must not lose. But "this had ignited nothing."

Because no one had a surefire solution to the Vietnam dilemma, advisers felt empowered to make the case for their viewpoint. It was as if the discussion about Vietnam had turned into a faith-based dispute with clashing egos. Each side was invested in its advice and uncertain about what would be effective; advisers felt free to urge their policy but perhaps more because they were mindful of the weakness in their opponents' arguments than from being confident of their prescription.

The inability of his advisers to reach a consensus discouraged Kennedy's hopes of finding an effective response to the Vietnam morass. The debate continued at a late afternoon meeting on September 10. McNamara, Taylor, and McCone argued for working with Diem to sideline Nhu and unite the country against the Viet Cong. They saw no alternative to Diem and feared chaos and defeat if he were removed from power. Harriman and Hilsman sharply disputed their conclusion. Diem could not win, and the only alternative for the United States was to find another leader who could defeat the communists. Hilsman acknowledged that this might require the use of American combat troops. Taylor opposed the introduction of combat forces either to oust Diem or fight the Viet Cong. Lodge, writing from Saigon, insisted that the time had come for the United States "to bring about the fall of the existing government."

Bobby Kennedy, who had been preoccupied with domestic struggles over civil rights, now joined or rejoined the conversation about Vietnam. At the September 10 meeting, the ongoing debate among the national advisers about how to win the war angered him. Mindful of his brother's frustration with a debate that seemed unending and unproductive, he pressed for a consensus: "All agreed that the war would go better without Nhu and Diem," he said. He insisted that they not burden the president with their differences. He wanted them to reach a consensus on Vietnam policy. But agreement remained beyond reach, and President Kennedy, hoping that they might yet find common ground, asked Forrestal to write a paper "recommending a delay in any decision for a sufficient time for the situation to ripen."

In the meantime, Kennedy, who knew that his advisers could not agree, wished to keep the argument about Vietnam out of the headlines. At the morning meeting on the tenth, he described himself as "disturbed at the tendency both in Washington and Saigon to fight out our own battles via the newspapers. . . . He wanted these different views fought out at this table and not indirectly through the newspapers." At the White House daily staff meeting the next day, the same day *New York Times* columnist James Reston published an article decrying the censorship of U.S. journalists in Saigon, Bundy raised the administration's press problem. He seemed "at a loss about what to do" about Vietnam in general and the press in particular. When told that Madame Nhu was coming to the United States, a visit Kennedy had made clear he opposed, Bundy, who was "already wobbly" and, according to the note taker, "close to the last blow," said, "This was the first time the world had been faced with collective madness in a ruling family since the days of the czars."

The continuing daily conversations about Vietnam at the State Department and the White House left everyone discouraged. Lodge kept pushing for a commitment to remove Nhu and Diem, while Rusk, McNamara, and Taylor maintained hopes of spurring Diem onto a fresh course. Rusk directed Lodge to have "frequent conversations" with Diem, but Lodge resisted, complaining that he had nothing new to bring up and saw "many better ways in which I can use my waking hours." Instead, he wanted Kennedy to send Lansdale to Saigon at once "to take charge, under my supervision, of all U.S. relationships with a change of government." But as Harriman explained to Lodge, differences of opinion were a deterrent to action.

On September 16, eighteen national security advisers debated the right course in Vietnam yet again. Rusk instructed Hilsman to draft two cables, one reflecting a "conciliatory approach" and the other the "pressure approach." The pressure policy aimed to force Diem into dropping Nhu and reforming his government, while the conciliation track assumed no change in the government and the rehabilitation of its leaders. "I think we have come to a position of stall in our attempts to develop a Washington consensus" on Vietnam, Forrestal told Bundy. The divide among the president's advisers was stimulating a war of leaks. "The longer we continue in an attitude of semi-public fluidity, the worse the leak problem becomes," he added. The only sure step Kennedy favored was putting a lid on the negative press stories on Vietnam. He wanted Lodge "to hush up the press in Saigon." Since he and his advisers had no good idea of how to ensure a victory in the fighting and end U.S. involvement in an unwanted war, Kennedy seemed to hope that matters would resolve themselves, but that would happen only if the press did not agitate the issue and pressure him into actions he was reluctant to take.

Because Kennedy saw no likelihood that Diem would be gone soon, he instructed Lodge to implement the pressure policy. In addition, he directed McNamara and Taylor to visit Vietnam once again to assess the state of the war and Diem's ability to defeat the communists. Advocates of dumping Diem were incensed at Kennedy's decision to send two of the most outspoken supporters of continued cooperation with Diem. Lodge immediately cabled his objection to a visit that "will be taken here as a sign that we have decided to forgive and forget and will be regarded as marking the end of our period of disapproval of oppressive measures." Lodge was furious. Why wasn't the White House listening to him? He was on the ground and believed he knew exactly what should be done. To appease him, Kennedy agreed to include Forrestal in the visiting team. Hilsman weighed in with a letter to Lodge asserting that "more and more of the town is coming around to our view and that if you in Saigon and we in the Department stick to our guns the rest will also come around. . . . A determined group here will back you all the way."

It was clear, however, that Kennedy simply didn't want to encourage a coup that would deepen U.S. commitments and increase the possibility of sending combat troops. To persuade Diem to follow America's lead, he directed McNamara and Taylor to shun any contact with coup generals and emphasize "the positive accomplishments of the last decade" that had resulted from U.S.-Vietnamese cooperation.

Predictably, the McNamara-Taylor visit solved nothing. A meeting with Diem was an exercise in futility. Having perfected the technique of speaking at length so as to limit what unwelcome visitors might say, Diem did most of the talking during the first two hours. It was a "virtual monologue" in which Diem simply repeated familiar observations about the fighting and the actions of his government. In the third hour, McNamara and Taylor made the case for reforms that could enhance the war effort and blunt criticism in the United States that threatened to reduce backing for Vietnam. Diem dismissed their complaints as unwarranted and generated by a hostile press corps attacking his government, him, and his family. McNamara and Taylor concluded that Diem was unmovable; he was indifferent to what they said.

In a report to Kennedy on their return to Washington, McNamara and Taylor reported significant progress in the fighting. They saw little likelihood of a successful coup and little prospect for government reforms. Nonetheless, they favored continuing pressure on Diem and Nhu and contacts with generals who might one day rise to the challenge and carry off a successful coup. They also predicted that "the major part of the U.S. military task can be completed by the end of 1965" and recommended that one thousand U.S. military advisers be withdrawn by the end of 1963. They gave no explanation for why the United States could leave Vietnam in a little over two years. In everything to do with Vietnam, wishful thinking won the day.

The White House then issued a press release that declared "the security of South Viet-Nam a major interest of the United States" and the determination of the administration to defeat the communist insurgency. The military support for the South Vietnamese was showing good progress and would be provided until the insurgency has been suppressed. "Secretary McNamara and General Taylor reported their judgment that the major part of the U.S. military task can be completed by the end of 1965," and one thousand U.S. military personnel could be withdrawn by the end of this year. Political tensions in Vietnam were "deeply serious," and the White House had "made clear its continuing opposition to any repressive actions in South Viet-Nam."

The public pronouncement was more an exercise in political posturing than a realistic assessment of current and future conditions in Vietnam. McNamara and Taylor understood that Kennedy opposed any expansion of U.S. military involvement in the war, and they were predicting communist defeat in the next two years that would allow the United States to withdraw its advisers. William H. Sullivan, an assistant to Averell Harriman who was part of the visiting group, told Taylor that the commitment to withdraw U.S. forces at the end of 1965 "would be considered a phony and a fraud and an effort to mollify the American public and just not be considered honest." But Kennedy was insisting on an end date to U.S. involvement in the war. When Bundy and members of the NSC questioned the wisdom of the announcement, McNamara and Taylor replied that they were "under orders." William Bundy thought "the words of the release on the military situation were extraordinarily unwise." Mindful of the questionable realism in placing limits on America's role in the fighting, Kennedy instructed that no formal announcement accompany the implementation of this decision. He did not wish to test the limits of public credulity.

Above all, now, he wanted to repress negative press accounts that he continued to think would make it difficult for him to limit, if not end, U.S. involvement in the conflict. Newspaper stories describing hostile State Department views of Diem, as well as Defense Department complaints that "inept diplomacy" was putting American interests at risk in Vietnam, angered him. As the White House prepared to release the statement summarizing the McNamara-Taylor findings, Kennedy told advisers, "Reports of disagreement do not help the war effort in Vietnam and do no good to the government as a whole. We must all sign on and with good heart set out to implement the actions decided upon." He insisted "that no one discuss with the press any measures that he may decide to undertake" on Vietnam. Bundy proposed that Kennedy instruct everyone not to say anything to the press that implied differences among policymakers. Bundy cabled Lodge: The president thinks it essential that the White House rather than the press inform the public about Vietnam. It was impermissible for the newspapers to describe the pressure Washington was putting on Diem. It would be better for the "press to consider us inactive than to trumpet a posture of 'major sanctions' and 'sweeping demands.'"

But Kennedy couldn't plug leaks or halt the flow of discouraging news coming from Saigon. In September and October, *New York Times* reports by Halberstam as well as critical columns and other negative headlines about American problems in Vietnam continued to irritate him. Despite a conscious effort by McNamara and Taylor to shun the press during their visit, Halberstam reported that the "U.S. mission is finding no easy solutions in Vietnam." Their tour underscored the "difficulty in assessing the impact of the political climate on the U.S.–aided war effort." On October 3, *New York Times* columnist Arthur Krock described "The Intra-Administration War in Vietnam." CIA operatives in Saigon were portrayed as at odds with Lodge, refusing on two occasions to carry out his orders. Halberstam also depicted Lodge as in disagreement with Harkins. "As you can appreciate, the story has caused concern in Washington," George Ball cabled Lodge, "since we have been making a serious effort in conjunction with McNamara-Taylor mission to achieve actual and visible unity" within the U.S. government. On the eighth, despite Kennedy's insistence on repressing news accounts of significant pressure on Saigon, Halberstam reported that the United States was halting some aid to Vietnam in hopes of forcing changes in Diem's government. On the seventeenth the *Times* reported that Nhu saw his country as losing faith in the United States.

Kennedy made Halberstam the focus of his campaign to restrain the press. It was not simply that Halberstam produced day-to-day headlines describing the vulnerabilities of Diem's government and its stumbling war effort; it was also that he had become an unspoken advocate of replacing the current leadership and winning the war. Halberstam did not think that the introduction of U.S. ground forces was the answer to the guerrilla insurgency. He had already concluded that it would trap the United States in a colonial war that would "parallel the French experience." So Kennedy and Halberstam partly agreed on the limits of U.S. involvement in Vietnam. But Kennedy did not want policy made or forced on him by unelected journalists. Specifically, he feared that Halberstam's hectoring was pressuring him into support of a coup that could further destabilize Vietnam and bring irresistible demands for intervention with American combat troops.

Three weeks into October, with conditions growing more uncertain in Saigon, Kennedy used a lunch meeting with Arthur
Sulzberger, who had recently become the publisher of the *New York
Times*, to ask that Halberstam be withdrawn from Saigon. Sulzberger refused Kennedy's assault on press freedom. The fact that
the president didn't like Halberstam's reports was insufficient to
compel his recall. Halberstam's reporting was a model of truth
telling. The *Times* had no desire to make policy with its lead stories,
however much Kennedy may have seen it that way. Halberstam was
providing an accurate portrait of an unpopular government and a
faltering civil war. It was up to Kennedy to face these realities and
not try to alter them by repressing the news out of Saigon.

But Kennedy could no more control press accounts than he could
his own advisers and events in Saigon. On October 9, after renewed
indications of coup planning reached Kennedy, he told Lodge
not to help "stimulate" a coup, but also not to discourage one if
it appeared likely to succeed and increase the effectiveness of the
military effort. While Lodge remained entirely supportive of the
generals now promising to oust Diem within a week after October
26, Harkins continued to advise the generals against toppling the
government and risking recent gains in the war. When Lodge and
Harkins conferred on the afternoon of October 23, they argued
about what the White House wanted them to do and the different
signals they were giving the generals. Speaking for the president,
Bundy instructed Lodge and Harkins to "stand back from any
non-essential involvement in these matters"—meaning that if there
were a coup, the White House wanted plausible deniability. When
Lodge responded that anything the United States did to thwart
a coup would be a mistake, Kennedy reiterated his concern that a
coup not "be laid at our door."

On the twenty-seventh, the divide among the advisers in Washington and Saigon grew more pronounced. Harriman and Hilsman convinced Ball, who was acting secretary of state while Rusk
was out of the country, to sign a "green light" cable to Lodge

telling the generals that Washington approved a coup. U. Alexis Johnson, who was excluded from their three-way exchange, believed that he was purposely kept out of the conversation because he opposed any such instruction. In telling Lodge to facilitate the coup, Harriman and Hilsman were taking advantage of Kennedy's ambivalence. He had neither approved nor opposed a coup, but simply said he didn't want it blamed on the United States. Kennedy's uncertainty about what to do about Vietnam allowed advisers to fill the policy vacuum.

On October 29, in an apparent reaction to the "green light" cable, Bundy told Kennedy that "all important separate instructions and reports made on any channel—State, CIA, DOD, USIA, and JCS—be sent over here during this next period for your personal information. . . . There is just no doubt at all that a good deal of our trouble in the last three months has come from difference of emphasis, at least in what we have said to the field." The instruction would allow him and Forrestal "to call to your attention any serious divergences. . . . I do not underestimate the sensitivity of this order." The Joint Chiefs, Defense Department, and CIA might object, "but your interest is not served by the uncritical acceptance" of their right to send unmonitored cables. It reflected Kennedy's feeling that he had lost control of policy.

The directive might have given Kennedy greater influence over future embassy actions in Saigon, but having encouraged the generals to act, Lodge believed that it was too late for the White House to pull back from a coup. He reported that a rebellion was "imminent" and that the United States was likely to be blamed, regardless of whether it succeeded or failed. Moreover, he saw no way to deter the generals from acting "short of informing Diem and Nhu with all the opprobrium that such an action would entail."

The limits of White House control became all too apparent on the afternoon of the twenty-ninth, when Kennedy met with all his principal advisers. Was the pending coup likely to succeed? he asked. Kennedy thought that the odds were against success, but his calculations were little more than guesswork. Bobby Kennedy, undoubtedly reflecting his brother's concerns, asserted that a coup would put America's position in Vietnam and all of Southeast Asia at risk. Rusk worried that if we opposed a coup, the generals would "turn against us and the war effort will drop off rapidly." Taylor sided with Bobby, warning that even "a successful coup would have a harmful effect on the war effort." McCone shared Taylor's view, but Rusk warned that if Diem remained in power, it would continue to jeopardize success against the Viet Cong. Harriman agreed. Kennedy now came out against the coup, saying that the opposing forces in Saigon were about equal, which made a coup "silly. If Lodge agrees with this point of view, then we should instruct him to discourage a coup." Caught between concerns that a failed coup would destroy U.S. ability to shape events in Vietnam and staying with a government that some believed was destined to lose the war, Kennedy abdicated control to Lodge, who had made his eagerness for a change of government clear.

At a subsequent meeting later that afternoon, Kennedy reiterated his eagerness to discourage the generals unless they were absolutely certain they could succeed. "We could lose our entire position in Southeast Asia overnight," he said. A cable to Lodge reiterated Kennedy's insistence on only supporting a coup that promised victory: The "burden of proof must be on coup group to show a substantial possibility of quick success. . . . A miscalculation could result in jeopardizing U.S. position in Southeast Asia."

But Lodge, who remained convinced that a coup was essential, dismissed Kennedy's demands for guaranteed success as beyond U.S. freedom to arrange. "Do not think we have the power to delay or discourage a coup," he responded. "We have very little influence on what is essentially a Vietnamese affair." He agreed "that a miscalculation could jeopardize position in Southeast Asia. We also run tremendous risks by doing nothing." Speaking for the president, Bundy replied, "We do not accept as a basis for U.S. policy that we have no power to delay or discourage a coup." He instructed him "to persuade" the generals "to desist at least until chances are better" if "there is not clearly a high prospect of success." In sum, the United States and Lodge specifically should only back a coup if they were sure it would succeed.

But the exchanges between the embassy and Washington had become irrelevant. On November 1, convinced by embassy indications that the United States would ensure their success, the generals overturned Diem's government and assassinated him and Nhu. Lodge was full of enthusiasm at the turn of events. He counseled prompt support and recognition of the new government: "We should, of course, give unmistakable signs of our satisfaction to the new leadership." He stressed "the very great popularity of this coup. . . . Every Vietnamese has a grin on his face today. Am told that the jubilation in the streets exceeds that which comes every New Year."

On learning that the coup was succeeding, Kennedy met with ten of his advisers to decide on whether to promptly recognize the new government. A few minutes into the meeting Forrestal brought in a cable reporting that Diem and Nhu had been killed. Kennedy was horrified. Taylor recorded: He "leaped to his feet and rushed from the room with a look of shock and dismay on his face which I had never seen before. He had always insisted that Diem must never suffer more than exile and had been led to believe or had persuaded himself that a change in government could be carried out without bloodshed." Shortly after, Schlesinger saw the president and remembered him as "somber and shaken." He could not believe that Diem and Nhu, good Catholics, had killed themselves, as the generals were alleging. He thought that they deserved a better fate. Moreover, the political repercussions of their deaths were not lost on him. As Bundy told Lodge, "there is danger that standing and reputation of incoming government may be significantly damaged if conviction spreads of their assassination."

In a recording on November 4, Kennedy ruminated about the coup, his part in encouraging it, and the prospects for the new government and the war. "I feel that we must bear a good deal of responsibility for it, beginning with our cable in early August in which we suggested the coup. In my judgment, that wire was badly drafted. It should never have been sent on a Saturday. I should not have given my consent to it without a roundtable conference at which McNamara and Taylor could have presented their views" against a coup.

Kennedy then described himself as "shocked by the death of Diem and Nhu." He recalled his contacts with Diem dating back many years and lamented his demise: "He'd held his country together, maintained its independence under very adverse conditions. The way he was killed made it particularly abhorrent." And so "the question now is whether the generals can stay together and build a stable government or whether Saigon will begin—whether public opinion in Saigon, the intellectuals, students etc.—will turn on this government as repressive and undemocratic in the not too distant future."

Regardless of what happened next, Kennedy was determined to separate the United States from Vietnam's future struggles. But having failed to bring Cuba, a much smaller island country in America's closest sphere of control, into line with administration goals, he doubted his capacity to dictate Vietnam's fate. His public posture was to do everything possible to ensure the autonomy of that country. "We must all intensify our efforts to help it [the new government] deal with its many hard problems," he told Lodge. Having encouraged the change in Saigon, "we thus have a responsibility to help this new government to be effective in every way that we can." He endorsed holding a conference in Honolulu on how to intensify the struggle against the Viet Cong. He also wanted the participants to discuss "how we can bring Americans out of there." Asked at a November 14 press conference if he still intended to bring home a thousand troops before the end of the year, he said it would be several hundred, but that he would wait to determine exact numbers until after the Honolulu meeting.

His eagerness to find a way out of Vietnam registered forcefully in a memo to Forrestal on November 21. As he was about to leave for a political fence-mending trip in Texas, where a division between Democratic Party conservatives and liberals threatened Kennedy's reelection prospects in 1964, he instructed Forrestal "to organize an in-depth study of every possible option we've got in Vietnam, including how to get out of there. We have to review the whole thing from the bottom to the top." In 1971, Forrestal told CBS that in an Oval Office conversation on November 21, Kennedy told him: "I want to start a complete and very profound review of how we got into this country, and what we thought we were doing, and what we think we can do. I even want to think about whether or not we should be there." It is impossible to say just what Kennedy would have done about Vietnam in a second term, if he had had one. But given the hesitation he showed about Vietnam during his thousand-day administration, it is entirely plausible that he would have found a way out of the conflict or at least not to expand the war to the extent Lyndon Johnson did.

"What He Is Slated to Become Depends on Us"

Kennedy's assassination on November 22, 1963, was a national trauma that continues to haunt Americans. Although solid evidence points to Lee Harvey Oswald as the lone gunman, some 70 percent of the country holds to the belief that a conspiracy cannot be ruled out. It is difficult for most people to accept that someone as inconsequential as Oswald—a dysfunctional, erratic character, who had an undistinguished period of service in the Marines and lived for two years in the Soviet Union before coming to Dallas, and working a menial job in the Texas Book Depository—could have killed someone as closely guarded as the president of the United States. Yet Kennedy himself—who had a keen sense of irony, the unpredictability of events, and the tragic nature of human affairs—would have been more accepting of the single gunman narrative. He would not have been surprised that the back brace he wore to help him get through the day without agonizing discomfort made a difference in ending his life: Had he not been wearing it, Oswald's first shot, which passed through his neck, would have toppled him and prevented a second fatal bullet from striking him in the back of the head.

Kennedy's interactions with his ministry of talent throw new light on his presidential performance as well as the agony of decision-making agitated by the uncertainties every adviser faced in trying to shape a better future. The retrospective judgments of some of Kennedy's associates on his leadership and what they believed he intended give us additional insights into his presidency. But they also expand our understanding of the advisers, whose reflections on the past tell us as much about them as the history they recount.

For those who saw themselves as best able to describe and defend Kennedy's presidency—his wife, brother, Arthur Schlesinger, and Ted Sorensen—the public's elevated opinion of him was justified by his actions. Devastated and anguished by Kennedy's assassination, they launched a campaign to promote a romanticized picture of a heroic leader selflessly serving the nation's best interests. Quoting the poet W. H. Auden, Sorensen said, "What he was he was; what he is slated to become depends on us."

On November 29, 1963, only a week after the president was slain, Jackie Kennedy led the way, sitting down with the journalist Theodore White to recount her husband's accomplishments. The interview, which appeared a week later in *Life* magazine, famously compared Kennedy's White House to King Arthur's Court, "the one brief shining moment" known as Camelot. She preserved the president's memory by lighting an eternal flame at his grave and renaming Florida's Cape Canaveral as Cape Kennedy; the manned spacecraft to the moon would be launched from that site in 1969. In December 1963, grieving New Yorkers renamed Queens's Idlewild Airport as John F. Kennedy International Airport.

In March 1964, Jackie Kennedy expanded on her campaign to memorialize her husband in a series of interviews with Arthur Schlesinger. Concerned that part of her remarks would offend some of the president's associates, and principally aiming to shape historical judgments, she instructed that they be closed until fifty years after her death. In 2011, however, her daughter, Caroline, published the interviews in a three-hundred-page book to mark the fiftieth anniversary of the Kennedy presidency. Despite some interesting revelations about Kennedy and the men around him, the book is mostly a continuation of the Camelot romance. Caroline Kennedy was not unmindful of the interviews' exaggerated, but understandable, regard for her father. In a foreword, she described the book as the product of "a young widow in the extreme stages of grief" and asserted that were her mother alive, she would have revised some of what she said in 1964.

Although she never retracted anything she told Schlesinger, during the thirty years after 1964 Jackie Kennedy became much more than John Kennedy's widow. In 1968, she married the Greek shipping magnate Aristotle Onassis, and spent considerable time abroad until Onassis died at the age of sixty-nine, in 1975. She then took up a career as a book editor, first at Viking and then at Doubleday. In 1980 she began a relationship with Maurice Tempelsman, a wealthy businessman, with whom she shared an interest in the arts and architecture. In 1994, she passed away from cancer at the age of sixty-four.

Jackie's recollections could be seen as a stand-in for what Kennedy himself would have done in a volume of memoirs: defend his historical record. In 1964, since she had no intention of publishing her memories in her lifetime and feared that Kennedy's standing would wane with the passage of time, she urged Schlesinger to write a book describing Kennedy's hopes of being a great president. She believed that Schlesinger's recollections and history would not only preserve Kennedy's memory but also advance the causes he believed in.

Schlesinger was more than happy to oblige. Although he was never a principal adviser and was mostly on the fringe of Kennedy's administration, mainly helping with speeches, as a professional historian he, along with Sorensen, was the White House official most capable of writing about JFK's presidency.

Within days of Lyndon Johnson's succession, Schlesinger concluded that Johnson saw him as a Kennedy devotee and was not keen to have him at the White House. As 1963 came to an end,

Schlesinger recorded in a diary, "I have not had a single communication from the President or his staff for the last month—not a request
to do anything, or an invitation to a meeting, or an instruction, or a
suggestion. . . . It seems clear that they are prepared to have me fade
away, which is OK by me." On January 27, 1964, Schlesinger sub-
mitted his resignation, which was accepted with "alacrity."

Johnson was eager to separate himself as much as possible from
the Kennedys. Although he saw initial political advantages from
a close association with the martyred president, at least until he
could win and hold the office in his own right, Johnson was determined to establish his administration as distinct from and superior to John Kennedy's. In January 1964, he began charting his
own legacy by announcing a War on Poverty, and in May he described his administration not as continuing the New Frontier but
as building a Great Society.

As soon as Schlesinger left the White House, he began working on a book about Kennedy's presidency. While he had been in
the habit of keeping a diary, his part in the 1960 campaign and
appointment as special assistant to the president had persuaded
him to become a more fastidious recorder of daily events. Jackie
Kennedy's suggestion to him confirmed his own intentions, and
within a year, excerpts of a Kennedy book, *A Thousand Days*, began
appearing in *Life*. Schlesinger described the book he published in
1966 as more "a personal memoir" that presented "only a partial
view" rather than "a comprehensive history." He predicted that it
would be some distant time in the future before a historian will
"immerse himself in the flood of papers in the Kennedy Library"
and write a more definitive, less subjective account of the Kennedy term. He hoped, however, that his book would contribute
to a positive assessment of Kennedy's leadership. Although many
books have been written about Kennedy and various aspects of his
administration, some of them highly critical of his temporizing on
civil rights, obsessive womanizing, and a hidden medical history
that might have cost him the 1960 election if known, Schlesinger's

volume still commands a significant readership and continues to shape judgments about Kennedy's place in history.

Like Jackie Kennedy and Schlesinger, Ted Sorensen struggled to make sense of Kennedy's death and committed himself to advancing his historical reputation. Although, as the record of Kennedy's interactions with his advisers shows, Sorensen played a limited role in policymaking, he was Kennedy's principal wordsmith. But Johnson viewed him as among the White House officials most closely identified with Kennedy and believed that his continuing presence as a speechwriter, at least for a time, would help to preserve the country's sense of continuity. Johnson pressed Sorensen to remain in his job and convinced him to help write a post-assassination speech for delivery on November 27. In the address, Sorensen had Johnson declare, "All I have I would have given gladly not to be standing here today." Johnson, however, deleted Sorensen's opening statement: "I who cannot fill his shoes must occupy his desk." There were limits to how far Johnson would go in paying tribute to the fallen president. Sorensen resented the deletion at the time and was eager to begin work toward the goal he set for himself of advancing Kennedy's "ideals and objectives." Vowing to "do all I could to keep John F. Kennedy's legacy alive," Sorensen persuaded Johnson to let him resign at the end of February 1964.

At once, he began writing *Kennedy*, which was published the following year to much acclaim. While he announced his determination not to produce a eulogy, noting Kennedy's acknowledgment of "imperfections and ignorance in many areas," the biography was a celebration of Kennedy's many personal and political attributes. Sorensen decried those who spoke more of Kennedy's "style than of his substance." Yes, his "style *was* special—the grace, the wit, the elegance, the youthful looks will rightly be long remembered. But what mattered most to him, and what in my opinion will matter most to history, was the substance—the strength of his ideas and ideals, his courage and judgment."

Forty-three years later, in 2008, two years before his own death at the age of eighty-two, Sorensen returned to the subject of his years with Kennedy in a memoir, *Counselor: A Life at the Edge of History.* "For eleven years," he wrote, "it had been my full-time job to advance his interests, invoke his name, and articulate his message in the struggles for justice at home and peace around the world. For the succeeding forty-plus years, I have made it my part-time mission to do the same." And in *Counselor,* he continued his efforts to secure Kennedy's historical reputation.

No one was more determined to carry John Kennedy's legacy forward than brother Bobby. His presence in the Johnson administration as attorney general was a source of mutual tension; their interactions during Kennedy's thousand days had intensified their reciprocal antagonism dating from the fifties. Bobby was angry at what he saw as Johnson's excessive haste in taking control of Air Force One and the Oval Office following the assassination, while Bobby angered Johnson when he ran past him to comfort Jackie after she had returned to Washington with the president's body. Momentarily, however, each saw a need to mute their differences for the sake of the country. At two meetings in the days immediately after Kennedy's death, Johnson told Bobby, "I need you more than the President needed you."

But the truce could last only so long. Tension between the two remained palpable. During Johnson's speech to the Joint Session of Congress on November 27, Bobby sat "pale, somber, and inscrutable, applauding faithfully, but his face set and his lips compressed." It impressed Schlesinger as "a particularly unbearable moment." Conflict over Bobby's possible candidacy for the vice presidency emerged during the first half of 1964. Johnson didn't want him, believing history would say that Bobby's presence on the ticket elected him president. Even though he told Bobby that he was out of the running and described it as part of a decision not to take anyone in the cabinet, Johnson couldn't free himself from the overdrawn fear that Bobby would steamroll the Democratic convention in August into selecting him anyway. That summer, after Johnson won passage of JFK's civil rights bill, Bobby decided to resign and run for a New York U.S. Senate seat. It was a relief to both him and Johnson.

After Bobby won his Senate election in November, he spent the next four years promoting the causes he associated with his brother's agenda. He wrote *Thirteen Days: A Memoir of the Cuban Missile Crisis*, which celebrated John's masterful resolution of the potentially disastrous confrontation with Moscow. He also sat for a number of oral history interviews with *New York Times* columnist Anthony Lewis; John Bartlow Martin, a journalist and JFK ambassador to the Dominican Republic; Burke Marshall, an attorney and head of the Kennedy Justice Department's Civil Rights Division; and Schlesinger. Covering everything from civil rights, with which Bobby was most intimately involved, to Kennedy's decision to name Johnson vice president, to his choice of advisers and cabinet officers, and the administration's Cuban and Vietnam policies, Bobby, like Schlesinger and Sorensen, recounted events as Kennedy might have described them in a memoir. Like Jackie Kennedy's reflections, some of Bobby's descriptions of people and events were seen as too candid to release while he remained actively involved in politics. In 1988, twenty years after he had run for president and been assassinated at the age of forty-two, Bobby Kennedy's recollections, which had become less controversial with the passage of time, appeared in print.

Bobby's own unrealized potential and his brother's unfinished presidency have given both of them an enduring hold on the public's imagination as heroic leaders who could have spared the country from missteps at home and abroad. They answer yearnings for better leadership in a more harmonious nation and world.

McGeorge Bundy was one member of a quartet of advisers who hoped to advance Kennedy's legacy by remaining at their jobs. The day after the assassination, Bundy told Schlesinger that "he intended to stay on as long as Johnson wanted him." Bundy was particularly intent on making sure that Vietnam was not lost to the communists, seeing this as a fixed Kennedy aim. When Johnson began expanding the war in March 1965 with Rolling Thunder, the bombing campaign against North Vietnam, Bundy warmly supported the decision. In the spring, he was eager to debate anti-war opponents, whom he saw as undermining the public backing he believed essential to a sustained war effort.

Although Bundy had serious doubts about the wisdom of dispatching large numbers of ground forces to Vietnam, he did not raise them directly with Johnson. He believed that "an effort had to be made" to save Vietnam "even if the odds favored defeat." Bundy's focus remained on the need to educate the public about the necessity of making the commitment to prevent a communist victory. But he and Johnson fell into conflict over building a consensus for the war effort. Bundy was critical of Johnson's decision to announce the first troop escalation in Vietnam in July 1965 at a noon press conference, "when no one was watching TV." Johnson dismissed Bundy's pressure for greater openness, saying, "If your mother-in-law . . . has only one eye, and it happens to be right in the middle of her forehead, then the best place for her is not in the livin' room with all of the company!" Mindful that Johnson "doesn't pay any attention to what I'm telling him," Bundy decided to resign in February 1966. He compared his last days in the White House to advisers in the Kremlin who were ignored and abused by Stalin.

After he left the government, Bundy served as president of the Ford Foundation for thirteen years, until 1979, when he became a professor of history at New York University, where he taught for ten years, followed by six years as a scholar-in-residence at the Carnegie Corporation. He died in 1996. In 1988, although he would publish *Danger and Survival: Choices About the Bomb in the First Fifty Years*, reflections on nuclear weapons, it was the failed Vietnam War that cast a constant shadow over his historical reputation. As Gordon Goldstein made clear in his 1998 book, *Lessons in Disaster*, a cooperative study with Bundy, Bundy struggled to make sense of the terrible misjudgments that led to the failed conflict. Although he knew it would seem self-serving, Bundy believed that the war was principally the result not of what the Kennedy and Johnson advisers, including the military chiefs, told them, but of what Kennedy and Johnson chose to do. It is perhaps more accurate to say that a close reading of the records shows Kennedy's responsibility to have been less the product of active commitments to fight a large war in Vietnam than his ambivalence: his eagerness to prevent a communist victory in Vietnam matched by his reluctance, indeed refusal, to turn the conflict into America's war, which risked Saigon's collapse. His unwillingness to come down decisively on one side or the other of these competing policies opened the way to Johnson's unequivocal determination to use U.S. power to preserve South Vietnam's autonomy, arguing that this is what Kennedy would have done.

In his pursuit of this goal, Rusk, McNamara, and Rostow aided and abetted Johnson. Rusk's nondescript posture under Kennedy, "his Buddha-like face and half-smile," Schlesinger called it, joined to "a montage of platitudes" in a soft-spoken Georgia drawl, made him something of a nonentity in Kennedy's circle of high-powered opinionated advisers like McNamara, Harriman, Rostow, and LeMay. His deference to Kennedy annoyed the president, who complained that Rusk "never gives me anything to chew on. . . . You never know what he is thinking." Rusk's courtly manner and deferential regard for higher authority perfectly suited Johnson. Regular foreign policy briefings for the vice president, in which they shared an enthusiasm for the Cold War clichés of the day like the defense of "the free world," gave Rusk an immediate place at the center of Johnson's administration. It did not hurt that they were both southerners who agreed on the compelling need for a civil rights revolution that would end segregation and disarm African American anger by giving blacks the chance to vote and compete on level ground with whites for a better life.

Rusk's determination not to allow communist control of South Vietnam, which he feared would lead to other acts of aggression and touch off a new round of recrimination against loyal public officials, echoed Johnson's openly stated pronouncement to Rusk, McNamara, Bundy, McCone, and Lodge on the third day of his presidency: "I am not going to lose Vietnam." Through all the turmoil over the next five years—the bombing of North Vietnam punctuated by pauses in hopes of inducing peace talks, the dispatch of more than 500,000 U.S. troops, with the deaths of more than 30,000 by 1968, and the eruption of antiwar protests that moved a French travel agent to advertise, "See America while it lasts"—Rusk backed Johnson's escalation and direction of the conflict at every turn, believing that a communist victory would be an impermissible blow to U.S. national security.

Like his many years in government, Rusk's post–State Department career was publicly muted. A professor of international law at the University of Georgia from 1970 to 1984, he did not publish a memoir, *As I Saw It*, until 1990, four years before he died at the age of eighty-five. It was an uncontroversial account with few recriminations, leaving it to history to render an independent verdict on his career. Unfortunately for Rusk, like Johnson he is doomed to be remembered not for any great advances in foreign affairs but as one of the principal architects of America's disastrous losing war in Vietnam.

Robert McNamara's historical reputation bears an even heavier share of the burden. Staying on as secretary of defense until February 1968, McNamara was even more instrumental than Rusk in encouraging first Kennedy and then Johnson to fight and win the war. Although he made an impressive mark as an industrial leader on the Ford Motor Company before becoming defense secretary; won plaudits for proposing Kennedy's quarantine of Cuba, which contributed so much to the peaceful resolution of the missile crisis; wisely backed the test ban treaty and, subsequently, nuclear disarmament; and provided well-regarded leadership at the World Bank between 1968 and 1981, he is largely remembered for his unyielding support of the decisions on bombing and troop deployments that went so wrong in Vietnam.

By 1968, McNamara understood how mistaken he, Johnson, Rusk, and the Joint Chiefs had been in their assumptions about

Vietnam. He became overtly morose about the war and began pressing Johnson to do whatever possible to end the conflict as soon as possible. After leaving the government, he remained largely silent about the war until publishing a mea culpa in 1995, *In Retrospect: The Tragedy and Lessons of Vietnam.* The book was a confession of sorts that brought him more criticism than praise for having finally owned up to the terrible miscalculations he did so much to produce. McNamara was no more successful in disarming critics when he sat for a series of interviews for a 2003 documentary, *Fog of War.* It was another attempt to win forgiveness for his unforgivable errors in Vietnam. McNamara passed away in 2009 at the age of ninety-three. His reputation as the longest-serving secretary of defense in U.S. history will be part of an endless argument about the triumphs and defeats of the Kennedy and Johnson presidencies.

Walt W. Rostow was the most unrepentant of all the Kennedy-Johnson architects of the war. In 1966, when Bundy resigned, Johnson made Rostow his national security adviser. As with Kennedy, Rostow urged Johnson to expand U.S. involvement in Vietnam to prevent a communist victory. He never regretted that advice or, unlike McNamara and Bundy, saw any reason to apologize for it. On the contrary, to the end of his life in 2003, at the age of eighty-six, he argued that the war may not have saved South Vietnam from communism but it gave the rest of Southeast Asia time to build its defenses. Like General William Westmoreland, who served as Johnson's top commander in Vietnam, Rostow believed that the United States did not lose the war but gave up the fight because of public weariness over the conflict. He asserted that those who lost loved ones in the war could take satisfaction from knowing that the United States stopped the dominoes from falling.

John McCone and Maxwell Taylor, two other administration hawks, did not stay on for very long with Johnson. McCone left in April 1964, and Taylor resigned as chairman of the Joint Chiefs of Staff in July 1964, when he became ambassador to Saigon for a year. It is doubtful that either one would have remained through a second Kennedy term. By November 1963, Kennedy had lost confidence in their respective judgments. Both had been hawkish during the missile crisis and much more committed to military interventions in Cuba and Vietnam than Kennedy. As for other members of the Joint Chiefs, they quietly retired in time, except for LeMay, who was forced to step down in February 1965. He remained in the public eye for more than three years after, declaring that the United States should bomb North Vietnam back to the Stone Age or at least, as he claimed he said, America had the capability to do it. The remark haunted him in 1968 when he agreed to become Alabama governor George Wallace's vice presidential running mate on the failed American Independent Party ticket. Unfairly identified with Wallace's segregationist views, LeMay retreated into private life, living in relative obscurity for twenty-two years before his death in 1990.

Paul Volcker, the chairman of the Federal Reserve Board from 1979 to 1987, believes that a president without advisers "is crippled in developing, defending and administering his policies." Kennedy's experience suggests a more complicated result. Through his thousand days in the White House he learned that even the brightest and most well meaning of advisers misjudge a situation and offer poor counsel. De Gaulle's guidance about gathering a variety of opinions on big policy questions and then following your own judgment resonated forcefully with Kennedy after two years as president. By November 1963, seeing how limited the expertise of the so-called experts was had made him a wiser decision-maker. While there would have been stumbles and reassessments during a second term, it is impossible to say exactly how the experience of his first four years would have played after his likely reelection. The initial hard lessons of his first term undoubtedly would have made him a more effective president in a second go-round. His tragic assassination in Dallas, however, deprived us of the chance to judge a second-term performance. It is easy nonetheless to believe that his premature death opened the way to events—the expanded Vietnam War, Nixon's election, Watergate—that changed America and the world for the worse.

Kennedy's death leaves us with unanswered questions: Would he have won reelection in 1964? And assuming that he did, would his health have held out in a second term? Would his womanizing have caught up with him and jeopardized his presidency? Would he have persuaded Congress to pass his four major legislative initiatives? Would he have reestablished relations with Cuba and found a way out of Vietnam? Would he have moved toward détente with the Soviet Union and possibly China?

Had Kennedy had a chance to write about his administration, he undoubtedly would have reflected on the cloud of uncertainty that hovered over everything they did and might have done in a second term. His time in the White House underscored for him that there are no experts in public policy—only men and women, with the best of intentions, guessing at what would work. The principal lesson of any presidential term, he would surely have acknowledged, is the anguish of choosing between imperfect options and having to take responsibility for lives lost and money wasted when fallible advisers and chiefs take wrong turns.

Whatever the outcome of a second Kennedy term might have been, speculation about it as better than what Johnson gave the country serves Kennedy's historical reputation, which remains extraordinarily high. The affection for him generated by his persona and the tragedy of his assassination have encouraged positive assessments of his leadership. Fifty years after his death, there is no sign that Kennedy's hold on Americans is anywhere in retreat.

Notes

○ ○ ○ ○

Chapter I: John F. Kennedy: Prelude to a Presidency

1 **Small wonder:** Schlesinger related his conversation to me in 2001. Also see, Arthur M. Schlesinger, Jr., *Robert Kennedy and His Times* (Boston: Houghton Mifflin, 1978), 228.

2 **Immediately after:** For JFK's medical history, see the Dr. Janet Travell medical records at the John F. Kennedy Library (JFKL). Also, Janet Travell Oral History at the JFKL. For LBJ's attack on JFK, see Robert Dallek, *Lone Star Rising: Lyndon Johnson and His Times, 1908–1960* (New York: Oxford University Press, 1991), 571–73. The medical bag is discussed in Abraham Ribicoff Oral History (all oral histories hereafter abbreviated OH), Columbia Universirty. On Nixon and the break-ins, see Robert Dallek, *An Unfinished Life: John F. Kennedy, 1917–1963* (New York: Little, Brown, 2003), 286, and the note on 755; also 299–300 on post-election questions about JFK's health; Theodore C. Sorensen, *Kennedy* (New York: Bantam Books, 1966), 268.

3 **Kennedy echoed:** Kenneth P. O'Donnell and David F. Powers, *Johnny, We Hardly Knew Ye* (Boston: Little, Brown, 1970), 234.

3 **Kennedy's route:** Dallek, *Unfinished Life*, 6–13, 19, 157–58.

6 **No one, however, contributed more:** Ibid., 14–25, 53, 112. Also Doris Kearns Goodwin, *The Fitzgeralds and the Kennedys: An American Saga* (New York: Simon & Schuster, 1987), 446–48, 498–99.

7 **Joe's reach:** Robert Dallek, *Franklin D. Roosevelt and American Foreign Policy, 1932–1945* (New York: Oxford University Press, 1995), 164; Michael R. Beschloss, *Kennedy and Roosevelt: The Uneasy Alliance* (New York: Norton, 1980), 123–28, and chap. 6; Dallek, *Unfinished Life*, 53–54.

9 **The fall in public:** Goodwin, *The Fitzgeralds and the Kennedys*, 600–04, 621–23.

9 **Joe's hopes for:** John H. Davis, *The Kennedys: Dynasty and Disaster, 1848–1983* (New York: McGraw-Hill, 1984), 104–06; U.S. Air Force Report, Aug. 14, 1944, JFKL; William G. Penny to JFKL, Aug. 14, 2001, JFKL.

10 **But not for long:** Dallek, *Unfinished Life*, 27, 33–35, 73–75.

11 **His medical ordeal:** Ibid., 76–77.

11 **Kennedy's medical issues:** Ibid., 51–68, 111–17.

12 **In 1945–46:** JFK Tape Recording 39, October 1960, JFKL; Joan and Clay Blair, Jr., *The Search for JFK* (New York: Berkley, 1974), 356.

13 **Once committed:** James McGregor Burns, *John Kennedy* (New York: Harcourt, Brace, 1959), 99; Twain quote is in Anthony Jay, ed., *The Oxford Dictionary of Political Quotations* (New York: Oxford University Press, 1996), 372; Charles Spaulding, OH, JFKL; JFK interview with James McGregor Burns, March 22, 1959, JFKL; Victor Lasky, *J.F.K.* (New York: Macmillan, 1963), 117.

14 **The Senate was:** Dallek, *Unfinished Life*, chap. 6, especially 177–78, 189–92, 226; Ted Widmer, *Listening In: The Secret White House Recordings of John F. Kennedy* (New York: Hyperion, 2012), 30–31.

15 **Journalists and party leaders:** William V. Shannon, *New York Post*, Nov. 11, 1957; James Reston, *New York Times*, Oct. 10, Nov. 10, 1958; Peter Lisagor, OH, JFKL; Widmer, *Listening In*, 287.

15 **Many in the Democratic Party:** Newton Minow, OH; William Attwood, OH; William Benton, OH; and William McCormick Blair, Jr., OH, Columbia University; Peter Collier and David Horowitz, *The Kennedys: An American Drama* (New York: Summit Books, 1984), 294.

16 **In running for:** Harris Wofford, *Of Kennedys and Kings* (New York: Farrar, Straus & Giroux, 1980), 36–37.

17 **"I claim not":** Quoted in David Donald, *Lincoln Reconsidered* (New York: Vintage Books, 1961), 138.

17 **His fight for:** Dallek, *Unfinished Life*, 252–58.

18 **The question shadowed:** O'Donnell and Powers, *Johnny, We Hardly Knew Ye*, 205–208; Sept. 12, 1960, speech, Box 1061, Pre-Presidential Papers (PPP), JFKL.

19 **It did not:** On the vote, see Louis Harris to Joseph Alsop, Nov. 16, 1960, "An Analysis of the 1960 Election for President," Joseph and Stewart Alsop Papers, Library of Congress (LOC); Theodore H. White, *The Making of the President, 1960* (New York: Atheneum, 1961), 350–65; JFK's New Frontier speech, July 15, 1960, Box 1027, PPP; and John Kenneth Galbraith, *Letters to Kennedy*, ed. James Goodman (Cambridge, MA: Harvard University Press, 1998), 10.

19 **Kennedy and his advisers:** On Nixon, see Robert Dallek, *Nixon and Kissinger: Partners in Power* (New York: HarperCollins, 2007), chap. 1.

20 **The importance of:** White, *Making of the President, 1960*, 286–94; Galbraith, *Letters to Kennedy*, 14.

21 **Kennedy won:** White, *Making of the President, 1960*, 350–65; Sorensen, *Kennedy*, 238–51; O'Donnell and Powers, *Johnny, We Hardly Knew Ye*, 229; Arthur Schlesinger, Jr., *A Thousand Days: John F. Kennedy in the White House* (Boston: Houghton Mifflin, 1965), 125.

22 **The same week:** See *New York Times*, Nov. 15, 1960; O'Donnell and Powers, *Johnny, We Hardly Knew Ye*, 229.

22 **Two meetings with:** Robert F. Kennedy, OH; Charles Spaulding, OH, JFKL; for the Dec. 5, 1960, and Jan. 19, 1961, meetings with Eisen-

hower, see the documents in Box 29A, President's Office Files (POF), JFKL, which list the topics for discussion.

24 **Kennedy, however, had no:** Clark Clifford, *Counsel to the President: A Memoir* (New York: Random House, 1991), 319–20.

24 **At the same time, Kennedy:** Richard Neustadt, "Organizing the Transition," Sept. 15, 1960, POF, JFKL; Richard Neustadt, OH, Columbia University; Schlesinger, *A Thousand Days*, 122–23.

25 **Neither Clifford nor:** The quote is in Carl M. Brauer, *Presidential Transitions: Eisenhower Through Reagan* (New York: Oxford University Press, 1986), 64.

26 **Kennedy saw:** The Adams and Wilson quotes are in Robert Dallek, "Presidential 'Disability': An American Dilemma," University Lecture, Boston University, 1999. For the Eisenhower comment, see Stephen E. Ambrose, *Eisenhower: The President* (New York: Simon & Schuster, 1984), 599–600.

26 **As the former Senate:** Dallek, *Lone Star Rising* on LBJ's pre–vice presidential career; Robert Dallek, *Flawed Giant: Lyndon Johnson and His Times, 1961–1973* (New York: Oxford University Press, 1998), 8–9; Jacqueline Kennedy, *Historic Conversations on Life with John F. Kennedy* (New York: Hyperion, 2011), 56; Arthur Schlesinger, Jr., *The Age of Roosevelt* (Boston: Houghton Mifflin, 1957–1960); Brauer, *Presidential Transitions*, 65–66.

28 **But whatever Kennedy:** George F. Kennan, OH, JFKL.

29 **Yet however wise:** Interview with Priscilla Johnson; Dallek, *Unfinished Life*, 151–52, 192–95.

31 **Kennedy's affinity for:** Mimi Alford, *Once Upon a Secret: My Affair with President John F. Kennedy and Its Aftermath* (New York: Random House, 2012), see esp. 82–83, 101–105, 109, 124–25, 127.

Chapter 2: Robert Kennedy: Adviser-in-Chief

35 **As Kennedy searched:** The Parnell quote is in Antony Jay, ed., *The Oxford Dictionary of Political Quotations* (New York: Oxford University Press, 2001), 284.

35 **Kennedy saw:** See my discussion of Joe Kennedy's views on his family in Dallek, *Unfinished Life*.

35 **When Jack first entered:** Ibid., 118–19, 121–26.

37 **Whatever Jack's limitations:** Widmer, *Listening In*, 32; Ralph G. Martin and Ed Plaut, *Front Runner, Dark Horse* (Garden City, NY: Doubleday, 1960), 133; Collier and Horowitz, *The Kennedys*, 183; Blair and Blair, Jr., *The Search for JFK*, 478–79, 495.

38 **Once in office:** Arthur Krock, OH, JFKL; Lasky, *J.F.K.*, 117.

38 **The alternative:** Schlesinger, *Robert Kennedy*, 42–66. Also see James W. Hilty, *Robert Kennedy: Brother Protector* (Philadelphia: Temple University Press, 1998), chaps. 1–3.

40 **While Jack served:** Dallek, *Unfinished Life*, 67–81.

41 **In the fall:** Schlesinger, *Robert Kennedy*, 60–93; RFK Diary in Folder "Trips 1951, Mid & Far East," Box 24, RFK Papers, JFKL.

42 **The moment came in 1952**: Thomas J. Whalen, "Evening the Score: John F. Kennedy, Henry Cabot Lodge, Jr., and the 1952 Massachusetts Senate Race," Ph.D. Dissertation, History Department, Boston College, 1998, 243–55, 285; Martin and Plaut, *Front Runner, Dark Horse*, 164; Schlesinger, *Robert Kennedy*, 94–98; Dallek, *Unfinished Life*, 168–76.

44 **In January 1953**: Schlesinger, *Robert Kennedy*, 99–109.

45 **Perhaps Bobby's**: Ibid., 109–29. Also Dallek, *Lone Star Rising*, 451–59.

46 **In 1956**: Schlesinger, *Robert Kennedy*, 130–33; Dallek, *Lone Star Rising*, 489–91, 502–504; Dallek, *Unfinished Life*, 203–208. "Hi, sonny" is in Evan Thomas, *Robert Kennedy: His Life* (New York: Simon & Schuster, 2000), 96.

49 **Despite his assessment**: Schlesinger, *Robert Kennedy*, 133–36; for JFK's speeches between September and November 1956, see Compilation of Speeches, JFKL; Herbert Parmet, *Jack: The Struggles of John F. Kennedy* (New York: Dial Press, 1980), 384–86; Goodwin, *Fitzgeralds and Kennedys*, 791–92.

50 **The 1956 ventures**: Schlesinger, *Robert Kennedy*, chaps. 8–9; Goodwin, *Fitzgeralds and Kennedys*, 787–88.

51 **By October**: Sorensen, *Kennedy*, 35, 117; Handwritten Notes: "Oct. 28, 1959—RFK House—Hyannis Port," JFKL; Paul B. Fay, Jr., *The Pleasure of His Company* (New York: Harper & Row, 1966), 76–77.

51 **Bobby's initial field assignment**: Dallek, *Lone Star*, 559; Jeff Shesol, *Mutual Contempt: Lyndon Johnson, Robert Kennedy, and the Feud that Defined a Decade* (New York: Norton, 1997), 10–11.

53 **And for Bobby Kennedy**: Schlesinger, *Robert Kennedy*, 194–97; Dallek, *Unfinished Life*, 239–51.

54 **West Virginia**: White, *Making of the President, 1960*, 96–114; Hubert H. Humphrey, *The Education of a Public Man* (Garden City, NY: Doubleday, 1976), 208, 216–17, 475; Schlesinger, *Robert Kennedy*, 198–204; Thomas, *Robert Kennedy*, 93–96; Dallek, *Unfinished Life*, 252–58.

56 **Jack and Bobby**: Schlesinger, *Robert Kennedy*, 204–206; Dallek, *Lone Star Rising*, 569–73; Earl Mazo, OH, Columbia University; Bobby Baker, *Wheeling and Dealing: Confessions of a Capitol Hill Operator* (New York: Norton, 1978), 118.

57 **It was an empty**: Schlesinger, *Robert Kennedy*, 206–11; Dallek, *Unfinished Life*, 267–74.

58 **Once Jack had**: Schlesinger, *Robert Kennedy*, 211–19; Thomas, *Robert Kennedy*, 100–08; Earl Mazo, OH, Columbia University; Abraham Ribicoff, OH, Columbia University; Parmet, *Jack*, 34; Eleanor Roosevelt to JFK, Aug. 16, 1960; Eleanor Roosevelt to Mary Lasker, Aug. 15, 1960, Box 32, POF.

61 **With Jack exhausted**: *Newsweek*, Nov. 21, 1960; *New York Times*, Nov. 23, 1960; Schlesinger, *Robert Kennedy*, 228–36; Thomas, *Robert Kennedy*, 109–11; John Seigenthaler, OH, JFKL; RFK to Drew Pearson, Dec. 15, 1960, Box 23, RFK Papers, JFKL; Dallek, *Unfinished Life*, 316–20.

65 **Bobby's appointment**: C. Douglas Dillon, OH, JFKL.

Chapter 3: "A Ministry of Talent"

67 **The day after:** Richard Reeves, *President Kennedy: Profile of Power* (New York: Simon & Schuster, 1993), 25.

68 **Compounding Kennedy's worries:** Dallek, *Unfinished Life*, 92–94.

68 **The Soviets might have:** L. James Binder, *Lemnitzer: A Soldier for His Time* (Washington, D.C.: Brassey's, 1997), chaps. I–17, especially pp. 2–12 and 281–83; Arthur M. Schlesinger, Jr., *Journals, 1952–2000* (New York: Penguin Press, 2007), 126.

71 **Admiral Arleigh Burke:** Schlesinger, *Thousand Days*, 200–201; Arthur Sylvester, OH, JFKL; *New York Times*, Jan. 28, 1961; Arleigh Burke, OH, JFKL; and Dallek, *Unfinished Life*, note for p. 337 on p. 762.

72 **Kennedy's biggest worry:** McGeorge Bundy to JFK, Jan. 30, 1961, Box 313, National Security File, JFKL; Roswell Gilpatric, OH; Memorandum, Feb. 24, 1961; JFK conference with military chiefs, Feb. 27, 1961, Box 345, National Security File; Binder, *Lemnitzer*, 315–16; "Thomas S. Power," Wikipedia; and Warren Kozak, *The Life and Wars of Curtis LeMay* (Chicago: Regnery, 2009), which paints a sympathetic portrait. Fred Kaplan, *The Wizards of Armageddon* (New York: Simon & Schuster, 1983), 43–44, 256, quotes LeMay; Sorensen is quoted on p. ix of Kozak.

74 **When Kennedy's national security adviser:** Kai Bird, *The Color of Truth: McGeorge Bundy and William Bundy: Brothers in Arms* (New York: Simon & Schuster, 1998), 208–10.

75 **At the time:** Schlesinger, *Thousand Days*, 912.

76 **No one on his staff:** Obituary, *New York Times*, Oct. 31, 2010.

76 **Sorensen was born:** Sorensen, *Counselor*, chaps. I–8, especially, pp. 93, 97, 100–102, 192–93; Schlesinger, *Thousand Days*, 17–19; Dallek, *Unfinished Life*, 179–80. I leave it to readers of my biography of Kennedy to see the extent to which JFK, his family, and staff misled voters in 1960 about his medical history and judge whether it would have changed the outcome of the race. Sorensen thinks fuller revelations would not have changed the result.

80 **During his lifetime:** Sorensen, *Counselor*, 129; Schlesinger, *Journals*, 143–44.

80 **Sorensen's importance:** *Counselor*, 195–96, 198–99.

81 **As Sorensen would eventually find out:** Ibid., 237–40.

81 **Arthur Schlesinger, Jr. was:** Sorensen, *Kennedy*, 296–97; Schlesinger, *Thousand Days*, chaps. I–3, especially pp. 143 and 162; Schlesinger, *Journals*, 63–93, 446–47; Edwin O. Guthman and Jeffrey Shulman, eds., *Robert Kennedy: In His Own Words* (New York: Bantam, 1988), 419.

83 **While Sorensen and Schlesinger:** David Halberstam, *Best and Brightest*, 3–10; Schlesinger, *Thousand Days*, 129.

85 **With Lovett out of the picture:** Robert S. McNamara, *In Retrospect: The Tragedy and Lessons of Vietnam* (New York: Times Books, 1995), chaps. I–4.

88 **The Kennedys didn't care:** Halberstam, *Best and Brightest*, 10; Schlesinger, *Thousand Days*, 166–67; Deborah Shapley, *Promise and Power: The Life and Times of Robert McNamara* (Boston: Little, Brown, 1993), xv–xvi, 11, 21, 88, 234, 270, 539–40.

89 **Kennedy was more focused:** Schlesinger, *Thousand Days*, 208–10; Halberstam, *Best and Brightest*, 44–46, 56–81; Bird, *Color of Truth*, 13–14, 135, 108, 151–53, 190, 192–93.

91 **Because Kennedy intended:** Schlesinger, *Thousand Days*, 150; Halberstam, *Best and Brightest*, 43; Dallek, *Flawed Giant*, 296; see the Walt W. Rostow File at JFKLibrary.org, Digital Archive, especially Rostow to JFK, Aug. 8, 1960; author interview with Rostow, July 27, 1992.

92 **Kennedy had initially:** Halberstam, *Best and Brightest*, 43–44; Bird, *Color of Truth*, 185–86.

93 **Bundy recruited:** Bird, *Color of Truth*, 186–89.

93 **Adlai Stevenson had wanted:** Halberstam, *Best and Brightest*, 26–28, 316; Guthman and Shulman, eds., *Robert Kennedy*, 6. The JFK-Stevenson relationship in 1960 is clearly documented in the 1960 Stevenson file, in POF, JFKL. Also see Abraham Ribicoff, OH; William Atwood, OH; John Sharon, OH; William McCormick Blair, OH, all at Columbia University; Reeves, *President Kennedy*, 25.

95 **Although Kennedy:** See Stevenson's report in his 1960 file, POF; also see, John Sharon, OH.

95 **Kennedy's thoughts:** On Fulbright's early years, see Randall B. Woods, *Fulbright: A Biography* (Cambridge, U.K.: Cambridge University Press, 1995); on his relations with JFK, see Fulbright's January–August 1961 file in POF, JFKL; also Fulbright's oral history in JFKL. On Fulbright's appointment as secretary, see Schlesinger, *Thousand Days*, 139–40; Guthman and Shulman, eds., *Robert Kennedy*, 36–37; and Halberstam, *Best and Brightest*, 29–30.

97 **With Fulbright eliminated:** Guthman and Shulman, eds., *Robert Kennedy*, 5, 37–38.

97 **It was not as if:** On Rusk's career up to 1960, see Warren I. Cohen, *Dean Rusk* (Totowa, NJ: Cooper Square, 1980), chaps. 1–5.

98 **When Kennedy offered Rusk:** Schlesinger, *Thousand Days*, 140–41; Halberstam, *Best and Brightest*, 32–37; Schlesinger, *Robert Kennedy*, 223; Brauer, *Presidential Transitions*, 88–89; Schlesinger, *Journals*, 98; Dean Rusk, *As I Saw It* (New York: Norton, 1990), 201–204.

100 **Rusk's caution:** Rusk, *As I Saw It*, 197–98.

101 **Yet Rusk was never:** Halberstam, *Best and Brightest*, 32–33, 63, 196–97.

101 **Once he made Rusk secretary:** Ibid., 11–24; Walter Isaacson and Evan Thomas, *The Wise Men: Six Friends and the World They Made* (New York: Simon & Schuster, 1986), 583; Howard B. Schaffer, *Chester Bowles: New Dealer in the Cold War* (Cambridge, MA: Harvard University Press, 1993), chap. 10, especially pp. 169–70.

103 **Kennedy's appointment of George Ball:** George W. Ball, *The Past Has Another Pattern: Memoirs* (New York: Norton, 1982), 157–62; James A. Bill, *George Ball: Behind the Scenes in U.S. Foreign Policy* (New Haven, CT: Yale University Press, 1997), 56–60.

104 **With his national security:** JFK's Inaugural Address is available at http://www.presidency.ucsb.edu/ws/?pid=8032; Galbraith, *Letters to Kennedy*, 11.

106 **Anyone listening:** George Gallup, *The Gallup Poll: Public Opinion, 1935–1971* (New York: Random House, 1972), 1676, 1691; Galbraith, *Letters to Kennedy*, 7.

106 **Yet, however much:** William E. Leuchtenburg, *A Troubled Feast: American Society Since 1945* (Boston: Little, Brown, 1973), 111, 113; Alan Ehrenhalt, "Are We as Happy as We Think?" *New York Times*, May 7, 2000; on poverty in the United States, see Michael Harrington, *The Other America* (New York: Macmillan, 1962); also see Scott Stossel, *Sarge: The Life and Times of Sargent Shriver* (Washington, DC: Smithsonian Books, 2004), chap. 25, especially pp. 336–38.

107 **Kennedy understood:** Dallek, *Unfinished Life*, 291–93.

109 **Powers was the first:** Pierre Salinger, *With Kennedy* (Garden City, NY: Doubleday, 1966), 71–72, 90; O'Donnell and Powers, *Johnny, We Hardly Knew Ye*, vii–ix, 52–55; Rose Fitzgerald Kennedy, *Times to Remember* (Garden City, NY: Doubleday, 1974), 310; Dallek, *Unfinished Life*, 127, 307, 476, note about JFK's womanizing on p. 779.

110 **Kenneth O'Donnell was:** Schlesinger, *Thousand Days*, 93; Salinger, *With Kennedy*, 64–65; O'Donnell and Powers, *Johnny, We Hardly Knew Ye*, 81ff, but especially 81, 252–57. RFK's committee was the Senate Select Committee on Improper Activities in the Labor or Management Field.

111 **Kennedy's objective:** Schlesinger, *Thousand Days*, 678–86, especially 685.

112 **Larry O'Brien:** Lawrence F. O'Brien, *No Final Victories: A Life in Politics— from John F. Kennedy to Watergate* (Garden City, NY: Doubleday, 1974), chaps. 1–6, especially pp. 100–101; Guthman and Shulman, eds., *Robert Kennedy*, 48–49; Dallek, *Flawed Giant*, 8–11.

114 **Pierre Salinger's selection:** Salinger, *With Kennedy*, chaps. 1–3 and pp. 49–59. Schlesinger, *Thousand Days*, 716–17; Sorensen, *Counselor*, 341.

116 **Kennedy's limited focus:** Schlesinger, *Thousand Days*, 133–36; Schlesinger, *Journals*, 93–96; Guthman and Shulman, eds., *Robert Kennedy*, 39–40; *New York Times* obituary of Dillon, Jan. 12, 2003; Halberstam, *Best and Brightest*, 435–36.

118 **For both substantive:** Schlesinger, *Thousand Days*, 137; Walter Heller, OH; Heller, "Meeting with the President-elect," Dec. 23, 1960; "Recollections of early Meetings with Kennedy," Jan. 12, 1964, Box 5, Walter Heller Papers, JFKL.

119 **Before Kennedy selected:** Wofford, *Of Kennedys and Kings*, 71–72; Brauer, *Presidential Transitions*, 78–79.

119 **A last consideration for Kennedy:** "Wofford, Harris," Biographical Directory of the United States Congress, Jan. 29, 2006; Wofford, *Of Kennedys and Kings*, 35ff., especially 36, 40, 58, 63–64, 67, 130–34.

122 **Despite Kennedy's directive:** Ibid., note on 133; Guthman and Shulman, eds., *Robert Kennedy*, 57, 77–79.

122 **Kennedy's choice:** Burke Marshall obituary, *New York Times*, June 3, 2003; Schlesinger, *Robert Kennedy*, 288–89.

123 **Executive action:** Dallek, *Flawed Giant*, 10–11.

123 **The president-elect also asked:** Dallek, *Lone Star Rising*, 529–32.

124 **The person most notably:** Jacqueline Kennedy, *Historic Conversations*, 319, 347–48, 201–203. The tour can be accessed on YouTube.

Chapter 4: "Never Rely on the Experts"

127 **Freezing weather:** President's Appointments, January 21, 1961; Charles Bartlett, OH, JFKL.

127 **He believed that the combined:** *Public Papers of the Presidents: John F. Kennedy, 1961* (Washington, DC: U.S. Government Printing Office, 1962), 5, 10, 15–16, 18.

128 **Kennedy's inspiration:** "Report to the President on the Peace Corps," February 1961; "Conversation between the President and Eleanor Roosevelt Discussing the Peace Corps," March 1, 1961, POF, JFKL. Also see Stossel, *Sarge*, 169–72, 198–208; Theodore Sorensen, *Counselor: A Life at the Edge of History*, 329–33; Peace Corps Online.

130 **The Alliance for Progress:** Schlesinger, *Thousand Days*, 193, 223–26; JFK's Inaugural Address; *Robert Kennedy in His Own Words*, 49.

130 **Goodwin was a brilliant:** Schlesinger, *Thousand Days*, 192–94.

131 **When Kennedy announced:** Stephen G. Rabe, "John F. Kennedy and Latin America," *Diplomatic History* (Summer 1999); Thomas Mann, OH, JFKL; Schlesinger to JFK, Feb. 6, 1961, POF; Schlesinger, *Thousand Days*, 205.

131 **The need for wise counsel:** *Foreign Relations of the United States: Cuba, 1961–1962* (Washington, DC: U.S. Government Printing Office, 1997), p. 5, n. 3; pp. 25, 44. (*Foreign Relations of the United States* hereafter abbreviated *FRUS.*)

132 **Because promises of a new day:** Ibid., 46–57.

133 **The great question then for Kennedy:** Ibid., 61–69, 89–90, 92–93. "Be landed gradually" is on 90. Howard Jones, *The Bay of Pigs* (New York: Oxford University Press, 2008); Bird, *Color of Truth*, 198; Dallek, *Unfinished Life*, 359–61; Robert Dallek, "The Untold Story of the Bay of Pigs," *Newsweek*, Aug. 22 and 29, 2011, 26, 28; Gordon M. Goldstein, *Lessons in Disaster: McGeorge Bundy and the Path to War in Vietnam* (New York: Times Books/Henry Holt, 2008), 38–40; O'Donnell and Powers, *Johnny, We Hardly Knew Ye*, 274.

137 **Part of Schlesinger's problem:** Bird, *Color of Truth*, 198–99.

137 **During February and March:** *FRUS: Cuba, 1961–62*, 107–108, 118–20, 143–45, 156–60, 177; *Robert Kennedy in His Own Words*, 246–47; Schlesinger, *Thousand Days*, 251.

139 **Undersecretary Chet Bowles:** *FRUS: Cuba, 1961–1962*, 178–81, 185–89; Richard Goodwin, *Remembering America: A Voice from the Sixties* (Boston: Little, Brown, 1988), 176–77; Jackie Kennedy, *Historic Conversations*, 112–13, 312–13; Schlesinger, *Journals*, 109; Schlesinger, *Thousand Days*, 250–51, 258–59; JFK speech to the American Newspaper Publishers Association, April 27, 1961, POF; *Robert Kennedy in His Own Words*, 242.

142 **Adlai Stevenson was yet another skeptic:** Schlesinger, *Thousand Days*, 271–72; *FRUS: Cuba*, 230–31.

143 **In the final days before the attack:** Ibid., 191–93, 200.

143 **Despite all precautions:** Jacqueline Kennedy, *Historic Conversations*, 182–83.

143 **The operation was a miserable failure:** Sorensen, *Kennedy,* 309; for the hidden CIA history, see Dallek, "The Untold Story of the Bay of Pigs," 26, 28; Jacqueline Kennedy, *Historic Conversations,* 185–86; *FRUS: Cuba, 1961–1962,* note on 221 about General Maxwell Taylor being assigned to chair a committee studying the failure, also 304–06; Schlesinger, *Journals,* 109–110; *Robert Kennedy in His Own Words,* 246–47; Dallek, *Unfinished Life,* 359–68; Jon Wiener, "Bay of Pigs Fifty Years Later: The Lessons Kennedy Never Learned," *Nation,* April 18, 2011.

147 **In a later conversation:** Jacqueline Kennedy, *Historic Conversations,* 190; McNamara, *In Retrospect,* 26–27; Bird, *Color of Truth,* 197–98; Goldstein, *Lessons in Disaster,* 41–43; PPP: JFK, 1961, 312–13; Schlesinger, *Thousand Days,* 289–90.

148 **Not only was it smart politics:** Schlesinger, *Thousand Days,* 258, 296; Goldstein, *Lessons in Disaster,* 41.

149 **Kennedy's public response:** Thomas E. Ricks, *The Generals: American Military Command from World War II to Today* (New York: Penguin Press, 2012), 220; Schlesinger, *Thousand Days,* 295–96; Jacqueline Kennedy, *Historic Conversations,* 183; "CIA: Maker of Policy or Toll?" *New York Times,* April 25, 1966; Thomas Powers, *The Man Who Kept the Secrets: Richard Helms and the CIA* (New York: Knopf, 1979), 115; Goodwin, *Remembering America,* 181. On Bissell, see Robert F. Kennedy Notes, April 22, 1961, 6–22: Cuba: Personal Notes, RFK Confidential Files, JFKL.

150 **While Kennedy sat on his anger:** Schlesinger, *Thousand Days,* 289; Goodwin, *Remembering America,* 183.

150 **The principal fall guy:** PPP: JFK, 1961, 307–308.

151 **Conflicting memos:** *FRUS: Cuba, 1961–1962,* 295–97, 302–304.

152 **Kennedy shared Bobby's:** Salinger, *With Kennedy,* 169–70; Wofford, *Of Kennedys and Kings,* 341–42.

153 **The Bowles onslaught:** *FRUS: Cuba, 1961–1962,* 304–306, 313–14; Goodwin, 187.

154 **Bobby Kennedy thought his attack:** *Robert Kennedy in His Own Words,* 264–65; Schlesinger, *Robert Kennedy,* 472.

155 **Bowles's open dissent:** PPP: JFK, 1961, 518–29; Schaffer, *Chester Bowles,* 220–30.

156 **The principal consequence:** *FRUS: Cuba, 1961–1962,* 306–307; also 309–310; Nancy Gibbs and Michael Duffy, *The Presidents Club: Inside the World's Most Exclusive Fraternity* (New York: Simon & Schuster, 2012), 186; RFK to JFK, June 1, 1961, quoted in Schlesinger, *Robert Kennedy,* 446–47.

157 **Kennedy now wanted a new voice:** *New York Times,* April 21, 1987; John M. Taylor, *An American Soldier: The Wars of General Maxwell Taylor* (Novato, CA: Presidio, 1989); Schlesinger, *Thousand Days,* 309–10; Schlesinger, *Robert Kennedy,* 448.

158 **In the first weeks of his term:** Schlesinger, *Thousand Days,* 338; *Robert Kennedy in His Own Words,* 247–48.

159 **Kennedy was not indifferent:** JFK, "Imperialism—The Enemy of Freedom," July 2, 1957, Compilation of Speeches, JFKL; Dallek, *Unfinished Life,* 350–53.

161 **At the end of January 1961:** *Foreign Relations of the United States: Vietnam,*
     1961 (Washington, DC: U.S. Government Printing Office, 1988),
     12–19.

162 **Because political change:** Walter S. Poole, *History of the Joint Chiefs of Staff:*
     The Joint Chiefs of Staff and National Policy, 1961–1964 (Washington, DC:
     U.S. Government Printing Office, 2011), 23–24; *FRUS: Vietnam, 1961,*
     28, 40, 46–47, 58–60; Halberstam, *Best and Brightest,* 110, 127–32.

164 **Among Kennedy's White House advisers:** *FRUS: Vietnam, 1961,* 68–69,
     82–86, 131; Walt W. Rostow, *The Stages of Economic Growth: A Non-*
     Communist Manifesto (Cambridge, UK: Cambridge University Press, 1960);
     Jacqueline Kennedy, *Historic Conversations,* 315–16; Halberstam, *Best and*
     Brightest, 128–31; Galbraith, *Letters to Kennedy,* 62.

166 **For all these doubts:** Schlesinger, *Thousand Days,* 302–303; *Foreign Relations*
     of the United States: Soviet Union, 1961–1963 (Washington, DC: U.S. Govern-
     ment Printing Office, 1998), 128; *FRUS: Vietnam, 1961,* 126–27.

167 **A Johnson trip:** George A. Smathers, OH, Senate Historical Office,
     Washington, D.C.; Dallek, *Flawed Giant,* 12–17; *FRUS: Vietnam, 1961,* 143.

168 **Yet as Johnson told Kennedy:** Dallek, *Flawed Giant,* 17–18; *FRUS: Vietnam,*
     1961, 149–57; Halberstam, *Best and Brightest,* 133–35.

170 **Walt Rostow didn't wait:** *FRUS: Vietnam, 1961,* 157–58, 166.

171 **Despite Kennedy's directive:** Ibid., 172–74.

171 **As it was, White House and Pentagon:** Ibid., 195–96, 198–200.

172 **With the Cuban failure:** "Off the Record Briefing with the President,"
     Dec. 31, 1961, Box WH 66, Arthur Schlesinger Papers, JFKL.

Chapter 5: "Roughest Thing in My Life"

173 **During his first months in office:** Schlesinger, *Thousand Days,* 315; Bayard
     Rustin, OH, Columbia University; Louis Martin to Sorensen, May 10,
     1961, Box 66, RFK Papers, JFKL.

174 **Kennedy hoped:** Taylor Branch, *Parting the Waters: America in the King Years,*
     1954–1963 (New York: Simon & Schuster, 1988), 405.

174 **An executive order:** Dallek, *Flawed Giant,* 23–29, 32–35.

175 **The White House came in for additional criticism:** Dallek, *Unfinished*
     Life, 383–88; Branch, *Parting the Waters,* 414–16.

176 **The administration's travails:** PPP: JFK, 1961, 396–403; Sorensen,
     Counselor, 334–35.

177 **On April 20:** JFK to LBJ, April 20, 1961, POF, JFKL.

178 **Kennedy largely knew:** Dallek, *Flawed Giant,* 21.

179 **McNamara and Rusk agreed:** Michael R. Beschloss, "Kennedy and
     the Decision to Go to the Moon," in Roger D. Launius and Howard
     E. McCurdy, eds., *Spaceflight and the Myth of Presidential Leadership* (Urbana:
     University of Illinois Press, 1997), 57–58.

179 **Kennedy was less inclined:** PPP: JFK, 1961, 403–405.

180 **Kennedy saw serious risks:** Dallek, *Flawed Giant,* 21–22. See Hugh L.
     Dryden, OH, March 24, 1964, JFKL, who recounts how little discussion
     there was with JFK about the moon venture. Everything went through
     Johnson. Also, Alan B. Shepard, Jr., OH, June 12, 1964, JFKL.

180 **Kennedy's speech also came:** *Foreign Relations of the United States: Berlin Crisis, 1961–1962* (Washington, DC: U.S. Government Printing Office, 1993), 3–9.

181 **Kennedy was keen to avoid:** Ibid., 21–22, 25–30; McGeorge Bundy, OH, #1, JFKL.

182 **Brandt's pessimism:** *FRUS: Berlin Crisis, 1961–1962*, 30–44; also Dallek, *Nixon and Kissinger*, 54–56; Dean Acheson, OH, JFKL; Schlesinger, *Thousand Days*, 380–81, 406, 413; Michael Fullilove, *Rendezvous with Destiny: How Franklin D. Roosevelt and Five Extraordinary Men Took America into the War and into the World* (New York: Penguin Press, 2013), 293.

184 **Conversations with Adenauer and:** *FRUS: Berlin Crisis, 1961–1962*, 45–51, 56.

184 **In 1961, in the immediate aftermath:** Ibid., 61–63, 66–69.

185 **Still, unless one side or the other:** Ibid., 77–79.

186 **The recommendations left Kennedy:** Ibid., 80–86; Dallek, *Unfinished Life*, 394–97.

186 **At an initial May 31 meeting:** Sir Alec Douglas-Home, OH; Isaiah Berlin, OH; Charles Bohlen, OH; Nicholas Wahl to McGeorge Bundy, May 1961, Box 331, National Security File; "President's Visit to de Gaulle," May 27, 1961, Box 116A, POF, all in JFKL; Beschloss, *Crisis Years*, 183; *Time*, June 9, 1961.

189 **As he prepared to meet:** Taubman, *Khrushchev: The Man and His Era* (New York: Norton, 2003), xviii–xx, 492–94.

190 **The State Department:** *FRUS: Soviet Union, 1961–1963*, 153–60, 164–70.

192 **Llewellyn Thompson also had doubts:** Ibid., 163–64.

192 **Averell Harriman:** Schlesinger, *Thousand Days*, 149–50; Halberstam, *Best and Brightest*, 73–75.

194 **Charles Bohlen doubted:** Charles Bohlen, OH, JFKL.

194 **But Khrushchev would concede no weakness:** Simon S. Montefiore, *Stalin: The Court of the Red Tsar* (New York: Vintage Books, 2005), 152–57; Taubman, *Khrushchev*, 213–15.

195 **Khrushchev learned the art:** Taubman, *Khrushchev*, xi, 427–28, 474–76.

196 **Khrushchev came to Vienna:** *New York Times*, June 4, 1961; Kenneth O'Donnell Tapes, Tape 51, JFKL; Beschloss, *Crisis Years*, 191–92; Jacqueline Kennedy, *Conversations*, 198.

197 **The initial formal discussions:** *FRUS: Soviet Union, 1961–1963*, 172–78; Beschloss, *Crisis Years*, 196, 234; O'Donnell and Powers, *Johnny, We Hardly Knew Ye*, 195; Taubman, *Khrushchev*, 493–500; *Robert Kennedy in His Own Words*, 28–29; Dallek, *Unfinished Life*, 404–14.

199 **What had particularly agitated:** *FRUS: Berlin Crisis, 1961–1962*, 87–98.

200 **Kennedy stopped in London:** Henry Brandon Diaries, June 9, 1961, Library of Congress; Harold Macmillan, *Pointing the Way, 1959–1961* (New York: Harper, 1972), 355–59.

201 **On returning to Washington:** *FRUS: Soviet Union, 1961–1963*, 232–37; PPP: JFK, 1961, 441–46.

202 **Kennedy now:** Ibid., 104–105, 107–109; National Action Security Memo No. 55, June 28, 1961, National Security File, JFKL; *The Gallup Poll, 1959–1971,* 1726, 1729; *President Kennedy,* 188–89; PPP: JFK, 1961, 476–77, 481.

203 **On the twenty-eighth, Kennedy:** Schlesinger, *Thousand Days,* 380; McGeorge Bundy, *Danger and Survival: Choices About the Bomb in the First Fifty Years* (New York: Random House, 1990), 371–75; *FRUS: Berlin Crisis, 1961–1962,* 138–41, 160–62; Dean Acheson, OH, April 27, 1964, JFKL; Douglas Brinkley, *Dean Acheson: The Cold War Years, 1953–1971* (New Haven, CT: Yale University Press, 1992), 125, 196; Robert L. Beisner, *Dean Acheson: A Life in the Cold War* (New York: Oxford University Press, 2006), 629–30; *Robert Kennedy in His Own Words,* 19; Jacqueline Kennedy, *Historic Conversations,* 30–31.

Chapter 6: Advice and Dissent

207 **Despite all the difficulties:** *PPP: JFK, 1961,* 481; Schlesinger, *Journals,* 122.

207 **Kennedy's initial problems:** Paul Boyer, *By the Bomb's Early Light* (New York: Pantheon Books, 1985), 352–55; Bundy, *Danger and Survival,* 324.

208 **In 1957–58, the Soviet Union:** Glenn T. Seaborg, *Kennedy, Khrushchev, and the Test Ban* (Berkeley: University of California Press, 1981), chaps. 1–2 and pp. 30, 32–34.

209 **In March 1961:** Ibid., pp. 45–48, 63–66; *Foreign Relations of the United States: Arms Control and Disarmament, 1961–1963* (Washington, DC: U.S. Government Printing Office, 1995), 38–41, 53–57, 69–71, 81–83.

210 **The exchanges at the conference:** *FRUS: Arms Control,* 83–92; Seaborg, *Kennedy, Khrushchev, and the Test Ban,* 66–68.

210 **After Vienna, Moscow:** Seaborg, *Kennedy, Khrushchev, and the Test Ban,* 68–78; Schlesinger, *Thousands Days,* 398; Reeves, *President Kennedy,* 223; PPP: JFK, 1961, 486–87; *FRUS: Arms Control,* 150–56.

211 **At the same time, Kennedy:** PPP: JFK, 1961, 618–26.

212 **But the Russians seemed impervious:** Seaborg, *Kennedy, Khrushchev, and the Test Ban,* 84, 111; *FRUS: Arms Control,* 217–21.

213 **Kennedy's ambivalence:** Seaborg, *Kennedy, Khrushchev, and the Test Ban,* 126–31; *FRUS: Arms Control,* 282–88.

213 **With little success in turning:** *FRUS: Cuba, 1961–1962,* 640–47, 654–55, 657.

215 **At the same time, however:** Ibid., 641, 657.

215 **But White House discussions:** Ibid., 664–68; Bird, *Color of Truth,* 199; Thomas, *Robert Kennedy,* 145–48.

217 **Yet the administration:** *FRUS: Cuba, 1961–1962,* 684–89; Thomas, *Robert Kennedy,* 151–52.

217 **Lansdale was the president's and Bobby's:** *Robert Kennedy in His Own Words,* 378; Schlesinger, *Robert Kennedy,* 461–62, 466–67; Thomas, *Robert Kennedy,* 148–50; *New York Times* obituary, Feb. 24, 1987; Jonathan Nashel, *Edward Lansdale's Cold War* (Amherst: University of Massachusetts Press, 2006).

219 **As much as the Kennedys:** *Robert Kennedy in His Own Words,* 378; Schlesinger, *Robert Kennedy,* 478–85; Thomas, *Robert Kennedy,* 151–52.

220 **The Joint Chiefs also got into:** The documents on Operation Northwoods are available at the National Archives website.

220 **Unhappily for the president:** Schlesinger, *Robert Kennedy,* 477–78.

221 **While nuclear talks and Cuba:** *Newsweek,* July 3, 1961.

221 **The *Newsweek* article:** *FRUS: Berlin Crisis, 1961–1962,* 187–202, 209–22; Acheson's comment is in Honoré M. Catudal, *Kennedy and the Berlin Wall Crisis* (Berlin: Berlin-Verlag, 1980), 182; McGeorge Bundy, OH, JFKL; Acheson Speech, July 25, 1961, POF, JFKL.

222 **To counter Acheson's assault:** PPP: JFK, 1961, 513–21.

223 **The questions to Kennedy:** Schlesinger, *Thousand Days,* 390–91.

223 **Kennedy saw the speech:** Dallek, *Unfinished Life,* 422–24; PPP: JFK, 1961, 533–40.

224 **Kennedy's speech had its desired effect:** Taubman, *Khrushchev,* 503–506.

225 **The Wall touched off:** Bird, *Color of Truth,* 212; Shapley, *Promise and Power,* 121; Poole, *History of the Joint Chiefs,* 148–49; *FRUS: Berlin Crisis, 1961–1962,* 339–42.

226 **Kennedy and his White House advisers:** O'Donnell and Powers, *Johnny, We Hardly Knew Ye,* 303; *FRUS: Berlin Crisis, 1961–1962,* 330–32.

226 **The pressure on Kennedy:** O'Donnell and Powers, *Johnny, We Hardly Knew Ye,* 299; Dallek, *Flawed Giant,* 19–20.

227 **But uncertainties remained:** *FRUS: Berlin Crisis, 1961–1962,* 359–60, 377–78, 392, 397–98, 435–37.

228 **The clash of opinions:** Ibid., 444–55; Salinger, *With Kennedy,* chapter 12.

229 **With the opening of negotiations:** Schlesinger, *Thousand Days,* 400.

229 **Kennedy's decision in early July:** *FRUS: Vietnam, 1961,* 195–96, 198–200, 205–207.

230 **Rostow was not alone:** Halberstam, *Best and Brightest,* 147; *FRUS: Vietnam, 1961,* 216–20.

231 **Others in the Kennedy administration:** Ibid., 234–36, 243–44, 248–51.

232 **Three days after his July 25:** Ibid., 252–56; Galbraith, *Letters to Kennedy,* 70–71; PPP: JFK, 1961, 624.

234 **Mindful that Kennedy:** Galbraith, *Letters to Kennedy,* 76–77.

234 **Galbraith thought:** *FRUS: Vietnam, 1961,* 256–57, 267–69, 292; William C. Gibbons, *The U.S. Government and the Vietnam War: Executive and Legislative Roles and Relationships, 1961–1964* (Washington, DC: U.S. Government Printing Office, 1984), 60–61; Halberstam, *Best and Brightest,* 150; White to JFK, Oct. 11, 1961, POF, JFKL.

235 **With the Berlin Wall:** *FRUS: Vietnam, 1961,* 283; Gibbons, *The U.S. Government and the Vietnam War,* 61–63.

236 **But no one could deny:** *FRUS: Vietnam, 1961,* 336–46; Gibbons, *The U.S. Government and the Vietnam War,* 66–71.

237 **In choosing the team:** Halberstam, *Best and Brightest,* 165–66; *FRUS: Vietnam, 1961,* 359, 362–63.

238 **On October 16:** *FRUS: Vietnam, 1961,* 381–82, 443.

238 **Despite Kennedy's directive:** Ibid., 456–57, 467–70.

239 **Mansfield's memo:** Gibbons, *The U.S. Government and the Vietnam War*, 72, n.2.

239 **Taylor and everyone on the mission:** *FRUS: Vietnam, 1961*, 427–30, 477–81, 489–94; Gibbons, *The U.S. Government and the Vietnam War*, 72–78.

240 **After seeing President Kennedy:** *FRUS: Vietnam, 1961*, 532–34, 538–40, 543–44, 547–48, 550–52, 572–73; Ball, *Past Has Another Pattern*, 363, 365; McNamara, *In Retrospect*, 38–39; Schlesinger, *Thousand Days*, 547.

241 **The memo gave Kennedy support:** *FRUS: Vietnam, 1961*, 576–78; Gibbons, *The U.S. Government and the Vietnam War*, 91–92.

242 **A fierce argument:** *FRUS: Vietnam, 1961*, 580–83, 601–603, 605–607; Gibbons, *The U.S. Government and the Vietnam War*, 101–102; Halberstam, *Best and Brightest*, 154.

244 **Bundy's advice resonated:** *FRUS: Vietnam, 1961*, 607–10, 603–604.

245 **Yet for all his skepticism:** Ibid., 591–94, 604, 615–18.

245 **McNamara later asserted:** Ibid., 615–18, 636–39, 664.

246 **In retrospect, McNamara saw:** McNamara, *In Retrospect*, 39–40.

247 **Writing thirty-five years after:** Arthur Krock, *Memoirs: Fifty Years on the Firing Line* (New York: Funk & Wagnalls, 1968), 332–33; Wofford, *Of Kennedys and Kings*, 379.

248 **In November 1961:** *FRUS: Vietnam, 1961*, 672, 678–79.

248 **As 1961 came to an end:** Steinbeck is quoted in T. D. Schellhardt, "Do We Expect Too Much?" *Wall Street Journal*, July 10, 1979; "JFK on Presidency," Box 23, David Powers Papers, JFKL; RFK interview, Oct. 23, 1961, Box 2, RFK Personal Papers, JFKL; PPP: JFK, 1962, 276.

Chapter 7: "The Greatest Adventure of Our Century"

251 **At the start of 1962:** Schlesinger, *Journals*, 141–42. Both State of the Union speeches can be viewed online. JFK's January 15, 1962, press conference is also online. Roy Wilkins, OH, Columbia University; Schlesinger, *Robert Kennedy*, 307–10, 376–78; *Robert Kennedy: In His Own Words*, 108–109, 112; Dallek, *Unfinished Life*, 492–96.

253 **In Kennedy's view, the greatest danger:** *FRUS: Cuba, 1961–1962*, 710, 720–21, 745–46, 771, 785.

254 **Yet the administration's:** Schlesinger, *Robert Kennedy*, 513; William Manchester, *The Death of a President* (New York: Harper & Row, 1967), 48–49, 85; *Robert Kennedy in His Own Words*, 27–28, 311; *FRUS: Berlin Crisis, 1961–1962*, 308–310.

255 **Khrushchev's aggressiveness:** Seaborg, *Kennedy, Khrushchev, and the Test Ban*, 126–31; *FRUS: Arms Control*, 295–97, 331–32, 411, 414–15, 439–42.

256 **Yet Kennedy was unwilling:** *FRUS: Arms Control*, 357–58, 372, 384, 447–48, 450, 456–59 .

256 **It also had the advantage:** Ibid., 487–88, 410–14; Seaborg, *Kennedy, Khrushchev, and the Test Ban*, 62–66; Philip Zelikow, Ernest May, and Timothy Naftali, eds., *The Presidential Recordings: John F. Kennedy: The Great Crises, July 30–October 28, 1962*, 3 vols. (New York: Norton, 2001), vol. I, 47–50.

258 **The resurfacing of the civil rights struggle:** Memorandum in the March 21–31, 1962, Folder, Box 50, POF, JFKL; Wofford, *Of Kennedys and Kings*, 159–61; *Robert Kennedy in His Own Words*, 157–58, 162–65; Berl Bernhard, OH, JFKL.

258 **Kennedy took comfort:** *The Gallup Poll, 1959–1971*, 1751, 1755, 1764–65, 1769, 1771.

259 **In the spring of 1962:** Dallek, *Unfinished Life*, 483–87.

260 **The struggle to find:** *FRUS: Vietnam, 1962*, 3–4, 14–16, 32; *The Pentagon Papers: The Defense Department History of United States Decision Making on Vietnam* (Boston: Bantam Books, 1971), vol. 2, 662–66.

261 **But Kennedy remained:** PPP: JFK, 1962, 17; Gibbons, *The U.S. Government and the Vietnam War*, 108.

262 **As the American role:** *FRUS, Vietnam, 1962*, 54, 73–90, 92–93, 95, 98, 124–25.

263 **On February 14:** *New York Times*, Feb. 14, 1962; PPP: JFK, 1962, 136–37.

264 **But the reporters in Vietnam:** Montague Kern, Patricia W. Levering, and Ralph B. Levering, *The Kennedy Crises: The Press, the Presidency, and Foreign Policy* (Chapel Hill: University of North Carolina Press, 1983), 5; *FRUS: Vietnam, 1962*, 129–32, 139, 156, 158–60.

265 **In February, as the U.S. effort:** *FRUS: Vietnam, 1962*, 171, 176–82, 186–87, 194–97; on Bobby Kennedy, see *New York Times*, Feb. 19, 1962, and *FRUS: Vietnam, 1962*, 230.

267 **While Kennedy battled:** McNamara, *In Retrospect*, 41–45.

268 **In the first half of 1962:** *FRUS: Vietnam, 1962*, 209, 216–19, 222, 233–34, 244–45, 273, 283.

268 **Kennedy had a sense of urgency:** *New York Times*, Feb. 8, 9, 10, 12, 17, 24, 28, March 8, April 1, 1962; *FRUS: Vietnam, 1962*, 206–207, 276, 305–306; Ball, *Past Has Another Pattern*, 367.

269 **Kennedy's eagerness:** Galbraith, *Letters to Kennedy*, 98–103.

271 **The same day Galbraith:** *FRUS: Vietnam, 1962*, 299–303; Isaacson and Thomas, *The Wise Men*, 583.

271 **By contrast, on April 6:** *FRUS: Vietnam, 1962*, 309–10, 317–18, 324–27; Halberstam, *Best and Brightest*, 35; Chester Bowles, *Promises to Keep: My Years in Public Life, 1941–1969* (New York: Harper & Row, 1971), 410–14.

273 **On May 1:** *FRUS: Vietnam, 1962*, 366–67.

273 **At the same time:** Ibid., 375–76, 399.

274 **Administration resistance:** Gibbons, *The U.S. Government and the Vietnam War*, 121–22; *FRUS: Vietnam, 1962*, 337–39, 353, 358, 364, 373, 379, 386–87; Shapley, *Promise and Power*, 146–51.

275 **Journalists who trailed McNamara:** Shapley, *Promise and Power*, 151–52; *New York Times*, May 17, July 25, 29, 1962; April 17, 1991; *FRUS: Vietnam, 1962*, 403–404, 489; Gibbons, *The U.S. Government and the Vietnam War*, 122–23.

277 **Because serious negotiations:** *FRUS: Vietnam, 1962*, 418–19, 425–26, 432–33, 437; Gibbons, *The U.S. Government and the Vietnam War*, 124–25.

Chapter 8: "If We Listen to Them, None of Us Will Be Alive"

279   And Schlesinger in particular: Schlesinger, *Journals*, 156–61.

280   In the summer of 1962: *FRUS: Vietnam, 1962*, 484–97, 506–10, 541.

280   Kennedy was reluctant: Ibid., 543–46; Gibbons, *The U.S. Government and the Vietnam War*, 12–21.

280   At the end of July: *FRUS: Vietnam, 1962*, 546–56; Shapley, *Promise and Power*, 160–61.

281   In August, the State Department: *FRUS: Vietnam, 1962*, 581, 583–84.

282   In September, Kennedy sent: Dallek, *Unfinished Life*, 495–500, 506–17.

283   Aside from conversations: Branch, *Parting the Waters*, 656–70; Schlesinger, *Robert Kennedy*, 315–16; Dallek, *Unfinished Life*, 514–18.

283   While Kennedy temporarily fixed: *FRUS: Vietnam, 1962*, 636–41, 660; Zelikow, May, and Naftali, eds., *Presidential Recordings*, vol. 2, 165, 169.

284   No one close to Kennedy: *New York Times*, Aug. 19, 28, 1962; Oct. 9, 1962; *FRUS: Vietnam, 1962*, 596–601.

285   Joe Mendenhall: *FRUS: Vietnam, 1962*, 649–50, 661–62, 671–72.

286   At the center of the administration's: Ibid., 679, 687.

287   While the administration struggled: *FRUS: Cuba, 1961–1962*, 947–49; Poole, *History of the Joint Chiefs*, 159–60.

287   John McCone: Halberstam, *Best and Brightest*, 152–53.

288   The minute McCone saw evidence: *FRUS: Cuba, 1961–1962*, 947, 950, 955, 957.

289   Roger Hilsman: Halberstam, *Best and Brightest*, 123, 190; *FRUS: Cuba, 1961–1962*, 963–66, 968, 1045, n. 1; Michael Forrestal, OH, JFKL; Zelikow, May, and Naftali, eds., *Presidential Recordings*, vol. 1, 130–31.

290   In trying to mute speculation: *FRUS: Cuba*, 1004, 1052, 1070–71.

290   When news of the Soviet buildup: *FRUS: Cuba*, 1002–1003; JFK Press Conference, Sept. 13, 1962, available online.

290   As with CIA and military: See Matthias Uhl and Vladimir I. Ivkin, "'Operation Atom': The Soviet Union's Stationing of Nuclear Missiles in the German Democratic Republic, 1959," *Bulletin: Cold War History Project* (Fall/Winter, 2001): 299–306. Also see Dallek, *Unfinished Life*, note for p. 537 on p. 787, recounting my conversation with Raymond L. Garthoff, March 19, 2002; *FRUS: Cuba*, 1083–84.

292   At the beginning of October: *Foreign Relations of the United States: Cuban Missile Crisis and Aftermath* (Washington, DC: U.S. Government Printing Office, 1996), 13–15; Zelikow, May, and Naftali, eds., *Presidential Recordings*, vol. 2, 393–95; McGeorge Bundy, OH, JFKL; Widmer, *Listening In*, 77.

293   Kennedy was convinced: Dallek, *Unfinished Life*, 538, 544; *FRUS: Cuba, 1961–1962*, 1047; *FRUS: Cuban Missile Crisis*, 28.

293   The result shocked Kennedy: Bird, *Color of Truth*, 226–27; Goldstein, *Lessons on Disaster*, 72–73; McNamara, *In Retrospect*, 32, 117; Schlesinger, *Journals*, 171–72; Thomas, *Robert Kennedy*, 209.

295   The presence of the missiles: *Robert Kennedy in His Own Words*, 14–16.

296   Shortly before noon: Zelikow, May, and Naftali, eds., *Presidential Recordings*, vol. 2, 397, n. 10, 409–11, 413–14; Jacqueline Kennedy, *Historic Conversations*, 273, n. 73, 274, 276, 278.

297 **Bundy, acting CIA director:** *Presidential Recordings*, Vol. 2, 399–402, 407–11, 413, 423.

299 **Listening to the discussion:** Ibid., 404–407, 411–13, 416, 421; *Robert Kennedy in His Own Words*, 38, 44; Aleksandr Fursenko and Timothy Naftali, *"One Hell of a Gamble": Khrushchev, Castro, and Kennedy, 1958–1964* (New York: Norton, 1997), 214.

300 **During the meeting, Bobby:** Zelikow, May, and Naftali, eds., *Presidential Recordings*, vol. 2, 416, 425.

301 **In the five hours before the group:** Ibid., 427–28; *FRUS: Cuban Missile Crisis*, 45–47, 49, n., 100.

302 **The same group of advisers:** Zelikow, May, and Naftali, eds., *Presidential Recordings*, vol. 2, 429–33.

302 **Rusk and Martin now weighed in:** Ibid., 433–35.

303 **Unwilling to decide:** Ibid., 435–39. On McNamara and nuclear weapons, see Shapley, *Promise and Power*, 119–20; McNamara, *In Retrospect*, 345; Robert S. McNamara, *Argument Without End: In Search of Answers to the Vietnam Tragedy* (New York: PublicAffairs, 1999), 158–59.

304 **Kennedy's hard line:** Zelikow, May, and Naftali, eds., *Presidential Recordings*, vol. 2, 439–43.

305 **Yet he could not discount:** Ibid., vol. 2, 444–47.

305 **McNamara didn't think the timing:** Ibid, vol. 2, 448–50, 468–69. On RFK and Bolshakov, see Fursenko and Naftali, *"One Hell of a Gamble,"* 109–14; Robert F. Kennedy, *Thirteen Days: A Memoir of the Cuban Missile Crisis* (New York: Norton, 1968), 19–22, 26–27; Isaiah Berlin, OH, JFKL.

307 **After Kennedy left:** Zelikow, May, and Naftali, eds., *Presidential Recordings*, vol. 2, 463–65.

308 **The recorded conversations:** For Stevenson's letter, see *FRUS: Cuban Missile Crisis*, 101–102; Dallek, *Unfinished Life*, 576.

309 **By Thursday morning:** Zelikow, May, and Naftali, eds., *Presidential Recordings*, vol. 2, 512–15.

310 **As the advisers convened:** Ibid., vol. 2, 516, 521–24.

310 **McNamara was more supportive:** Ibid., vol. 2, 525–29.

311 **Kennedy was not convinced:** Ibid., vol. 2, 528–29, 541, 550, 552.

311 **If Kennedy needed support:** For a portrait of Bohlen, see Isaacson and Thomas, *The Wise Men*; Zelikow, May, and Naftali, eds., *Presidential Recordings*, vol. 2, 524–25; *FRUS: Cuban Missile Crisis*, 96–97, 107.

312 **Bohlen's departure:** Zelikow, May, and Naftali, eds., *Presidential Recordings*, vol. 2, 515, n. 20; *Robert Kennedy in His Own Words*, 18.

313 **Kennedy was also content to have:** On Thompson, see Taubman, *Khrushchev*, 397, 449–50, 458; Zelikow, May, and Naftali, eds., *Presidential Recordings*, vol. 2, 532–35, 539, 547–49.

314 **After two days of discussion:** Zelikow, May, and Naftali, eds., *Presidential Recordings*, vol. 2, 557, 563, 565, 567–68.

314 **A series of evening meetings:** Ibid., vol. 2, 572–77.

315 **It is striking that Kennedy:** Dallek, *Unfinished Life*, 92–94, 517; Reeves, *President Kennedy*, 363.

315 **The meeting confirmed his assumption:** Zelikow, May, and Naftali, eds., *Presidential Recordings*, vol. 2, 580–98.

317 **Kennedy was also angry:** Schlesinger, *Robert Kennedy*, 511.

317 **While Kennedy had concluded:** Sorensen, *Kennedy*, 692; *FRUS: Cuban Missile Crisis*, 116–22.

318 **Despite his show of confidence:** Zelikow, May, and Naftali, eds., *Presidential Recordings*, vol. 2, 600–01; *FRUS: Cuban Missile Crisis*, 126–36, 141–51, 162–63.

319 **On Monday, October 22:** *FRUS: Cuban Missile Crisis*, 153, 157–63.

319 **The initial Soviet response:** Ibid., 170–71, 174–75.

319 **That evening, at the end:** Ibid., 177; Robert Kennedy, *Thirteen Days*, 49, 98.

320 **October 24 was a day:** *FRUS: Cuban Missile Crisis*, 177; Zelikow, May, and Naftali, eds., *Presidential Recordings*, vol. 3, 183–85; Fursenko and Naftali, *"One Hell of a Gamble,"* 258; Robert Kennedy, *Thirteen Days*, 52; Schlesinger, *Robert Kennedy*, 514.

321 **There were also hopeful signs:** Zelikow, May, and Naftali, eds., *Presidential Recordings*, vol. 3, 188, 191, 196–97, 209–10.

321 **Yet the crisis was far from over:** Ibid., 197; Shapley, *Promise and Power*, 176–78.

322 **But even if Kennedy:** *FRUS: Cuban Missile Crisis*, 174–75, 185–87.

322 **But Schlesinger passed along:** Ibid., 187–88, 198.

323 **Kennedy sent his reply:** Zelikow, May, and Naftali, eds., *Presidential Recordings*, vol. 3, 232–81; *FRUS: Cuban Missile Crisis*, 210–12, 224–26, 232.

324 **On Friday morning:** Zelikow, May, and Naftali, eds., *Presidential Recordings*, vol. 3, 286, 297–302, 309–10, 313, 317, 321, 323, 328.

324 **If he had to resort:** Ibid., 346–48.

325 **Meanwhile, all this talk:** Ibid., 331–36; Fursenko and Naftali, *"One Hell of a Gamble,"* 259–60, 263–65.

325 **And then at about nine:** *FRUS: Cuban Missile Crisis*, 235–40.

326 **Although an end to the crisis:** Zelikow, May, and Naftali, eds., *Presidential Recordings*, vol. 3, 356–87.

327 **The discussion continued:** Ibid., vol. 3, 387–483, especially 387–400 and 427–28. For JFK to NK, Oct. 27, 1962, *FRUS: Cuban Missile Crisis*, 268–69.

328 **The letter was to be hand-delivered:** Zelikow, May, and Naftali, eds., *Presidential Recordings*, vol. 3, 483–88.

328 **Kennedy had ample reason:** *FRUS: Cuban Missile Crisis*, 275.

329 **Kennedy was spared:** Fursenko and Naftali, *"One Hell of a Gamble,"* 283–87; Zelikow, May, and Naftali, eds., *Presidential Recordings*, vol. 3, 512–17.

329 **Kennedy and his civilian advisers:** Zelikow, May, and Naftali, eds., *Presidential Recordings*, vol. 3, 517–18; Mimi Alford, *Once Upon a Secret*, 93–96; Jacqueline Kennedy, *Historic Conversations*, 236, 262–63; Isaiah Berlin, OH, JFKL; Michael Beschloss, *The Crisis Years: Kennedy and Khrushchev, 1960–1963* (New York: HarperCollins, 1991), 544. JFK's remark to Galbraith is quoted by Sheldon Stern, "Noam Chomsky and the Cuban

Missile Crisis," Oct. 18, 2012, History News Network, online. JFK to
McNamara, Nov. 5, 1962, Box 274, National Security File, JFKL; Poole,
History of the Joint Chiefs, 183–85.

331 **What lessons did Kennedy:** Schlesinger, *Robert Kennedy*, 507; Schlesinger,
Thousand Days, 831; Jacqueline Kennedy, *Historic Conversations*, 254, 271;
Zelikow, May, and Naftali, eds., *Presidential Recordings*, vol. 3, 518; *Robert
Kennedy in His Own Words*, 18, 420. Also see James G. Blight, Bruce J.
Allyn, and David A. Welch, *Cuba on the Brink: Castro, the Missile Crisis, and the
Soviet Collapse* (New York: Pantheon Books, 1993), 249, 352–56, on the
likelihood of a nuclear exchange.

333 **Bobby should have included:** Anatoly Dobrynin, *In Confidence: Moscow's
Ambassador to America's Six Cold War Presidents (1962–1986)* (New York:
Times Books/Random House, 1995), 78–79, 84–93.

Chapter 9: "Mankind Must Put an End to War"

335 **In December 1962:** Gallup, vol. 3: *The Gallup Poll, 1959–1971*, 1793,
1796, 1798–99, 1810; Meyer Feldman to JFK, Aug. 15, 1963, POF;
O'Donnell and Powers, *Johnny, We Hardly Knew Ye*, 13; PPP: JFK, 1963,
828; Schlesinger, *Journals*, 185.

336 **Only Vietnam cast a shadow:** *FRUS: Vietnam, 1962*, 750–51, 757–58,
761, 763–65, 789–96.

337 **After visiting Vietnam:** Ibid., 779–87, 797–98; Halberstam, *Best and
Brightest*, 205–08; Gibbons, *The U.S. Government and the Vietnam War*, 131.

338 **A front-page *New York Times* story:** Gibbons, *The U.S. Government and the
Vietnam War*, 132–34; O'Donnell and Powers, *Johnny, We Hardly Knew Ye*,
15; David Kaiser, *American Tragedy: Kennedy, Johnson and the Origins of the Vietnam War* (Cambridge, MA: Harvard University Press, 2000), 180; *New
York Times*, Dec. 3, 1962.

339 **In public, he tried to maintain:** JFK Press Conference, Dec. 12, 1962,
JFKL.

339 **Events in January:** *FRUS: Vietnam, Jan.–Aug., 1963* (Washington, DC:
U.S. Government Printing Office, 1991), 1–3; Roger Hilsman, *To Move a
Nation: The Politics of Foreign Policy in the Administration of John F. Kennedy* (New
York: Dell, 1968), 447–49.

340 **Because Mansfield's report:** Hilsman, *To Move a Nation*, 453–54; *FRUS:
Vietnam, 1963*, 3–4, n. 1, 5, 49–50, 52, 60.

341 **Kennedy at once read:** Ibid., 63, 73, 89–91, 94–95, 97–98. On the tensions between the civilian and military advisers, see John M. Newman,
JFK and Vietnam (New York: Warner Books, 1992), 305, 312–13.

342 **For the next three months:** PPP: JFK, 1963, 11, 20, 34, 243–44.

343 **While Kennedy stood aside, the debate:** David Halberstam, *The Making of
a Quagmire* (New York: Rowman & Littlefield, 2008), chapters 6 and 7.
For the Wheeler quote, see Hilsman, *To Move a Nation*, 426.

344 **The argument in Washington and Saigon:** *FRUS: Vietnam, January–August
1963*, 105–06, 126–28, 132–33.

345 **The one thing policymakers:** Halberstam, *Making of a Quagmire*, 90.

345 **The U.S. military:** Ibid., 89–90.

346 **There were also dissenting voices:** Ibid., 93–101. Also see Neil Sheehan, *A Bright Shining Lie: America in Vietnam* (New York: Random House, 1988).

346 **The State Department also downplayed:** *FRUS: Vietnam, January–August 1963,* 161–62, 169–73, 189.

347 **Robert G. K. Thompson:** Ibid., 193, 198–200, 205, 207–13.

348 **Still, Diem and the Nhus:** Ibid., 207–13, 222–25, 232–35, 243–46.

349 **By the spring of 1963:** Ibid., 256–58, 261–62, 265, 268, 270; taped conversation, May 7, 1963, JFKL.

349 **In May, if Kennedy needed:** *FRUS: Vietnam, January–August 1963,* 277–78, 283, 294–300, 303–05.

350 **Kennedy was caught between:** *New York Times,* May 5, 1963; JFK Press Conference, May 22, 1963; O'Donnell and Powers, *Johnny, We Hardly Knew Ye,* 16. Also see Newman, *JFK and Vietnam,* 319–25, who argues that Kennedy was playing a double game: talking in public about winning the war, while using it as a cover to withdraw under the guise of victory.

351 **Because Kennedy had lost:** *FRUS: Vietnam, January–August 1963,* 316–24.

351 **Pressure to dump Diem:** *New York Times,* June 9, 14, 16, 1963; *FRUS: Vietnam, January–August 1963,* 362–64, 366–69, 374, 377–78, 381–83.

352 **Rusk's warning to Diem:** Dallek, *Unfinished Life,* 594–606.

355 **Despite Kennedy's injunction:** *FRUS: Vietnam, January–August 1963,* 386–87, n. 5, 393–95, 405–09, 413–15; *New York Times,* June 14, 1963.

356 **Kennedy's eagerness:** *FRUS: Arms Control,* 599–601; *Foreign Relations of the United States: Kennedy-Khrushchev Exchanges, 1961–1963* (Washington, DC: U.S. Government Printing Office, 1996), 204–05, 234–36, 238–40.

356 **Khrushchev's response:** *FRUS: Kennedy-Khrushchev Exchanges,* 238–39, 247–49, 253–55, 262–65; Seaborg, *Kennedy, Khrushchev, and the Test Ban,* 178–81, 184–85.

357 **Although Khrushchev clearly:** Seaborg, *Kennedy, Khrushchev, and the Test Ban,* 193–95; PPP: JFK, 1963, 80.

357 **Prime Minister Macmillan:** *FRUS: Arms Control,* 655–58, 676–78, 683–87, 693–99.

358 **On May 30, Kennedy:** Ibid., 707–08; *Saturday Review of Literature,* Nov. 7, 1964; Norman Cousins, *The Improbable Triumvirate: John F. Kennedy, Pope John, Nikita Khrushchev* (New York: Norton, 1972), 111–20.

359 **Cousins's urgings:** Sorensen, *Kennedy,* 821–26; PPP: JFK, 1963, 459–64; *FRUS: Arms Control,* 710–14; Dallek, *Unfinished Life,* 618–21.

360 **But Castro:** The documents covering the discussions in November about the IL-28s are in *FRUS: Kennedy-Khrushchev Exchanges;* JFK's quote is on 223; and *FRUS: Cuban Missile Crisis.* For an excellent summary of the dispute, see Fursenko and Naftali, *"One Hell of a Gamble,"* 290–310.

361 **Yet Kennedy could not ignore:** *FRUS: Cuban Missile Crisis,* 379–80, 394, 499, 548, 574, 582–83.

362 **For many of Kennedy's advisers:** Ibid., 587–89, 597, 608. See Russell D. Hoffman, "The Effects of Nuclear Weapons," online, 1999, who cites the 1962 Air Force pamphlet; *Robert Kennedy in His Own Words,* 247–48; Schlesinger, *Thousand Days,* 337–38; taped conversation, Dec. 5, 1962, POF, JFKL.

363 **On December 27:** JFK Conference with Joint Chiefs, Dec. 27, 1962, Box 345, National Security File, JFKL.

364 **Two days later, Kennedy spoke:** Richard Goodwin, "President Kennedy's Plan for Peace with Cuba," *New York Times*, July 5, 2000; *FRUS: Cuban Missile Crisis*, 635–36; O'Donnell and Powers, *Johnny, We Hardly Knew Ye*, 276–77; Schlesinger, *Robert Kennedy*, 535–38; Lawrence Freedman, *Kennedy's Wars: Berlin, Cuba, Laos, and Vietnam* (New York: Oxford University Press, 2000), 225–26; Jacqueline Kennedy, *Historic Conversations*, 193–94.

365 **There were good reasons:** *FRUS: Cuban Missile Crisis*, 648–51.

365 **Kennedy liked Bundy's plan:** Ibid., 658–62, 665.

366 **Kennedy wanted:** Ibid., 666–68, 670–75; and Schlesinger, *Robert Kennedy*, 538.

366 **Kennedy was torn:** *FRUS: Cuban Missile Crisis*, 668–69, 681–88.

367 **At an NSC meeting:** Ibid., 681–94, 698–99, 708–10.

369 **By March, Kennedy:** Ibid., 713–18; RFK to JFK, March 14, 1963, Box 35, Theodore Sorensen Papers, JFKL; Schlesinger, *Robert Kennedy*, 538–39; Thomas, *Robert Kennedy*, 239.

370 **But he had limited control:** *FRUS: Cuban Missile Crisis*, 728, 732–34.

371 **The raids opened a new round:** Ibid., 739–43.

372 **Under pressure from the CIA and the Florida:** Ibid., 748–54. Also see Taubman, *Khrushchev*, 578–81.

373 **Had Kennedy's advisers:** Gallup, *The Gallup Poll, 1959–1971*, 1993, 1800, 1807, 1811, 1815.

373 **In response to all the cross pressures:** *FRUS: Cuban Missile Crisis*, 754–56, 759–60; Bundy, *Danger and Survival*, 462.

374 **On the same day Kennedy wrote:** *FRUS: Cuban Missile Crisis*, 761–64.

375 **The discussion of how to deal with Castro:** Ibid., 780, 791, 795–97, 802–04, 814, 821–23.

376 **And so the CIA just continued:** Ibid., 822, 828–34, 837–38, 842–45, 851. McCone's warning is in recently released material in the RFK Papers, JFKL.

Chapter 10: "The Two of You Did Visit the Same Country, Didn't You?"

379 **In the second half of 1963:** Widmer, *Listening In*, 86, 208–09.

380 **Kennedy sent a stellar delegation:** Schlesinger, *Thousand Days*, 902–09.

380 **The more difficult battle:** Taubman, *Khrushchev*, 603.

381 **Kennedy's principal problem:** Poole, *History of the Joint Chiefs*, 102–04; JFK–Mike Mansfield telephone conversation, Aug. 12, 1963, JFKL; Bernard J. Firestone, "Kennedy and the Test Ban: Presidential Leadership and Arms Control," in Douglas Brinkley and Richard T. Griffiths, eds., *John F. Kennedy and Europe* (Baton Rouge: Louisiana State University Press, 1999), 82–85.

382 **Bringing the Chiefs to support:** Poole, *History of the Joint Chiefs*, 104–05; telephone conversations: JFK-LeMay, July 19, 1963; JFK-Rusk, July 24, 1963; JFK-Truman, July 26, 1963; JFK-Fulbright, Aug. 23, 1963, JFKL; Firestone, "Kennedy and the Test Ban," 88–93.

383 **As he was winning his fight:** *FRUS: Cuban Missile Crisis*, 837–38, 848, 853.

383 **Still, no matter how much:** Ibid., 861–63.

384 **Despite Kennedy's plea:** Ibid., 864–65.

384 **Kennedy, supported by Bundy and McNamara:** Schlesinger, *Robert Kennedy*, 551–52; *FRUS: Cuban Missile Crisis*, 868–70.

385 **The CIA immediately countered:** Ibid., 871–73.

385 **At the same time, Kennedy saw:** Ibid., 875–77; Poole, *History of the Joint Chiefs of Staff, 1961–1964*, 184.

386 **Ending agitation about Cuba:** *FRUS: Cuban Missile Crisis*, 877.

386 **Still, the president, Bobby:** Schlesinger, *Robert Kennedy*, 552–53; *Robert Kennedy in His Own Words*, 376.

387 **On October 31, Attwood:** *FRUS: Cuban Missile Crisis*, 879–83, especially 882.

388 **Like Castro, Kennedy found:** Schlesinger, *Robert Kennedy*, 553–54; *FRUS: Cuban Missile Crisis*, 883–89.

389 **Castro was not:** *Public Papers of the Presidents: John F. Kennedy, 1963* (Washington, DC: U.S. Government Printing Office 1964), 872–77; Schlesinger, *Robert Kennedy*, 554–56; Dallek *Unfinished Life*, 662–63.

391 **Pressure on Diem to settle:** *FRUS: Vietnam, January–August 1963*, 432, 447–49, 451–53.

392 **What was Kennedy supposed to believe:** *New York Times*, July 3, 1963; *FRUS: Vietnam*, 455–56, 465–66, 470–78, 481.

393 **At a July 17 press conference:** JFK Press Conference, July 17, 1963.

393 **At the end of July:** *FRUS: Vietnam, January–August 1963*, 541–43.

394 **On August 5, when the press:** Ibid., 553–55, 557–60.

394 **On August 15, Kennedy met:** Ibid., 567; JFK–Henry Cabot Lodge taped conversation, Aug. 15, 1963, Tape 104/A40, JFKL.

395 **Kennedy, of course, could not control:** *New York Times*, Aug. 15, 1963; *FRUS: Vietnam, January–August 1963*, 584, 589.

395 **For all his antagonism:** *Robert Kennedy in His Own Words*, 161–65; Reeves, *President Kennedy*, 363; *FRUS: Vietnam*, 585–88, 590–91.

395 **On August 21, when reports:** Tape 106/A41, Aug. 21, 1963, JFKL; *FRUS: Vietnam*, 598–602, 604–05.

396 **Embassy dispatches:** *FRUS: Vietnam*, 611–14, 620–25.

397 **On August 24, Forrestal:** Ibid., 625–26, 628–31. On the controversy, see Gibbons, *The U.S. Government and the Vietnam War*, 148–50.

397 **On August 26, Kennedy met:** Tape 107/A42, Aug. 26, 1963, JFKL; *FRUS: Vietnam*, 638–41.

398 **While the White House debated:** Ibid., 647, 649.

399 **But the problem of what to do:** Ibid., 650, n. 6, 653, 658–59.

399 **The 4 P.M. meeting:** Ibid., 659–65; tape 107/A42, Aug. 27, 1963, JFKL.

400 **The struggle over how to proceed:** *FRUS: Vietnam, January–August 1963*, 668, 670–72.

400 **Meanwhile, back in Washington:** *FRUS: Vietnam, August–December 1963* (Washington, DC: U.S. Government Printing Office, 1991), 1–9; tapes 107/A42, 108, and 108/A43, all Aug. 28, 1963; Schlesinger, *Robert Kennedy*, 713–14.

401 **At the evening meeting:** *FRUS: Vietnam, August–December 1963*, 12–17.

402 **During all this debate:** Schlesinger, *Thousand Days*, 968–73.

403 **The pressure on Kennedy:** *FRUS: Vietnam, August–December 1963*, 20–31, 35.

403 **The problem, however:** Ibid., 38, 53, 55, 63–64, 66, 76, 78–79.

404 **Later that morning, at another:** Gibbons, *The U.S. Government and the Vietnam War*, 160–61; *FRUS: Vietnam, Aug–Dec. 1963*, 69–74. The mea culpas of McNamara and Bundy are respectively in McNamara, *In Retrospect*, and Goldstein, *Lessons in Disaster*. Interview with McGeorge Bundy, Sept. 25, 1993.

405 **Lodge was now instructed:** *FRUS*, 75–79; JFK Interview with Cronkite, Sept. 2, 1963, online; Gibbons, *The U.S. Government and the Vietnam War*, 163.

406 **In suggesting that it was Vietnam's:** *FRUS: Vietnam, August–December 1963*, 98–104, 111–12.

406 **Nonetheless, Vietnam remained:** Ibid., 146, 161–63, 165; tape 109, Sept. 10, 1963, JFKL.

407 **The inability of his advisers:** *FRUS: Vietnam, August–December 1963*, 169–71, 176.

408 **In the meantime, Kennedy:** Ibid., 166, 174–75.

409 **The continuing daily conversations:** Ibid., 192–93, 195, 203, 205, 208–209, 212, 218–20, 235; Gibbons, *The U.S. Government and the Vietnam War*, 177–80.

409 **Because Kennedy saw no likelihood:** *FRUS: Vietnam, August–December 1963*, 252–56, 278–79, 282–83.

410 **Predictably, the McNamara-Taylor visit:** Ibid., 311, 317, 321, 336–39, 346, 353.

411 **The White House then issued:** PPP: JFK, 1963, 759–60.

411 **The public pronouncement:** Gibbons, *The U.S. Government and the Vietnam War*, 186, 188.

412 **Above all, now, he wanted to repress:** *FRUS: Vietnam, August–December 1963*, 166, 350–52; Gibbons, *The U.S. Government and the Vietnam War*, 188–89.

412 **But Kennedy couldn't plug leaks:** *New York Times*, Sept. 29, Oct. 3, 8, 17, 1963; *FRUS: Vietnam, August–December 1963*, 364–65.

413 **Kennedy made Halberstam the focus:** Halberstam, *Making of a Quagmire*, 201–03; Susan E. Tifft and Alex S. Jones, *The Trust: The Private and Powerful Family Behind the New York Times* (Boston: Little, Brown, 1999), 388–89.

414 **But Kennedy could no more control:** *FRUS: Vietnam, August–December 1963*, 393, 423–24, n. 5, 427, 429.

414 **On the twenty-seventh, the divide:** Gibbons, *The U.S. Government and the Vietnam War*, 196; *FRUS*, 453–55, 467. The details about the cable on the twenty-seventh are noticeably absent from the *FRUS* volume; and the Kennedy Library, as Gibbons, the author of *The U.S. Government and the Vietnam War*, points out, has no record of this directive. We only have the recollections of Ball, Hilsman, and Johnson that such a cable was sent, but its date is in dispute.

415 **The limits of White House control:** *FRUS: Vietnam, August–December, 1963,* 468–75.

416 **But Lodge, who remained:** Ibid., 484–85, 488, 500–502.

417 **But the exchanges:** Ibid., 525–26, 533, 537.

417 **In a recording on November 4:** JFK tape, Nov. 4, 1963, JFKL.

418 **Regardless of what happened next:** *FRUS: Vietnam, August–December 1963,* 579–80; JFK Press Conference, Nov. 14, 1963; William J. Rust, *Kennedy in Vietnam* (New York: Da Capo Press, 1985), 3–5; Forrestal's quote can be found on the Internet under JFK Nov. 21, 1963, discussion of Vietnam with Forrestal.

Epilogue: "What He Is Slated to Become Depends on Us"

422 **For those who saw themselves:** Sorensen, *Counselor,* 372; Jacqueline Kennedy, *Historic Conversations,* xiii; Glen Johnson, "Camelot Revisited," Associated Press, 1995.

422 **In March 1964, Jackie Kennedy:** Jacqueline Kennedy, *Historic Conversations,* foreword and introduction, especially xv, xx–xxi.

423 **Schlesinger was more than happy:** Schlesinger, *Journals,* 218–19, 224.

424 **As soon as Schlesinger left:** Schlesinger, *Thousand Days,* ix, x.

425 **Like Jackie Kennedy and Schlesinger:** Sorensen, *Kennedy,* 3–8; Sorensen, *Counselor,* chaps. 27–28, especially 371–72, 382.

426 **No one was more determined:** Schlesinger, *Journals,* 210, 214–15, 228–30, 237; Dallek, *Flawed Giant,* 57, 137–43, 165.

427 **After Bobby won his Senate election:** Robert Kennedy, *Thirteen Days; Robert Kennedy: In His Own Words.*

427 **McGeorge Bundy was one member:** Schlesinger, *Journals,* 205, 238, 333; Bird, *Color of Truth,* 319–23, 348–49; Goldstein, *Lessons in Disaster,* 128–29, 193–201, 217–18.

428 **After he left the government:** Goldstein, *Lessons in Disaster,* 144–50.

429 **In his pursuit of this goal:** Schlesinger, *Thousand Days,* 432–37; Dallek, *Flawed Giant,* 87–88, 99; Rusk, *As I Saw It; Time,* Feb. 4, 1908.

430 **Robert McNamara's historical reputation:** Dallek, *Flawed Giant,* 494–95; McNamara, *In Retrospect.*

431 **Walt W. Rostow was the most unrepentant:** Dallek, *Flawed Giant,* 296–97; interview with Walt W. Rostow, July 27, 1992.

432 **Paul Volcker, the chairman:** Paul Volcker, "What the New President Should Consider," *New York Review of Books,* Dec. 6, 2012, 10.

Bibliography

o o o o

Alford, Mimi. *Once Upon a Secret: My Affair with President John F. Kennedy and Its Aftermath.* New York: Random House, 2012.

Ambrose, Stephen E. *Eisenhower: The President.* New York: Simon & Schuster, 1984.

Baker, Bobby. *Wheeling and Dealing: Confessions of a Capitol Hill Operator.* New York: Norton, 1978.

Ball, George W. *The Past Has Another Pattern: Memoirs.* New York: Norton, 1982.

Beisner, Robert L. *Dean Acheson: A Life in the Cold War.* New York: Oxford University Press, 2006.

Beschloss, Michael R. *The Crisis Years: Kennedy and Khrushchev, 1960–1963.* New York: HarperCollins, 1991.

———. *Kennedy and Roosevelt: The Uneasy Alliance.* New York: Norton, 1980.

Bierce, Ambrose. *Cynic's Word Book.* New York: Doubleday, Page, 1906.

Bill, James A. *George Ball: Behind the Scenes in U.S. Foreign Policy.* New Haven, CT: Yale University Press, 1997.

Binder, James. *Lemnitzer: A Soldier for His Time.* Washington, DC: Brassey's, 1997.

Bird, Kai. *The Color of Truth: McGeorge Bundy and William Bundy: Brothers in Arms.* New York: Simon & Schuster, 1998.

Blair, Clay, Jr., and Joan Blair. *The Search for JFK.* New York: Berkley, 1974.

Blight, James G., Bruce J. Allyn, and David A. Welch. *Cuba on the Brink: Castro, the Missile Crisis, and the Soviet Collapse.* New York: Pantheon Books, 1993.

Bowles, Chester. *Promises to Keep: My Years in Public Life, 1941–1969.* New York: Harper & Row, 1971.

Boyer, Paul. *By the Bomb's Early Light.* New York: Pantheon Books, 1985.

Brauer, Carl M. *Presidential Transitions: Eisenhower Through Reagan.* New York: Oxford University Press, 1986.

Brinkley, Douglas, *Dean Acheson: The Cold War Years, 1953–1971.* New Haven, CT: Yale University Press, 1992.

Brinkley, Douglas, and Richard T. Griffiths, eds. *John F. Kennedy and Europe.* Baton Rouge: Louisiana State University Press, 1999.

Bundy, McGeorge. *Danger and Survival: Choices About the Bomb in the First Fifty Years.* New York: Random House, 1988.

Burns, James McGregor. *John Kennedy.* New York: Harcourt, Brace, 1959.

Catudal, Honoré M. *Kennedy and the Berlin Wall Crisis.* Berlin: Berlin-Verlag, 1980.

Clifford, Clark. *Counsel to the President: A Memoir.* New York: Random House, 1991.

Cohen, Warren I. *Dean Rusk.* Totowa, NJ: Cooper Square, 1980.

Collier, Peter, and David Horowitz. *The Kennedys: An American Drama.* New York: Summit Books, 1984.

Cousins, Norman. *The Improbable Triumvirate: John F. Kennedy, Pope John, Nikita Khrushchev.* New York: Norton, 1972.

Dallek, Robert. *Flawed Giant: Lyndon Johnson and His Times, 1961–1973.* New York: Oxford University Press, 1998.

———. *Franklin D. Roosevelt and American Foreign Policy, 1932–1945.* New York: Oxford University Press, 1995.

———. *Lone Star Rising: Lyndon Johnson and His Times, 1908–1960.* New York: Oxford University Press, 1991.

———. *Nixon and Kissinger: Partners in Power.* New York: HarperCollins, 2007.

———. *An Unfinished Life: John F. Kennedy, 1917–1963.* New York: Little, Brown, 2003.

Davis, John H. *The Kennedys: Dynasty and Disaster, 1848–1983.* New York: McGraw-Hill, 1984.

Dobrynin, Anatoly. *In Confidence: Moscow's Ambassador to America's Six Cold War Presidents (1962–1986).* New York: Times Books/Random House, 1995.

Donald, David. *Lincoln Reconsidered.* New York: Vintage Books, 1961.

Fay, Paul B., Jr. *The Pleasure of His Company.* New York: Harper & Row, 1966.

Freedman, Lawrence. *Kennedy's Wars: Berlin, Cuba, Laos, and Vietnam.* New York: Oxford University Press, 2000.

Fullilove, Michael. *Rendezvous with Destiny: How Franklin D. Roosevelt and Five Extraordinary Men Took America into the War and into the World.* New York: Penguin Press, 2013.

Fursenko, Aleksandr, and Timothy Naftali. *"One Hell of a Gamble": Khrushchev, Castro, and Kennedy, 1958–1964.* New York: Norton, 1997.

Galbraith, John Kenneth. *Ambassador's Journal: A Personal Account of the Kennedy Years.* Boston: Houghton, Mifflin, 1969.

———. *Letters to Kennedy.* Edited by James Goodman. Cambridge, MA: Harvard University Press, 1998.

Gallup, George. *The Gallup Poll: Public Opinion, 1935–1971.* New York: Random House, 1972.

Gibbons, William C. *The U.S. Government and the Vietnam War: Executive and Legislative Roles and Relationships, 1961–1964.* Washington, DC: U.S. Government Printing Office, 1984.

Gibbs, Nancy, and Michael Duffy. *The Presidents Club: Inside the World's Most Exclusive Fraternity.* New York: Simon & Schuster, 2012.

Goldstein, Gordon M. *Lessons in Disaster: McGeorge Bundy and the Path to War in Vietnam.* New York: Times Books/Henry Holt, 2008.

Goodwin, Doris Kearns. *The Fitzgeralds and the Kennedys.* New York: Simon & Schuster, 1987.

———. *Team of Rivals: The Political Genius of Abraham Lincoln.* New York: Simon & Schuster, 2012.

Goodwin, Richard. *Remembering America: A Voice from the Sixties.* Boston: Little, Brown, 1988.

Guthman, Edwin O., and Jeffrey Shulman, eds. *Robert Kennedy: In His Own Words.* New York: Bantam Books, 1988.

Halberstam, David. *The Best and the Brightest.* New York: Ballantine Books, 1993.

———. *The Making of a Quagmire.* New York: Rowman & Littlefield, 2008.

Harrington, Michael. *The Other America: Poverty in the United States.* New York: Macmillan, 1962.

Hilsman, Roger. *To Move a Nation: The Politics of Foreign Policy in the Administration of John F. Kennedy.* New York: Dell, 1968.

Hilty, James W. *Robert Kennedy: Brother Protector.* Philadelphia: Temple University Press, 1998.

Humphrey, Hubert H. *The Education of a Public Man.* Garden City, NY: Doubleday, 1976.

Isaacson, Walter, and Evan Thomas. *The Wise Men: Six Friends and the World They Made.* New York: Simon & Schuster, 1986.

Jay, Anthony, ed. *The Oxford Dictionary of Political Quotations.* New York: Oxford University Press, 1996.

Jones, Howard. *The Bay of Pigs.* New York: Oxford University Press, 2008.

Kaiser, David. *American Tragedy: Kennedy, Johnson, and the Origins of the Vietnam War.* Cambridge, MA: Harvard University Press, 2000.

Kaplan, Fred. *The Wizards of Armageddon.* New York: Simon & Schuster, 1983.

Kennedy, Jacqueline. *Historic Conversations on Life with John F. Kennedy.* New York: Hyperion, 2011.

Kennedy, Robert F. *Thirteen Days: A Memoir of the Cuban Missile Crisis.* New York: Norton, 1968.

Kennedy, Rose Fitzgerald. *Times to Remember.* Garden City, NY: Doubleday, 1974.

Kern, Montague, Patricia W. Levering, and Ralph B. Levering. *The Kennedy Crises: The Press, the Presidency, and Foreign Policy.* Chapel Hill: University of North Carolina Press, 1983.

Kozak, Warren. *The Life and Wars of Curtis LeMay.* Chicago: Regnery, 2009.

Krock, Arthur. *Memoirs: Fifty Years on the Firing Line.* New York: Funk & Wagnalls, 1968.

Lasky, Victor. *J.F.K.* New York: Macmillan, 1963.

Launius, Roger D., and Howard E. McCurdy. *Spaceflight and the Myth of Presidential Leadership.* Urbana: University of Illinois Press, 1997.

Leuchtenburg, William E. *A Troubled Feast: American Society Since 1945.* Boston: Little, Brown, 1973.

Manchester, William. *The Death of a President.* New York: Harper & Row, 1967.

Martin, Ralph G., and Ed Plaut. *Front Runner, Dark Horse.* Garden City, NY: Doubleday, 1960.

McNamara, Robert S. *Argument Without End: In Search of Answers to the Vietnam Tragedy.* New York: PublicAffairs, 1999.

———. *In Retrospect: The Tragedy and Lessons of Vietnam.* New York: Times Books, 1995.

Montefiore, Simon S. *Stalin: The Court of the Red Tsar.* New York: Vintage Books, 2005.

Nashel, Jonathan. *Edward Lansdale's Cold War.* Amherst: University of Massachusetts Press, 2006.

Newman, John N. *JFK and Vietnam.* New York: Warner Books, 1992.

Oberdorfer, Don. *Senator Mansfield: The Extraordinary Life of a Great American Statesman and Diplomat.* Washington, DC: Smithsonian Books, 2003.

O'Brien, Lawrence F. *No Final Victories: A Life in Politics—from John F. Kennedy to Watergate.* Garden City, NY: Doubleday, 1974.

O'Donnell, Kenneth P., and David F. Powers. *Johnny, We Hardly Knew Ye.* Boston: Little, Brown, 1970.

Olson, James S. *Where the Domino Fell.* New York: Wiley-Blackwell, 1996.

Parker, Richard. *John Kenneth Galbraith: His Life, His Politics, His Economics.* New York: Farrar, Straus & Giroux, 2005.

Parmet, Herbert. *Jack: The Struggles of John F. Kennedy.* New York: Dial Press, 1980.

The Pentagon Papers: The Defense Department History of United States Decisionmaking on Vietnam. Boston: Bantam Books, 1971.

Poole, Walter S. *History of the Joint Chiefs of Staff: The Joint Chiefs of Staff and National Policy, 1961–1964.* Washington, DC: U.S. Government Printing Office, 2011.

Powers, Thomas. *The Man Who Kept the Secrets: Richard Helms and the CIA.* New York: Knopf, 1979.

Public Papers of the Presidents: John F. Kennedy, 1961–1963. Washington, DC: U.S. Government Printing Office, 1962–64.

Reeves, Richard. *President Kennedy: Profile of Power.* New York: Simon & Schuster, 1993.

Ricks, Thomas E. *The Generals: American Military Command from World War II to Today.* New York: Penguin Press, 2012.

Rostow, Walt W. *The Stages of Economic Growth: A Non-Communist Manifesto.* Cambridge, UK: Cambridge University Press, 1960.

Rusk, Dean. *As I Saw It.* New York: Norton, 1990.

Rust, William J. *Kennedy in Vietnam.* New York: Da Capo Press, 1985.

Salinger, Pierre. *With Kennedy.* Garden City, NY: Doubleday, 1966.

Schaffer, Howard B. *Chester Bowles: New Dealer in the Cold War.* Cambridge, MA: Harvard University Press, 1993.

Schlesinger, Arthur M. *The Age of Roosevelt.* Boston: Houghton Mifflin, 1957–60.

———. *Journals, 1952–2000.* New York: Penguin Press, 2007.

———. *Robert Kennedy and His Times.* Boston: Houghton Mifflin, 1978.

———. *A Thousand Days.* Boston: Houghton Mifflin, 1965.

Seaborg, Glenn T. *Kennedy, Khrushchev, and the Test Ban.* Berkeley: University of California Press, 1981.

Shapley, Deborah. *Promise and Power: The Life and Times of Robert McNamara.* Boston: Little, Brown, 1993.

Sheehan, Neil. *A Bright Shining Lie: America in Vietnam.* New York: Random House, 1988.

Shesol, Jeff. *Mutual Contempt: Lyndon Johnson, Robert Kennedy, and the Feud that Defined a Decade.* New York: Norton, 1997.

Sorensen, Theodore C. *Counselor: A Life at the Edge of History.* New York: HarperCollins, 2008.

———. *Kennedy.* New York: Bantam Books, 1966.

Stossel, Scott. *Sarge: The Life and Times of Sargent Shriver.* Washington, DC: Smithsonian Books, 2004.

Taubman, William. *Khrushchev: The Man and His Era.* New York: Norton, 2003.

Taylor, John M. *An American Soldier: The Wars of General Maxwell Taylor.* Novato, CA: Presidio Press, 1989.

Thomas, Evan. *Robert Kennedy: His Life.* New York: Simon & Schuster, 2000.

Tifft, Susan E., and Alex S. Jones. *The Trust: The Private and Powerful Family Behind the New York Times.* New York: Little, Brown, 1999.

U.S. Department of State. *Foreign Relations of the United States: Arms Control and Disarmament, 1961–1963.* Washington, DC: U.S. Government Printing Office, 1995.

———. *Foreign Relations of the United States: Berlin Crisis, 1961–1962.* Washington, DC: U.S. Government Printing Office, 1993.

———. *Foreign Relations of the United States: Cuba, 1961–1962.* Washington, DC: U.S. Government Printing Office, 1997.

———. *Foreign Relations of the United States: Cuban Missile Crisis and Aftermath.* Washington, DC: U.S. Government Printing Office, 1996.

———. *Foreign Relations of the United States: Kennedy-Khrushchev Exchanges, 1961–1963.* Washington, DC: U.S. Government Printing Office, 1996.

———. *Foreign Relations of the United States: Soviet Union, 1961–1963.* Washington, DC: U.S. Government Printing Office, 1998.

———. *Foreign Relations of the United States: Vietnam, 1961.* Washington, DC: U.S. Government Printing Office, 1988.

———. *Foreign Relations of the United States: Vietnam, 1962.* Washington, DC: U.S. Government Printing Office, 1990.

———. *Foreign Relations of the United States: Vietnam, August–December, 1963.* Washington, DC: U.S. Government Printing Office, 1991.

———. *Foreign Relations of the United States: Vietnam, January–August 1963.* Washington, DC: U.S. Government Printing Office, 1991.

White, Theodore H. *The Making of the President, 1960.* New York: Atheneum, 1961.

Widmer, Ted. *Listening In: The Secret White House Recordings of John F. Kennedy.* New York: Hyperion, 2012.

Wofford, Harris. *Of Kennedys and Kings.* New York: Farrar, Straus & Giroux, 1980.

Woods, Randall B. *Fulbright: A Biography.* New York: Cambridge University Press, 1995.

Zelikow, Philip, Ernest May, and Timothy Naftali, eds. *The Presidential Recordings: John F. Kennedy: The Great Crises, July 30–October 28, 1962.* 3 vols. New York: Norton, 2001.

Acknowledgments

o o o o

Every book is the product of a collective effort: archivists, editors, and colleagues generous enough to take time from their own work to read the manuscript.

John Wright, my agent and friend for twenty years, suggested this project and offered wise counsel on the broad scope of the book as well as on its details. Tim Duggan, my editor at Harper-Collins now for the third time, read three versions of the manuscript, each time pressing the case for revisions that have made a considerable difference in bringing Kennedy's many advisers into sharper focus and enriching our understanding of Kennedy the man and policymaker.

Geri Dallek, as usual, was unrelenting in reminding me that readers want to know the people you are describing—not simply as men trying to find answers to impossible questions about war and peace, but also as flesh and blood characters struggling with their own inner demons and reach for historical influence. Like Tim Duggan, she deserves a special shout-out for offering wise counsel. None of this, however, is meant to suggest that either she or Tim should share responsibility for whatever defects remain in the organization and composition of the text.

Several others have helped bring the book to life, including Emily Cunningham, associate editor at HarperCollins, whose keen eye for the right phrase and more economical use of quotations have made this a much more readable book.

Matthew Dallek, with whom I had the pleasure of teaching some courses at the University of California in Washington, UCDC, offered compelling advice in several discussions about what I was trying to do in the book. I had the chance to try out some of my ideas in both his class and my own at Stanford University in Washington, SIW, where I have been teaching for five years under the guidance of Adrienne Jamieson, the Center's superb director. Peter Kovler, who knows more about national politics than any of the so-called experts I have met over the years, has been another helpful sounding board on a book about presidential advisers.

I am also grateful to Tom Pitoniak for his excellent copyediting. He has saved me from numerous errors. Lydia Weaver, the production editor, applied her expertise to the publication of my third HarperCollins book. I am in her debt for making the process so relatively easy. She is a master of her craft.

Finally, I cannot resist saying thank you to President Barack Obama, who has graciously hosted four dinners for presidential historians, where I had a close-up look at what a president hoped he could learn from history. It provided a glimpse into how a president interacted with men and women trying to offer useful judgments on the not entirely different problems earlier presidents, including Kennedy, faced.

Index

o o o o

About the Author

ROBERT DALLEK is the author of *An Unfinished Life: John F. Kennedy, 1917–1963* and *Nixon and Kissinger: Partners in Power*, among other books. His writing has appeared in *The New York Times*, *The Washington Post*, *The Atlantic*, and *Vanity Fair*. He is an elected fellow of the American Academy of Arts and Sciences and of the Society of American Historians, for which he served as president in 2004–2005. He lives in Washington, D.C.

BOOKS BY ROBERT DALLEK

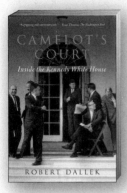

CAMELOT'S COURT
Inside the Kennedy White House
Available in Paperback and eBook

Fifty years after John F. Kennedy's assassination, presidential historian Robert Dallek, whom *The New York Times* calls "Kennedy's leading biographer," delivers a riveting portrait of this president and his inner circle of advisors—their rivalries, personality clashes, and political battles. In *Camelot's Court*, Dallek analyzes the brain trust whose contributions to the successes and failures of Kennedy's administration—including the Bay of Pigs, civil rights, the Cuban Missile Crisis, and Vietnam—were indelible.

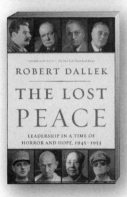

THE LOST PEACE
Leadership in a Time of Horror and Hope, 1945–1953
Available in Paperback and eBook

Provocative, illuminating, and based on a lifetime of research, *The Lost Peace* is a penetrating look at the misjudgments that caused enormous strife and suffering during a most critical period in history: from the closing months of World War II through the early years of the Cold War. The men who led the world—principally Churchill, Stalin, de Gaulle, Mao, Truman, Syngman Rhee, and Kim Il Sung—executed astonishingly unwise actions that propelled the nuclear arms race. The decisions of these great men, for better and often for worse, had profound consequences for the following decades.

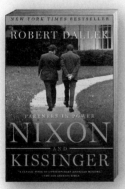

NIXON AND KISSINGER
Partners in Power
Available in Paperback and eBook

Working side by side in the White House, Richard Nixon and Henry Kissinger were two of the most compelling, contradictory, and powerful figures in America in the second half of the twentieth century. Tapping into a wealth of recently declassified archives, Robert Dallek uncovers fascinating details about Nixon and Kissinger's tumultuous personal relationship and brilliantly analyzes their shared roles in monumental historical events.

Visit HarperCollins.com for more information about your favorite HarperCollins books and authors.

Available wherever books are sold.